Rotations, Quaternions, and Double Groups

Rotations, Quaternions, and Double Groups

Simon L. Altmann

Brasenose College

Oxford

DOVER PUBLICATIONS, INC.

Mineola, New York

Copyright

Bibliographical Note

This Dover edition, first published in 2005, is a slightly corrected republication of the edition published by Oxford University Press, Oxford and New York, 1986.

Library of Congress Cataloging-in-Publication Data

Altmann, Simon L., 1924-
Rotations, quaternions, and double groups / Simon L. Altmann.
p. cm.
Originally published: Oxford : Clarendon Press ; New York : Oxford University Press, 1986.
ISBN 0-486-44518-6 (pbk.)
1. Rotation groups. 2. Quaternions. 3. Finite groups. 4. Representations of groups I. Title.

QC174.17.R65A48 2005
530.12—dc22

2005049630

Manufactured in the United States of America
Dover Publications, Inc., 31 East 2nd Street, Mineola, N.Y. 11501

On peut avoir trois principaux objets dans l'étude de la vérité: l'un, de la découvrir quand on la cherche; l'autre, de la démontrer quand on la possède; le dernier, de la discerner d'avec le faux quand on l'examine.

One can have three principal objects in the study of truth: the first, to discover it when one searches for it; the second, to prove it when one possesses it; the last one, to distinguish it from falsity when one examines it.

> Blaise Pascal (*ca.* 1711). ***Réflections sur la géometrie en géneral. De l'esprit géométrique et de l'art de persuader. Oeuvres complètes.*** Présentation et notes de L. Lafuma. Éditions du Seuil, Paris, 1963, p. 348.

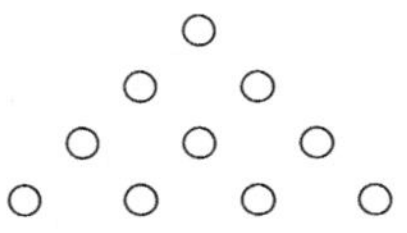

The Pythagorean Tetractys

PREFACE

In order to describe the scope of this book, I shall start from the remark that rotation operators are often obtained as by-products of the angular momentum operators used in quantum mechanics. Partly as a result of this approach, rotations are then parametrized by means of the well-known Euler angles, which suffer from three defects: they are not always unique, they are very cumbersome to determine in the finite rotation groups (point groups), and they do not provide a scheme for the multiplication of rotations.

An entirely different approach to rotations is possible, which was introduced by Olinde Rodrigues in 1840 but which has never been used. The rotation operators are obtained in this approach by an entirely geometric method, which not only provides a much better insight into their nature, but also leads most naturally to the parametrization of rotations by parameters that coincide with quaternions. These parameters are unique, exceedingly easy to determine, and – because they are quaternions – they provide an algebra that permits the multiplication of rotations in a very simple way. At the same time, and most importantly, these parameters determine unambiguously the phase factors that appear in the angular-momentum representations for half-integral quantum numbers. This result leads to a rigorous formulation of the representations of the rotation group, either as projective representations or by means of double groups.

The main object of this book is to present a consistent description of this geometric and quaternionic treatment of rotations, so far badly neglected. Because it was my purpose to plug this important gap, and because I hope to reach an audience, such as molecular and solid-state theorists and crystallographers, primarily interested in point groups and which may not be familiar with Lie-group theory, I have avoided the use of the latter: there are excellent treatments of rotations based on this method, and the very few ideas of continuous groups which are required in this book can be made, at the expense of generality, of course, reasonably self-contained.

I assume, on the other hand, some familiarity with simple notions of group theory. The relation between irreducible representations and energy levels, e.g., is discussed in many undergraduate courses in chemistry and some in physics. To have started from scratch in group theory would not only have made the book much longer but would also have reduced the impact of its main message. I have, however, provided a reasonably self-contained revision of group theory, matrices, and angular momentum, the first five chapters or so, in fact, containing material suitable for undergraduate work in these subjects. I hope in this way to bridge the gap between the elementary textbooks on group theory and angular momentum, and the more advanced treatises on the rotation group.

It is very difficult in a book of this type to deal with a wide range of applications, beyond those standard in molecular problems. Solid-state applications, or relativistic ones, would have required an extensive development of basic theories beyond the scope of this book. The important feature provided by the methods given here for those that require such applications, is that the phase factors of the representations which must be used in such cases are all consistently determined, and if this much is achieved then the purpose of this book is fulfilled.

In most elementary treatments of rotations, existential proofs are often given which guarantee that certain matrices are obtainable, but which do not provide them in a form acceptable to the user. This is so because it is now universally accepted in this subject that all bases of representations must satisfy the Condon and Shortley convention. All proofs in this book, instead, are designed so as to provide always matrices consistent with this convention. The reader must appreciate that there is a great difference between a purely existential proof and one that is constructive in the way described, which requires meticulous attention to often-ignored technicalities. This is all to the good, since a mastery of such technicalities is essential in order to obtain sound results in this subject.

There are two approaches which may be adopted in a book of this type. One is to produce first a complete account of all the theoretical topics that are required. The other is to produce an account of background material sufficient to make a start, and to develop then the new mathematical tools required when the need for them emerges. I have adopted largely this second approach because I felt that, for the audience I had in mind, it is important to establish a motivation for learning a new method. I am conscious, however, of the drawbacks of such an approach, in particular because some topics then spread out over various chapters, rather than being neatly packed in one pigeon hole. (To help connect the various strands, though, the index provides careful indications of the context of the various entries in it.)

Readers may notice that certain definitions and conventions are sometimes summarized or even repeated in various chapters. I had in mind that it is a rare reader these days who reads a book like this from cover to cover and, although it is impossible to make each chapter entirely self-contained, I have tried to provide enough clues for the reader who expects to gather some information from a later section of the book not having read the whole. Partly for this purpose, a complete set of the conventions and major definitions used is given in Chapter **0**, where notation is explained. Frequent reference to this chapter is strongly advised.

One feature of the book is that many of the numerous problems provide an outlet for some of the more technical proofs of the book, for which reason complete solutions are provided in Chapter **16**. This device often allows me to concentrate the work of a chapter into the more illuminating aspects of the discussion.

I have felt that, at this level, it was not necessary to provide a reading scheme for the book. The reader will realize that many topics in the first five chapters can be skipped and, if necessary, used later as a reference. Likewise, the more technical aspects of many chapters can safely be left for a second reading.

I shall not discuss further here the overall scheme of the book, since the reader will get, I hope, a fairly clear idea of the scope and intentions of this work by reading the introductory Chapter **1**. It must be appreciated, though, that this chapter contains numerous forward references to topics which will be fully developed later on, so that some of the more technical aspects of the discussion are likely to remain somewhat obscure on a first reading. I hope that, if this is the case, this material will be found much clearer after most of the book, Chapter **12** in particular, has been read.

I should like now to thank the many people that made this book possible. A set of lectures on double groups which I gave in Easter 1976 at the Instituto de Química Física Rocasolano, Madrid, at the invitation of Ramón Gancedo, helped me to crystallize my thoughts on the subject. A lot of the basic material for the book was prepared in 1979 at the Centre of Mathematical Physics, Austin, Texas, where I gave a lecture course on the subject at the invitation of Professors F. A. Matsen and A. Böhm. Various discussions with my friend, Dr Christopher Bradley, were most useful at several stages of the work, as well as the collaboration of Drs Fernando Palacio, Peter Herzig, and Rainer Dirl in the research from which much of this book grew.

Doctors I. Grattan-Guinness and J. Gray read Chapter **1** and provided very useful comments on various historical problems. I am grateful to Professor Ioan James for a discussion of the topological proof in §**10**-4. Professor John Ackrill gave me some advice on the Pythagorean school.

Doctor Peter Herzig read the whole typescript and provided numerous corrections. Professor Alun Morris contributed most generously a pure mathematician's insight in his reading of the text. I am also grateful to two anonymous readers of the Oxford University Press for their help in improving the text. Most of all, I owe the warmest thanks to Dr Brian Sutcliffe, who read the text with meticulous care and did most of the exercises. His extensive and penetrating comments have been most important in improving the book in all its aspects.

I am very grateful to Signor Italo Calvino for permission to use the quotation from *Palomar* in Chapter **11** and to Emecé Editores for the quotation from Señor Jorge Luis Borges in Chapter **4**.

Oxford S.L.A
January 1985

CONTENTS

0

NOTATION–CONVENTIONS–HOW TO USE THIS BOOK

CROSS REFERENCES AND REFERENCES

§**3**-5 §5	Chapters are numbered with bold-face numerals, but in all cross-references, formulae, etc., the number of the current chapter is dropped: §**3**-5 is section 5 of Chapter **3**; §5 is section 5 of the current chapter.
(**6**-3.5) (3.5) (5)	Equations, figures, and tables are numbered by the section number followed, after a dot, by a numeral. The examples shown are of eqn (5) of, respectively, Chapter **6**, §3; section 3 of current chapter; current section.
L3,R3	Left-hand and right-hand side of (3), respectively.
3 2,3 ,3 3|4	(*Left margins of displayed formulae*) '3' means that the equation on its right follows from (3). '2,3' means that on the line on its right eqns (2) and (3) are applied in that order. ',3' means that eqn (3) is applied, but not immediately, in the line on its right. '3|4' means that the line on its right follows by applying eqn (3) to eqn (4). The letters F, T, P preceding any of the above mean a reference to a Figure, Table, or Problem respectively.
Rodrigues (1840)	Identifies a paper of that date in the alphabetic list at the end of the book.

PROBLEMS

The page number on the right of the heading of a set of problems indicates the page where the solutions to that set of problems are given.

GENERAL MATHEMATICAL NOTATION

$=_{\text{def}}$	means that a definition is entailed.
$\equiv$	identically equal.
$a \Rightarrow b$	a entails b (if a is true then b is true).
$\Leftrightarrow$	entails and is entailed by.
$\exists$	there exists.
$\in$	belongs to.
$\forall$	all.
$\{g\}$	set of all elements g.
$\mathbb{R}^n, \mathbb{R}^3$	set of all n-tuples of real numbers. $\mathbb{R}^3$ is also referred to as 'configuration space'.
$\mathbb{C}^n$	set of all n-tuples of complex numbers.
$A \to B$	map of set A into set B.
$a \mapsto b$	map of element a (pre-image) onto element b (image).
$a \Leftrightarrow b$	one-to-one map.
i	imaginary unit, to be distinguished from the inversion $\boldsymbol{i}$.
$x\,y\,z$	components of a vector $\mathbf{r}$. Independent variables.
$\mathsf{x\,y\,z}$	dependent variables. (The function $f(x) = \mathsf{x}$, etc.) The distinction is often left implicit. See end of §**2**-4 and footnote §**5**-1.
$\mathbf{x\,y\,z}$	space-fixed right-handed orthonormal axes.
$\mathbf{i\,j\,k}$	right-handed orthonormal vectors fixed to the rotating configuration space. (See Major Conventions and Definitions below.) In Chapter **12**, $\mathbf{i}$, $\mathbf{j}$, $\mathbf{k}$ are used as the $\mathbf{x}$, $\mathbf{y}$, $\mathbf{z}$ above.

BRACKETS

Commas, without notational significance, may be inserted for the sake of clarity between elements of round and angular brackets.

$\langle a_1 a_2 \ldots a_n \vert$	row vector of elements $a_1, a_2, \ldots, a_n$. It may be abbreviated as $\langle a \vert$, §**2**–4.
$\vert a_1 a_2 \ldots a_n \rangle$	column vector of elements $a_1, a_2, \ldots, a_n$. It may be abbreviated as $\vert a \rangle$, §**2**-4.
(abc)	vector of components a, b, c along some stated axes. This symbol is used when the distinction between row and column vectors made by the angular brackets is not required.

$(\bar{a}bc)$, $\vert\bar{a}bc\rangle$	vectors of components $-a, b, c$. Do not confuse the bars here with those used for transformed vectors, as in the next symbol.
$\langle\bar{a}_1\bar{a}_2\ldots\bar{a}_n\vert$	the transform of the vector $\langle a_1a_2\ldots a_n\vert$ under some stated operation.
$\langle a\vert$, $\vert a\rangle$	Dirac's bra and ket respectively.
$[g_i, g_j]$	commutator: $g_ig_j - g_jg_i$.
$\mathbf{[}g_i, g_j\mathbf{]}$	projective factor.
$[\![a, \mathbf{A}]\!]$	quaternion.
$A_{[ij]}$	element ij (itself a matrix) of supermatrix A.

MATRIX NOTATION

See Table **2**-6.1

$A_{[ij]}$	matrix element ij (itself a matrix) of supermatrix A.
$A \otimes B$	direct product.

ANGULAR MOMENTUM, ROTATION GROUP GENERATORS, AND RELATED QUANTITIES

$\mathbf{I}$	angular momentum operator, (**4**-2.12).
$\boldsymbol{\nabla}$	$\mathbf{r}\times\nabla \equiv i\mathbf{I}$, (**4**-2.6).
$I_xI_yI_z$	angular momentum components.
$\hat{\mathbf{J}}$	vector with matrix components $\hat{J}_x, \hat{J}_y, \hat{J}_z$, (**3**-5.10), (**4**-3.3).
$\hat{\mathbf{I}}$	matrix corresponding to $\mathbf{I}$, (**4**-3.7).
$I_\pm$	shift operators, (**4**-5.1).
$\hat{I}_x\hat{I}_y\hat{I}_z$	matrix components of $\hat{\mathbf{I}}$ (**4**-9.5), (**4**-9.8).
$\boldsymbol{I}_x\boldsymbol{I}_y\boldsymbol{I}_z$	SU(2) matrices corresponding to binary rotations (**6**-4.1).
$\boldsymbol{\sigma}_x\boldsymbol{\sigma}_y\boldsymbol{\sigma}_z$	Pauli matrices, (**4**-9.9), ($\sigma_x, \sigma_y, \sigma_z$ are orthogonal mirrors, Problem **11**-9.1).

ROTATIONS AND ROTATION MATRICES

Commas, without notational significance, may be inserted for the sake of clarity inside the brackets of the rotation and rotation matrix symbols.

$R(\phi\mathbf{n})$	rotation around axis $\mathbf{n}$ by angle ϕ (distinguished from the polar coordinate, for which the symbol φ is used). §§**2**-1, **3**-2, **3**-6.

$R(\phi\mathbf{x})$	rotation about axis $\mathbf{x}$ (or $\mathbf{y}$, $\mathbf{z}$) by ϕ. In this case the axes $\mathbf{x}, \mathbf{y}, \mathbf{z}$ are orthonormal and fixed in space.
$R(\phi\Pi)$	rotation by ϕ with pole Π, §**9**-1.
$R(\lambda; \boldsymbol{\Lambda}) \equiv [\![\lambda, \boldsymbol{\Lambda}]\!]$	as $R(\phi\mathbf{n})$ (from which symbol it is distinguished by the semicolon) for $\lambda = \cos\frac{1}{2}\phi$, $\boldsymbol{\Lambda} = \sin\frac{1}{2}\phi\ \mathbf{n}$ (Euler–Rodrigues or quaternion parameters).
$R(ab)$	rotation in terms of the Cayley–Klein parameters, $a = \lambda - \mathrm{i}\Lambda_z$, $b = -\Lambda_y - \mathrm{i}\Lambda_x$.
$R(\rho\tau)$	rotation in terms of the quaternion complex parameters, $\rho = \lambda + \mathrm{i}\Lambda_z$, $\tau = \Lambda_x + \mathrm{i}\Lambda_y$.
$R(\alpha\beta\gamma)$	rotation in terms of the Euler angles (of which γ is the first and α the third), §**3**-1.
$\hat{R}^j(\ )$	rotation matrix of dimension $2j+1$ in any of the above parametrizations and in the canonical basis $\langle u_j^j \ldots u_{-j}^j\vert$ or its dual (see below). For j half-integral, it is understood that the caret ˆ is changed into ˇ.
$\hat{R}_{\mathbf{r}}^1(\ )$	as $R^1(\ )$ above, i.e. rotation matrix of dimension 3 but in the basis $\vert xyz\rangle \equiv \mathbf{r}$, where $\mathbf{r} \in \mathbb{R}^3$, or in its dual $\langle xyz\vert$, (**3**-1.15), (**3**-3.11).
A	SU(2) matrix, §**6**-2; Chapter **7**.
$X(\xi)$, $Y(\eta)$, $Z(\zeta)$	Rotations about orthogonal axes $\mathbf{x}, \mathbf{y}, \mathbf{z}$ by angles ξ, η, ζ respectively, §**3**-5.
$\hat{X}(\xi)$, $\hat{Y}(\eta)$, $\hat{Z}(\zeta)$	Corresponding matrices, §**3**-5.
$C_{2x}C_{2y}C_{2z}$	Binary rotations around $\mathbf{x}, \mathbf{y}, \mathbf{z}$ (BB).
$\sigma_x\sigma_y\sigma_z$	Mirrors perpendicular to $\mathbf{x}, \mathbf{y}, \mathbf{z}$ (OM), Problem **11**-9.1.

GROUP NOTATION

G, H	groups of elements $g_1, g_2, \ldots, h_1, h_2 \ldots$ respectively.
$\vert G\vert$	order of G.
$H \subset G$, $G \supset H$	H subgroup of G, §**2**-5.
$H \lhd G$	H invariant subgroup of G, §**2**-5.
$\hat{G}(g)$, $\check{G}(g)$	matrix representative of g in a vector and in a projective representation, respectively, of G.
g, σ, R	operators in configuration space ($\mathbb{R}^3$).
$\dot{g}$, $\dot{\sigma}$, $\dot{R}$	operators in function space. (The dot is dropped whenever possible, since function operands identify this type of operator.) §**2**-3.
g_ig_j	result (g_k say) of the product of g_i with g_j.
g_i, g_j	operator g_j followed by g_i, (**8**-2.3).
g_i^g	conjugate of g_i by g: the operation that results from the product gg_ig^{-1}, (**2**-5.2).

$(g_i)^g$	operation whose pole is the conjugate of the pole of g_i by g (that is, the pole obtained by acting with g on the pole of g_i). g_i^g and $(g_i)^g$ may or may not coincide. §**11**-6.
$G \otimes H$	direct product of G and H.
$g_i \oplus g_j$	direct sum of elements, i.e. the inclusion of elements g_i and g_j in the same set.
$\textstyle\bigoplus\!\sum_i g_i$	direct sum of all elements g_i.
E, e	identity elements.

DOUBLE-GROUP NOTATION

See (**13**-3.3) to (**13**-3.10)

BASES

$\langle u_j^j \ldots u_{-j}^j|$	basis of the $2j+1$ functions u_m^j, $j \geqslant m \geqslant -j$, which are the eigenfunctions of the z component of the angular momentum, I_z, in the Condon and Shortley phase convention. For integral j, they are the spherical harmonics $Y_l^m(\theta\varphi)$.
$|\mathrm{u}_1 \mathrm{u}_0 \mathrm{u}_{-1}\rangle$	dual of the basis $\langle u_1^1 u_0^1 u_{-1}^1| \equiv \langle Y_1^1 Y_1^0 Y_1^{-1}|$. The vector $\mathbf{r} \in \mathbb{C}^3$ with components $\mathrm{u}_1, \mathrm{u}_0, \mathrm{u}_{-1}$ is called a spherical vector, (§**5**-1).
$|xyz\rangle \equiv \mathbf{r}$	$\mathbf{r} \in \mathbb{R}^3$, a vector in configuration space.
$\langle \mathrm{xyz}|$	dual of the above. $\mathrm{x} \equiv \mathrm{x}(\theta\varphi)$ stands for the Cartesian coordinate x written as function of θ and φ, §**2**-4.
$\langle u| \otimes \langle v|$	direct-product basis, (**2**-5.22).
$\langle u| \bar{\otimes} \langle v|$	symmetrized direct-product basis, (**14**-3.3).

MAJOR CONVENTIONS AND DEFINITIONS

Cross-references to the numbers of paragraphs in this section are given in square brackets. These rules are valid for the projective representation method and have to be partly modified for the double-group method, as stated in [18] below.

1. *Axes.* The axes $\mathbf{x}, \mathbf{y}, \mathbf{z}$ are a set of orthogonal right-handed axes centred at O and fixed in space. The axes $\mathbf{i}, \mathbf{j}, \mathbf{k}$ are a set of orthonormal right-handed axes fixed to the (rotating) unit ball centred at O.
2. *Configuration space.* All points in $\mathbb{R}^3$ are referred to the axes $\mathbf{x}, \mathbf{y}, \mathbf{z}$.

3. *Configuration space operators.* They are always used in the *active picture* in which the whole of $\mathbb{R}^3$ (in solidarity with the axes $\mathbf{i}, \mathbf{j}, \mathbf{k}$) is rotated with respect to $\mathbf{x}, \mathbf{y}, \mathbf{z}$. For $\mathbf{r} \in \mathbb{R}^3$, its coordinates change from x, y, z to $\bar{x}, \bar{y}, \bar{z}$ after the transformation, these being the coordinates of the transformed vector, $\mathrm{g}\mathbf{r} \equiv \bar{\mathbf{r}}$. The matrix of the transformation is defined so that it transforms the components x, y, z written as a *column vector.*

4. *Function space operators.* They are always defined in the *active picture.* The transform $\dot{g}f$ of the function f under the function space operator $\dot{g}$ is defined so that

$$\dot{g}f(\mathrm{g}\mathbf{r}) = f(\mathbf{r}) \qquad \Leftrightarrow \qquad \dot{g}f(\mathbf{r}) = f(\mathrm{g}^{-1}\mathbf{r}).$$

5. *Positive hemisphere.* Disjoint sets $\mathfrak{H}$ (positive hemisphere) and $\bar{\mathfrak{H}}$ (negative hemisphere) are defined to cover the whole of the unit sphere. (See [8].)

6. *Rotation angles.* All rotations are classified as positive or negative with a rotation angle ϕ in the range

$$-\pi < \phi \leqslant \pi.$$

Positive and negative rotations are defined so that they are seen counterclockwise and clockwise, respectively, from outside the unit sphere.

7. *Rotation poles.* The pole of a proper rotation $\mathrm{g} \in G$ is the point of the unit sphere that is left invariant by the rotation g and such that, from it, the rotation is seen counterclockwise. The pole of an improper rotation $i\mathrm{g} \in G$ is the pole of the proper rotation g (not necessarily an operation of G).

8. *Standard set of poles.* From [6] and [7] it follows that for all proper and improper rotations of G a unique pole is defined, that belongs to $\mathfrak{H}$ for all positive rotations and all binary rotations and that belongs to $\bar{\mathfrak{H}}$ for all negative rotations.

While forming this standard set of poles, attention must be paid to the choice of $\mathfrak{H}$, which may be done as follows. Choose a pole $\Pi(\mathrm{g}_i)$ of a *positive* (conveniently so defined) rotation g_i. Form $\mathrm{g}\Pi(\mathrm{g}_i)$ for all $\mathrm{g} \in G$, disregarding antipoles as they appear. (This choice is not unique, but it may become so if a subgroup of G is chosen which generates the same number of positive poles.) The set of positive poles thus obtained, for one g_i in each class of G, will entail the choice of $\mathfrak{H}$. The rule stated ensures that the class of g_i is properly described and, if a subgroup of G has been used in the manner indicated, it also guarantees correct subduction to the subgroup.

9. *Standard $\phi\mathbf{n}$ parameters.* For each operation g its standard pole is associated with a unique positive rotation angle

$$0 \leqslant \phi \leqslant \pi,$$

and a unique axis $\mathbf{n}$ ($|\mathbf{n}|=1$) which belongs to $\mathfrak{H}$ for positive rotations and $\bar{\mathfrak{H}}$ for negative rotations.

10. *Standard quaternion set.* Given the standard $\phi\mathbf{n}$ parameters, a unique quaternion can be defined for every proper or improper rotation of G as follows:

$$g \leftrightarrow [\![\lambda, \mathbf{\Lambda}]\!], \qquad \lambda = \cos \tfrac{1}{2}\phi, \qquad \mathbf{\Lambda} = \sin \tfrac{1}{2}\phi \; \mathbf{n}.$$

The set of all such quaternions for each $g \in G$ is called the standard quaternion set.

11. *Quaternion multiplication rules.* Given

$$g_i \leftrightarrow [\![\lambda_i, \mathbf{\Lambda}_i]\!], \qquad g_j \leftrightarrow [\![\lambda_j, \mathbf{\Lambda}_j]\!],$$

where both quaternions belong to the standard set, then

$$[\![\lambda_i, \mathbf{\Lambda}_i]\!][\![\lambda_j, \mathbf{\Lambda}_j]\!] = [\![\lambda_i\lambda_j - \mathbf{\Lambda}_i \cdot \mathbf{\Lambda}_j, \lambda_i\mathbf{\Lambda}_j + \lambda_j\mathbf{\Lambda}_i + \mathbf{\Lambda}_i \times \mathbf{\Lambda}_j]\!] = \pm[\![\lambda_k, \mathbf{\Lambda}_k]\!],$$

where

$$[\![\lambda_k, \mathbf{\Lambda}_k]\!] \leftrightarrow g_k,$$

is a quaternion of the standard set.

12. *Multiplication rules and projective factors.* Given the results in [11], then

$$g_i g_j = g_k$$

and

$$[g_i, g_j] = \pm 1,$$

the latter result corresponding to the $\pm$ sign in the quaternion product in [11].

13. *Gauge.* The rule given in [12] for the projective factor entails the use of the Pauli gauge (§**11**-9) in the projective representations of improper point groups.

14. *Bases of the representations.* Unless the contrary is explicitly stated (by a suffix on the symbol $\hat{R}^j$ or $\check{R}^j$ of a representation matrix) the bases of the representations are canonical bases of functions u^j_m, eigenfunctions of the total angular momentum with angular momentum quantum number j and belonging to the eigenvalue m of the z component of the angular momentum operator. They are canonical in the sense that the set u^j_m, $j \geq m \geq -j$ satisfies the Condon and Shortley phase convention (§**4**-8).

15. *Form of the bases.* The canonical bases must always be written as row vectors and in the order from u^j_j in the left to u^j_{-j} in the right. Corresponding *dual bases* in two and three dimensions are defined by column vectors.

16. *Form of the representations.* It is

$$R(\lambda;\mathbf{\Lambda})\langle u^j_j u^j_{j-1}\ldots u^j_{-j}| = \langle u^j_j u^j_{j-1}\ldots u^j_{-j}|\,\hat{R}^j(\lambda;\mathbf{\Lambda}).$$

Here, the caret in $\hat{R}^j$ should be changed into $\check{R}^j$ for j half-integral and the parametrization of the rotation can be changed into other systems. The matrix $\hat{R}^j(\lambda;\mathbf{\Lambda})$ is given by (**14**-2.9) and its sign is always positive for the standard set quaternions $[\![\lambda, \mathbf{\Lambda}]\!]$. (Although it becomes undetermined in the Euler angle parametrization.) The above equation gives the transform of an individual function of the basis as follows:

$$R(\lambda;\mathbf{\Lambda})u^j_m = \sum_{m'=j}^{-j} u^j_{m'}\hat{R}^j(\lambda;\mathbf{\Lambda})_{m'm}.$$

The dual of the basis $\langle u^j_j\ldots u^j_{-j}|$ is defined so that it is a column-vector basis that transforms by precisely the same matrix $\hat{R}^j(\lambda;\mathbf{\Lambda})$ as $\langle u^j_j\ldots u^j_{-j}|$. For the dual basis, $\hat{R}^j(\lambda;\mathbf{\Lambda})$ must pre-multiply the basis.

17. *Euler angles* (if used). They are defined in the active picture in terms of rotations of $\mathbb{R}^3$ with respect to the fixed axes $\mathbf{x}, \mathbf{y}, \mathbf{z}$. $R(\alpha\beta\gamma)$ is a rotation by γ around $\mathbf{z}$, followed by a rotation by β around $\mathbf{y}$, followed by a rotation by α around $\mathbf{z}$. The ranges for the angles are given in (**3**-1.11).

18. *Changes required for double-group work.* Rules [1] to [9] are unchanged. In [10], take

$$g \leftrightarrow [\![\lambda, \mathbf{\Lambda}]\!], \qquad \tilde{g} \leftrightarrow -[\![\lambda, \mathbf{\Lambda}]\!],$$

thus defining an extended standard quaternion set. In [12] take

$$g_i g_j = g_k \qquad \text{or} \qquad g_i g_j = \tilde{g}_k,$$

depending on the quaternion that appears on the right of the quaternion product in [11]. Rule [13] is always used implicitly in double groups. [14] and [15] are unchanged but, in [16], $\hat{R}^j(\lambda;\mathbf{\Lambda})$ must be changed in sign when an operation $\tilde{g}$ is considered.

1

INTRODUCTION

> Quaternions came from Hamilton after his really good work had been done; and, though beautifully ingenious, have been an unmixed evil to those who have touched them in any way, including Clerk Maxwell.
>
> Lord Kelvin (1892). *Letter to Hayward* quoted by S. P. Thompson (1910), *The life of William Thomson, Baron Kelvin of Largs*. Macmillan, London, vol. II, p. 1070.

Such robust language as Lord Kelvin's is now largely forgotten, but the fact remains that the man in the street is strangely averse to using quaternions, despite the most sanguine early expectations for them in the highest quarters (see epigraph to Chapter **12**). Side by side with matrices and vectors, now the *lingua franca* of all physical scientists, quaternions appear to exude an air of nineteenth-century decay, as a rather unsuccessful species in the struggle-for-life of mathematical ideas. Mathematicians, admittedly, still keep a warm place in their hearts for the remarkable algebraic properties of quaternions but, alas, such enthusiasm means little to the harder-headed physical scientist.

This introduction will attempt to highlight certain problems of interpretation as regards quaternions which may seriously have affected their progress, and which might explain their present parlous status. For claims were made for quaternions that quaternions could not possibly fulfill, which made it difficult to grasp what quaternions are excellent at – and this is at doing the work of this book; that is, handling rotations and double groups. Admittedly, that quaternions play a rôle in the study of rotations has been known for as long as they have existed, if not longer, and yet their practical use in this context has been minimal in comparison with other, more cumbersome and less accurate, methods.

It is not possible to understand the problems that we face unless we go a little over the history of the subject; and the history of quaternions, more perhaps than that of any other nineteenth-century mathematical subject, is dominated by the extraordinary contrast of two personalities, the inventor of quaternions, Sir William Rowan Hamilton, Astronomer Royal of Ireland, and Olinde Rodrigues, one-time director of the Caisse Hypothécaire (a bank specializing in lending money on mortgages) at the Rue Neuve-Saint-Augustin in Paris (Booth 1871, p. 107).

Hamilton was a very great man indeed; his life is documented in minute detail in the three volumes of Graves (1882); and a whole issue in his honour was published in 1943, the centenary of quarternions, in the *Proceedings of the Royal Irish Academy*, vol. **A50** and, in 1944, in *Scripta Mathematica*, vol. **10.** There are also two excellent new biographies (Hankins 1980, O'Donnell 1983) and numerous individual articles (e.g. Lanczos 1967). Of Hamilton, we know the very minute of his birth, precisely midnight between 3 and 4 August 1805, in Dublin. Of Olinde Rodrigues, despite the excellent one-and-only published article on him by Jeremy Gray (1980), we know next to nothing. He is given a mere one-page entry in the Michaud *Biographie Universelle* (1843) as an 'economist and French reformer'. So little is he known, indeed, that Cartan (1938, p. 57) invented a non-existent collaborator of Rodrigues by the surname of Olinde, a mistake repeated by Temple (1960, p. 68). Booth (1871) calls him *Rodrigue* throughout his book, and Wilson (1941, p. 100) spells his name as *Rodriques.* Nothing that Rodrigues did on the rotation group – and he did more than any man before him, or than any one would do for several decades afterwards – brought him undivided credit; and for much of his work he received no credit at all. This Invisible Man of the rotation group was probably born in Bordeaux on 16 October 1794, the son of a Jewish banker, and he was named Benjamin Olinde, although he never used his first name in later life. The family is often said to have been of Spanish origin, but the spelling of the family name rather suggests Portuguese descent (as indeed asserted by the *Enciclopedia Universal Ilustrada Espasa-Calpe*). He studied mathematics at the École Normale, the École Polytechnique not being accessible to him owing to his Jewish extraction. He took his doctorate at the new University of Paris in 1816 with a thesis that contains the famous 'Rodrigues formula' for Legendre polynomials, for which he is mainly known (Grattan–Guinness 1983). The little that we know about him is only as a paranymph of Saint-Simon, the charismatic Utopian Socialist, whom he met in May 1823, two months after Saint-Simon's attempted suicide. So we read (Weill 1894, p. 30) that the banker Rodrigues helped him in his illness and destitution and supported him financially until his death in 1825. That Rodrigues must have been very well off we can surmise from Weill's reference to him as belonging to high banking circles, on a par with the wealthy Laffittes (1894, p. 238). When Saint-Simon died, Rodrigues shared with another disciple of Saint-Simon's, Prosper Enfantin, the headship of the movement, thus becoming *Père Olinde* for the acolytes; but in 1832 he repudiated Enfantin's extreme views of sexual freedom and he proclaimed himself the apostle of Saint-Simonism. In August that year he was charged with taking part in illegal meetings and outraging public morality and was fined fifty francs (Booth 1871). Neither Booth nor Weill even mention that Rodrigues was a mathematician: the single reference to this (Booth 1871, p. 100) is that in 1813 he was Enfantin's tutor in mathematics at the École Polytechnique. Indeed, all that we know about him in the year 1840 when he published his fundamental paper on the Euclidean and rotation groups, is that he was 'speculating at the Bourse'. (Booth 1871, p. 216.)

Besides his extensive writings on social and political matters, Rodrigues published several pamphlets on the theory of banking and was influential in the development of French railways. He died in Paris, however, almost forgotten (Michaud 1843). Even the date of his death is uncertain: 26 December 1850 according to the *Biographie Universelle*, or 17 December 1851 according to Larousse (1866). Sébastien Charléty (1936, pp. 26, 294), although hardly touching upon Rodrigues in his authoritative history of Saint-Simonism, gives 1851 as the year of Rodrigues's death, a date which most modern references seem to favour.

Hamilton survived Rodrigues by fourteen years and had the pleasure, three months before his death in 1865, to see his name ranked as that of the greatest living scientist in the roll of the newly elected Foreign Associates of the American National Academy of Sciences. And quite rightly so: his achievements had been immense by any standards. In comparison with Rodrigues, alas, he had been born with no more than a silver-plated spoon in his mouth: and the plating was tarnishing. At three, the family had to park various children with relatives and William was sent to his uncle, the Revd James Hamilton, who ran the diocesan school at Trim. That was an intellectually explosive association of a child prodigy and an eccentric pedant: at three, William was scribbling in Hebrew and at seven he was said by a don at Trinity College Dublin to have surpassed the standard in this language of many Fellowship candidates. At ten, he had mastered ten oriental languages, Chaldee, Syriac, and Sanscrit amongst them, plus, of course, Latin and Greek and various European languages. (The veracity of the reports of these linguistic feats is, however, disputed by O'Donnell 1983.) Mathematics – if one does not count mental arithmetic, at which he was prodigious – came late but with a bang, when at seventeen, reading on his own Laplace's *Mécanique Céleste,* he found a mistake in it which he communicated to the President of the Irish Academy. His mathematical career, in fact, was already set in 1823 when, still seventeen, he read a seminal paper on caustics before the Royal Irish Academy.

From then on Hamilton's career was meteoric: Astronomer Royal of Ireland at 22, when he still had to take two quarterly examinations as an undergraduate, knight at 30. Like Ørsted, the Copenhagen pharmacist who had stirred the world with his discovery of the electromagnetic interaction in 1820, Hamilton was a Kantian and a follower of the

Naturphilosophie movement then popular in Central Europe. For Hamilton 'The design of physical science is . . . to learn the language and interpret the oracles of the universe' (*Lecture on Astronomy*, 1831, see Graves 1882, vol. I, p. 501). He discusses in 1835, prophetically, (because of the later application of quaternions in relatively theory) 'Algebra as the Science of Pure Time'. He writes copiously both in prose and in stilted verse. He engages in a life-long friendship with Wordsworth, and goes to Highgate in the spring of 1832 to meet Coleridge, whom he visits and with whom he corresponds regularly in the next few years, the poet praising him for his understanding 'that *Science*, . . . , needs a *Baptism*, a regeneration in Philosophy' or Theosophy. (Graves 1882, vol. I, p. 546).

Hamilton had been interested in complex numbers since the early 1830s and he was the first to show, in 1833, that they form an algebra of couples. (See Hamilton 1931, vol. III.) Given the real and imaginary units, with the multiplication laws $1^2=1$, $\mathrm{i}^2=-1$, the elements of the algebra are the complex numbers of the form $a+\mathrm{i}\,b$, with a and b reals. Of course, to say that they form an algebra merely means that the formal rules of arithmetical operations are valid for the objects so defined. For over ten years Hamilton tried to extend this concept in order to define a triplet, with one real and two imaginary units, i and j (whatever they be!). This, however, not even he could do.

Then Monday, 16 October 1843 came, one of the best documented days in the history of mathematics. Hamilton's letter to his youngest son, of 5 August 1865 (Graves 1882, vol. II, p. 434), is almost too well known, but bears brief repetition. That morning Hamilton, accompanied by Lady Hamilton, was walking along the Royal Canal in Dublin towards the Royal Irish Academy, where Hamilton was to preside at a meeting. As he was walking past Broome Bridge (erroneously referred to as Brougham Bridge by Hamilton and called by this name ever since), Hamilton, in a flash of inspiration, realized that three, rather than two, imaginary units were needed, with the following properties:

$$\mathrm{i}^2=\mathrm{j}^2=\mathrm{k}^2=-1, \tag{1}$$

$$\mathrm{ij}=\mathrm{k}, \qquad \mathrm{ji}=-\mathrm{k}, \tag{2}$$

and cyclic permutation $\mathrm{i}\to\mathrm{j}\to\mathrm{k}\to\mathrm{i}$ of (2). As everyone knows, and as de Valera was to do almost a century later on his prison's wall, Hamilton carved these formulae on the stone of the bridge. (Lady Hamilton was undoubtedly a very patient lady.) Armed now with four units, Hamilton called the number

$$q=a+b\mathrm{i}+c\mathrm{j}+d\mathrm{k} \tag{3}$$

a *quaternion*. Thus were quaternions born and baptized: it was entered on the Council Books of the Academy for that day that Mr W. R. Hamilton was given leave to read a paper on quaternions at the First General Meeting of the Session, 13 November 1843.

One of the various falsities that have to be dispelled about quaternions is the origin of this name, since entirely unsupported sources are often quoted; in particular Milton, *Paradise Lost*, v. 181 (Mackay 1977, p. 70) and the *Vulgate*, *Acts* 12:4 (Temple 1981, p. 46). Of course, we know that Milton was a favourite poet of Hamilton at 24 (Graves 1882, vol. I, p. 321) and to suggest that he was not aware of Acts and the apprehension of Peter by a quaternion of soldiers would be absurd. But no one with the slightest acquaintance with Hamilton's thought would accept the obvious when the recondite will do. In his *Elements of Quaternions* (Hamilton 1899, p. 114) we find our first clue: 'As to the mere *word*, *quaternion*, it signifies primarily (as is well known), like its Latin original, "Quaternio" or the Greek noun τετρακτύς, a *Set of Four*'. The key word here is 'tetractys', and there is evidence for this coming from Hamilton's closest, perhaps his only real pupil, P. G. Tait, who writing in the *Encyclopaedia Britannica* (see article on Quaternions in the XIth Edition) says: 'Sir W. R. Hamilton was probably influenced by the recollection of its Greek equivalent the Pythagorean Tetractys . . . , the mystic source of all things . . . '. That Tait very much believed in this, is supported by the unattributed epigraph in Greek in the title page of his own treatise on quaternions (Tait 1890): these are verses 47 and 48 of *Carmen Aureum* (*Golden Song*), a hellenistic Pythagorean poem much in vogue in the Augustan era, the full text of which appears in Diehl (1940, p. 45). Of course, the concept of the tetractys embodying, as we shall see, multiple layers of meaning in a single word, must have attracted Hamilton: for Pythagoras, having discovered that the intervals of Greek music are given by the ratios 1:2, 3:2, 4:3, made it appear that *kosmos*, that is, order and beauty, flow from the first four digits, 1, 2, 3, 4, the sum of which gives the perfect number 10, and is symbolized by the tetractys, the sacred symbol depicted beneath the epigraph of the present book (p. v). (See Guthrie 1962, p. 225.) The Pythagoreans used to take an oath by the tetractys, as recorded by Sextus Empiricus (see Kirk, Raven, and Schofield 1983, p. 233): 'The Pythagoreans are accustomed sometimes to say "All things are like number" and sometimes to swear this most potent oath: Nay, by him that gave to us the *tetractys*, which contains the fount and root of ever-flowing nature'. That the tetractys exercised the imagination of Hamilton, there is no doubt: besides the cryptic footnote in the Treatise, already quoted, we find Augustus De Morgan (with whom Hamilton entertained a very copious correspondence) acknowledging on 27 December 1851 a sonnet from Hamilton (apparently lost) on the tetractys. It is tempting to speculate that Hamilton might have been introduced to the tetractys by Coleridge, who alludes to 'the adorable tetractys, or tetrad'. (Barfield 1972, p. 252).

Now back to 16 October 1843. That same evening Hamilton wrote a long and detailed draft of a letter to his friend John Graves, first published by A. J. McConnell (1943) and included in Hamilton (1931 vol. III). Next day, a final letter was written and sent, later published in the *Philosophical Magazine* (Hamilton 1844*a*). The November report to the Irish Academy was published almost at the same time (Hamilton 1844*b*). We can thus follow almost hour by hour Hamilton's first thoughts on quaternions. Although in the morning of the glorious day he had been led to the discovery through the algebra of the quaternions, by the evening (and in this he acknowledges the influence of Warren 1829), he had been able to recognize a relation between quaternions and what we now call rotations. And in this, sadly, we cannot but see the germ of the canker that eventually consumed the quaternion body. For here Hamilton identifies the scalar part of the quaternion (the coefficient a in eqn 3) with $\mu \cos \rho$, where he calls μ the *modulus* and ρ the *amplitude* of the quaternion, the latter purporting to be (but unfortunately not quite being) what we would now call the rotation angle of the quaternion. As regards the remaining three coefficients in (3) they all contained a factor $\mu \sin \rho$ multiplied by some functions of two angles, φ and ψ, that determine the orientation of the rotation axis. It is, of course, easy to understand why Hamilton was led to this parametrization, from which he never moved away, becoming later, as we shall see, even further entrenched in it.

Argand had observed in 1806 that the imaginary unit i rotates what we would now call a vector in the Argand plane by $\frac{1}{2}\pi$, which made it possible to visualize the relation $\mathrm{i}^2 = -1$. From that point of view, in fact, i^2 should be a rotation by π (a *binary rotation*) which, acting on a vector in that plane, multiplies it by the factor -1. If we want to identify a quaternion unit, say j, in (3), we must choose φ and ψ so that b and d vanish, take the modulus μ equal to 1 and the 'amplitude' ρ equal to $\frac{1}{2}\pi$ so that a (which is now $\cos \rho$) vanishes and c (which is $\sin \rho$) equals 1. Thus j, and in the same manner all quaternion units, are identified with 'quadrantal' rotations, in the language later introduced by Hamilton (1853, p. 64, art. 71), that is rotations by $\frac{1}{2}\pi$. Of course, this appears to be satisfactory insofar as it explains, *à l'*Argand, eqn (1). Truly enough, this is not even a compendium of Hamilton's thoughts in his first two papers mentioned: he was not given to put his thoughts in words of one syllable, for he always had a multi-layered universe in his mind's eye. Where the argument can be seen in all its beautiful clarity is in Clifford (1878, p. 351). The sad truth is that, appealing as this argument is, to identify the quaternion units with rotations by $\frac{1}{2}\pi$ is not only not right, but entirely unacceptable in the study of the rotation group, a point to which we shall return later.

We must move on towards the full geometrical identification of the quaternion, *à la* Hamilton, keeping, of course, to his parametrization of a as $\cos\rho$ (we can safely take μ as unity). If we take a in (3) as zero, that is ρ as $\frac{1}{2}\pi$, we are left with what Hamilton called a *pure quaternion* $b\mathrm{i}+c\mathrm{j}+d\mathrm{k}$, for which he coined the word *vector* (Hamilton 1846, p. 54). (This is the reason why to this day, i, j, and k are still used for the unit vectors.) I must say, of course, that the contribution of Hamilton to the later development of vectors in the hands of Heaviside and Gibbs was considerable if not crucial. Already in the very day of the discovery he had comprehended both the scalar and vector products. And I must also remind the reader that, as the inventor of the name, it was in his gift to call $b\mathrm{i}+c\mathrm{j}+d\mathrm{k}$ anything. The trouble is that, without having a name for them, people well before 1846 were already thinking about certain physical objects very much as we think today about vectors, and that the identification of Hamilton's 'vectors' with what they actually had in mind created a deal of confusion. To make matters worse, Hamilton, and to a much greater degree his followers, became dogmatic and intolerant (Klein 1926, p. 182) and a great deal of sterile discussion ensued. It must have been evident for decades that something was seriously wrong here, as Klein, one of the leading nineteenth-century geometers, implied himself (1926, p. 186). Yet, the earliest explicit statement that I have been able to find is one by Marcel Riesz (1958, p. 21): 'Hamilton and his school professed that the quaternions make the study of vectors in three-space unnecessary since every vector can be considered as the vectorial part... of a quaternion... this interpretation is grossly incorrect since the vectorial part of a quaternion behaves with respect to coordinate transformations like a bivector or "axial" vector and not like an ordinary or "polar" vector'. We shall see in §**12**-3 that the 'vector' part of a quaternion is indeed an axial vector, but we shall also see in §§**12**-8, **12**-9 that the uncritical identification of a pure quaternion with any vector at all is fraught with danger: for a pure quaternion is nothing else than a binary rotation. (§**12**-2.)

We must now go back a little and examine in more detail the relation between quaternions and rotations as it developed over the years in Hamilton's mind. We must stress that Hamilton's everlasting monument (see MacDuffee 1944) is his construction of objects that, except for commutativity, obey the same algebra as that of real and complex numbers: and Hamilton was aware of this – although he could not foresee that quaternions were to receive in 1878, at the hands of Frobenius, the supreme accolade of being proved to be the only possible objects with this property. Hamilton himself had proved that quaternions could always be multiplied and divided; that is, unless q_2 be zero q_1/q_2 is always another quaternion q. This, of course, must also be true if q_1 and q_2 are pure quaternions, that is vectors r_1 and r_2 in Hamilton's language. (We shall assume for simplicity that they are both of the same length.) Hamilton could very easily prove (see §**12**-9) that the quotient q of two vectors r_1 and r_2 separated by an angle ρ is a quaternion with axis normal to the plane (r_1, r_2) and with 'amplitude' ρ, the latter having the same value and meaning previously discussed. This, of course, fitted beautifully with the desired picture for the equation $\mathrm{i}^2 = \mathrm{j}^2 = \mathrm{k}^2 = -1$.

This picture of quaternions coming from their algebra impressed Hamilton so much that in his *Lectures on Quaternions* (1853, p. 122), and ever after, the primary definition of a quaternion that he used was 'The quotient of two vectors, or the operator which changes one vector into another' as later adopted by the *Oxford English Dictionary*. This definition became the core of the quaternion dogma; a most unfortunate choice, since it entails defining the rotation angle of the quaternion q as the angle between r_1 and r_2. We shall see in §**12**-9 that, because r_2 (like r_1) is not a true vector, qr_2 is not a rotation of r_2 into r_1, and that the angle between r_2 and r_1 happens to be precisely one-half of the rotation angle that must be associated with the quaternion q.

That there must be some trouble of this type can be seen when we pass from a *rectangular rotation* as just discussed (i.e. a rotation in which the rotation axis is normal to the vector) to a *conical rotation* in which the rotation axis is at an angle to the vector to be rotated, which will thus move on a cone around the rotation axis. Hamilton and others saw at once that when this is the relation between q and r_2 then qr_2 cannot be the transformed vector r_1, since it is easy to prove that qr_2 is not a pure quaternion at all and thus cannot be identified with a vector. It is perhaps significant that Hamilton solved the problem but did not publish the answer. Cayley (1845) was the first to go into print, although in his collected papers (Cayley 1888, vol. I, p. 586, note 20) he concedes priority to Hamilton. The result is this: qr_2q^{-1} turns out to be a pure quaternion r_1. When the components of r_1 are obtained they are 'precisely those given for such transformations by M. Olinde Rodrigues .... It would be an interesting question to account, *à priori* for the appearance of these coefficients here'. And in this note Cayley acknowledges that 'I was led to it [the above formula of transformation] by Rodrigues' formulae'. Let us see what Hamilton has to say about this: 'The SYMBOL OF OPERATION $q(\ )q^{-1}$, where q may be called (as before) the operator quaternion, while the symbol (suppose r) of the operand quaternion is conceived to occupy the place marked by the parentheses ... ' can be regarded as 'a conical rotation of the axis of the operand round the axis of the operator, through *double* the angle thereof'. (Hamilton 1853, p. 271, my itals). It is clear that there are two problems here: one is why do we have to have two different transformation rules for a vector, and the other why in the second case, the angle of rotation is 'doubled'. (See §§**12**-8, **12**-9.) Or, looking at the problem in another way, we have two ways in which we can associate an angle of rotation with a quaternion q, as being either α (given by the rectangular transformation) or β (equal to 2α) as the result of the conical transformation. Hamilton chose α to parametrize the rotation, whereas the true angle, as Rodrigues (1840) discovered, was β. And this is the paper to which Cayley referred, almost certainly never read by Hamilton, and never again quoted by any of the major quaternionists. Cayley's question was never answered – not until the last epigones of Hamilton were safely in their tombs. (It is not impossible, in fact, that a full answer was never given until §**12**-8 was written!)

E. T. Bell (1937, Chapter 19), labelled the last twenty years of Hamilton 'The Irish Tragedy'. Lanczos (1967) compares them with Einstein's fruitless search for a unified field theory in his own last two decades. The truth is probably somewhere between these two views. For Hamilton suffered the weight of his own greatness: it was not enough for him to have an algebra, it was not enough for him to have a geometry, he had to 'interpret the oracles of the Universe' and the oracles trailed in front of him false clues that no one was to unravel for another three-score years after his death.

The last years of Hamilton, despite his immense fame, were not without worries: Continentals were spreading rumours that the great Gauss had actually discovered quaternions but never bothered to publish. (They were right, as shown by his notes of 1819, published in 1900; see Gauss 1863, vol. VIII, pp. 357–62.) In letters to De Morgan of January 1852 (Graves 1882, vol. II, p. 490, vol. III, p. 330), Hamilton attacks these allegations. Curiously enough, of Rodrigues, who in 1840 not only had invented quaternions bar the name, but also published his formulae, never a word.

After Hamilton's death, a byzantine posture was adopted by his followers, intent on stopping the flood of rebellion from across the Atlantic. Thus P. G. Tait (1890, p. vi): 'Even Prof. Willard Gibbs must be ranked as one of the retarders of quaternion progress, in virtue of his pamphlet on *Vector Analysis*, a sort of hermaphrodite monster, compounded of the notations of Hamilton and of Grassmann.' Alexander Macfarlane, a former student of Tait teaching at Texas, became the leading force behind an International Association for Promoting the Study of Quaternions and Allied Systems of Mathematics, which actually published a Bulletin from 1900 to 1923. Its influence extended as far as Japan where Kimura in 1907 became one of the major influences on the Association, but it did not succeed in preventing the rise of vectors and the consequent decline of quaternions.

A trickle of work, however, revealed that quaternions were not quite dead and might yet find a place in life. In 1912 two papers appeared that would have warmed Hamilton's heart, since time actually entered quaternions in a presentation of special relativity (Silberstein 1912, Conway 1912). In 1923, J. D. Bernal, when still an undergraduate, wrote a brilliant Prize Essay for Emmanuel College, Cambridge, in which he applied quaternions to the structure of crystal lattices (Bernal 1923). Unfortunately, it was not published until 1981. A couple of papers appeared on the optics of mirrors (Tuckerman 1947, Wagner 1951), while Coxeter (1946) and Gormley (1947) dealt with geometry. Whittaker (1940) expressed a hope that quaternions might still become 'the most natural expression of the new physics' (of which more later), while Ehlers, Rindler, and Robinson (1966) and then Synge (1972) moved some way towards that goal. Hüller and Kroll (1975) applied quaternions to molecular rotations.

The Rodrigues programme

We have seen that Hamilton constructed his quaternions as an algebra, whence the elements of the algebra were given a dual rôle (as very clearly explained by Clifford 1878) as operators (rotations) and operands (vectors). It is of course in this respect that quaternions form a vastly more satisfying structure than Gibbs's vectors, which do not form an algebra. In order to identify the quaternion *qua* operator, Hamilton wrote the scalar part as $\mu \cos \rho$ (μ being a stretching factor, that he called the *tensor* part of the quaternion – nothing to do with modern tensors), ρ being identified with a rotation angle. When ρ equals $\frac{1}{2}\pi$ – as must be the case for a quaternion unit for which the scalar part vanishes – then the square of the operator changes the sign of the operand, providing a picture for the equation $i^2 = -1$.

It is obvious here that we are trying to do many things all at the same time; and that because rotations then become subservient to the algebra, their treatment is not satisfactory, as we have already seen.

Historically, however, a treatment of rotations and quaternions had been going on for many years before 1843, quite independently of Hamilton and taking a diametrically opposed view to his. This treatment was entirely geometrical, and because it tried to do a simple job in a simple way it was clear and it was precise, and it was entirely successful; but it was largely ignored by everyone.

Let us consider rotations of a unit sphere with fixed centre. The first problem we have to solve is to prove that if we apply one rotation after another the net result is a rotation of the sphere around some axis by some angle. (This, of course, is crucial in order to establish the group structure of the set of all possible rotations of the sphere.) Euler (1775*a*) solved a rather more general problem than this, since he considered the composition of two successive affine transformations (translation-rotations) and showed that the orientation of the final axes depends on six angular parameters, of which three can be eliminated algebraically, leaving three parameters only, thus determining a rotation. It must be made clear that Euler's approach is algebraic, not geometrical, and that it is not constructive. That is, he does not provide closed expressions to determine the angle and axis of the resultant rotation. Euler, however, is most often credited for the solution of the existential, geometric, and constructive problems regarding the composition of two rotations.

The problem just described is deeply related to a slightly more general one: to prove that the most general motion of a sphere with fixed centre is a rotation. (Clearly, all that is required, if the above result is used, is to prove that the most general motion of such a sphere can always be expressed as the composition or product of two rotations. See §**2**-1.) This result is often, appropriately, referred to as *Euler's theorem*. It is, in fact, a theorem which Euler (1775*a*, p. 202, and Fig. 2, Tab. II), explicitly enunciates as such, and which he then proves geometrically. His proof is a direct one in the sense that the product of two rotations is not explicitly employed.

It was the paper by Rodrigues (1840) that solved all three aspects of the first-mentioned Euler problem of the product of two rotations. In §8 he describes most clearly, without a figure, a geometrical construction that, given the angles and axes of two rotations, determines the orientation of the resultant axis of rotation and the geometrical value of its angle of rotation. This geometrical construction (of which more later) is called by Hilton (1903, p. 24) 'Euler's or Rodrigues's construction', and it has been called Euler's construction ever since (Buerger 1956, p. 36, McKie and McKie 1974, p. 40). This construction is fully described in §**9**-2 where, with deep apologies to Rodrigues, the putative attribution to Euler is still used, for the benefit of clarity. (It could admittedly be argued that the geometrical construction used by Euler in the proof of his above-mentioned Theorem contains embryonically the solution of the problem of the composition of two rotations.)

The 'Euler' construction, despite holding the geometrical key to the rotation group, has been assiduously ignored in treatments of the rotation group and angular momentum. I have not been able to find a single instance of its use or description in this context, not even in the recent monumental works of Biedenharn and Louck (1981, but see below) and Normand (1980). That it was ignored by the quaternionists, goes without saying. In 1848, eight years after Rodrigues's paper, we find G. G.

Stokes, (not a fool he: the famous Cambridge Professor of Mathematics), complaining that no geometrical method of composition of two rotations exists. Sylvester (1850) rises to the challenge and produces a pretty picture of the Rodrigues construction – the first ever published – with a very clear explanation and not a word of reference to Rodrigues, or Euler for that matter. Hamilton (1853, p. 328) rediscovered geometrically the results of the Rodrigues construction, but a very much related result of his appears to be better known (1853, p. 330). This result asserts that, given a spherical triangle, three successive rotations around its edges by double the corresponding dihedral angles equate the identity. This is the result actually quoted by Biedenharn and Louck (1981, p. 193) and by Whittaker (1937), although the original construction of Rodrigues provides a more direct solution to the important problem of determining the product of two rotations.

Having solved the existential and geometric problems as regards the product of two rotations, Rodrigues provides, in §19 of his paper, closed formulae for determining the resultant angle and axis of rotation. In order to do this, he parametrizes a rotation with four parameters. If ϕ is the angle of rotation and n_x, n_y, n_z the components of a unit vector denoting the axis of rotation, his parameters are:

$$\cos\tfrac{1}{2}\phi, \qquad \sin\tfrac{1}{2}\phi\ n_x, \qquad \sin\tfrac{1}{2}\phi\ n_y, \qquad \sin\tfrac{1}{2}\phi\ n_z. \tag{4}$$

If they are replaced into the quaternion (3), taking the place of the quaternion components a, b, c, d, then the formula for the multiplication of rotations that Rodrigues provides is precisely Hamilton's multiplication rule for quaternions. Felix Klein (1884, trans. p. 38) credits Rodrigues with the discovery of the rotation parameters and acknowledges that they were unknown to Euler: these parameters are universally called the Euler–Rodrigues parameters, a terminology that regretfully, for clarity, I shall keep in this book.

It is worthwhile discussing briefly how Euler came to be credited for these parameters. Klein actually asserted that Euler (1775*b*, p. 217) has used the *symmetric parameters* obtained by dividing the last three parameters in (4) by the first one. Euler, in fact, never used these symmetric parameters explicitly, although, after some algebra, his formulae can be made to yield them. Half angles, in particular, the appearance of which is essential in order to obtain the correct quaternionic multiplication rules for rotations, are not used by Euler. Klein's statement is so constructed, however, that it appears to imply that even the parameters (4) themselves, which he later called the *quaternion parameters* (Klein 1897, p. 5), were known to Euler. Klein and Sommerfeld (1897, p. 56), who discuss very clearly the relation between quaternions and rotations (pp. 55–68), as well as the conical transformation (vol. IV, pp. 939–44), give the parameters (4) without any attribution. Their formulae are quoted by Klein's collaborators, Schoenflies and Grübler (1902, p. 204), with an ambiguous attribution to Euler: 'These formulae are quite often treated in connection with Euler. We owe a first essential advance to O. Rodrigues . . . '.

We have already seen that Cayley (1845) recovered the results of Rodrigues's formulae, but rather than starting from a crystal-clear geometrical construction, he proceeded by way of the somewhat mysterious $q\,(\;)\,q^{-1}$ operator, and in so doing destroyed the splendid symmetry of Rodrigues' formulae in such a way that the parameters (4) became obscured.

It is most significant that, as a difference with Hamilton's approach to rotations, the 'Euler' construction (see §**9**-2) introduces most naturally the rotation angles via their halves, as is reflected in the structure of the Rodrigues quaternion ('Euler–Rodrigues' parameters). This did not worry Rodrigues in the least – he did not carry, like Hamilton, all the world's problems on his shoulders. We can understand, however, why such a parametrization was unacceptable to the quaternionists. We have seen that the scalar part of a quaternion unit must vanish, which in the Rodrigues quaternion (4) requires that ϕ be equal to π. (Then, on taking $n_y = n_z = 0$, say, we recover Hamilton's i, and so on.) This means that one has to identify the quaternion units with *binary rotations* (rotations by π), which is absurd, since everyone knows that the square of a binary rotation is a rotation by 2π, which is the identity and not -1, as i^2 is. Nature and history were playing games here with the quaternionists. How were they to know that the 'absurd' would become common place, that Cartan (1913) was going to discover objects (spinors) that are indeed multiplied by -1 under a rotation by 2π? (See §§**5**-3 and **6**-7.) Moreover, when the topology of the rotation group became understood through the work of Hermann Weyl in the 1920s (see Chapter **10**), it became natural to accept that the square of a binary rotation multiplies the identity by -1 and thus behaves like the quaternion units. Needless to say, it is this entirely geometrical and topological approach that we shall follow in this book. It will lead us to quaternions in a most natural way.

It must be stressed that the Rodrigues approach to rotations, by emphasizing their multiplication rules and by regarding them entirely as operators, fully reveals the group properties of the set of all proper rotations, the full orthogonal group SO(3), as it is now called. Rodrigues's analysis is astonishingly penetrating. He recognizes that rotations do not commute and then he goes on to define infinitesimal rotations, which commute and from which, he shows, all other rotations can be generated. He thus anticipates the idea of the infinitesimal group generators later fully developed by Sophus Lie. The whole theory of the rotation group is in fact contained in some form in the Rodrigues paper, as noticed by Gray (1980). The question of the infinitesimal group generators is of much importance to us, because we shall be able to prove (§**6**-4) that, under certain conditions, the infinitesimal group generators in the rotation group can be taken to coincide – surprisingly – with three orthogonal binary rotations. In the Hamilton parametrization, however, binary rotations are read as 'quadrantal', that is rotations by $\frac{1}{2}\pi$, and this shows that this parametrization affects the very essence of the rotation group. One of the major features of this book will be, in fact, the emphasis placed on the rôle of binary rotations, not only in realizing the quaternion units but also in determining the structure of the rotation group.

The 'Euler–Rodrigues' rotation parameters of eqn (4) have not in practice been much used, having been largely replaced by a set of three independent angular parameters, the Euler angles (§**3**-1). We shall see however, that the Euler–Rodrigues parameters (4) offer extraordinary advantages in working with the rotation group. In particular, we can use the whole apparatus of the quaternion algebra with them, which makes the work much easier than in any other parametrization. Substantial advantages of precision also appear when dealing with the spinor representations.

Before we leave this subject, and Rodrigues (1840), it should be said that his paper goes further than a mere treatment of the rotation group. For Rodrigues is mainly concerned with the Euclidean group, that is the group of all translations and rotations. He shows that translations commute amongst themselves but not with rotations and has the remarkably ingenious idea of realizing translations as infinitesimal rotations around infinitely distant rotation axes. As Gray (1980) points out, he establishes the structure of the Euclidean group as what we would now call a semidirect product of the translation group with the rotation group.

Rotations by 2π

Rotations by 2π have already raised their head in this discussion and they require a little analysis, since they involve points of doctrine. There are two schools of thought as regards the description of 2π rotations which, purely for convenience, I shall call the geometrical and the physical schools. I shall first describe the geometrical approach. In it, a symmetry operation is first and foremost a coordinate transformation that changes the position vector $\mathbf{r}$ of a point of the space into another position vector $\mathbf{r}'$. In order to specify a rotation (see §**2**-1) you need a unit sphere; you have to mark two points $\mathbf{r}_1$ and $\mathbf{r}_2$ on it, perform some transformation of the sphere (leaving its centre fixed) and then determine the new position vectors $\mathbf{r}_1'$ and $\mathbf{r}_2'$ of the two points marked. In order to determine uniquely and precisely the rotation effected, all that you need are the initial and final coordinates, $\mathbf{r}_i$, $\mathbf{r}'_i$ $(i = 1, 2)$: whatever has happened in between is entirely irrelevant. This is what a coordinate transformation means and if you do accept this, then you must also accept that it is impossible to rotate a sphere by an angle equal to or larger than 2π. Try this experiment: measure the $\mathbf{r}_i$'s, then get out of the room while somebody spins the sphere in any manner whatsoever until it stops. You then return to the room and find that the sphere has rotated, i.e. that the $\mathbf{r}_i$ have changed into $\mathbf{r}'_i$: it is just as likely that you will find a host of angels standing on the tip of a needle as it is that you will find any of the angular coordinates of $\mathbf{r}_i$ displaced by 2π or more. Notice that the crux of the matter is the difference between 'to spin' and 'to rotate'. The latter is a coordinate transformation and thus it has no history and no time, whereas to spin an object is a process realized in time and it thus has a history.

All this is very well, but we know that when we rotate an electron by 2π its wave function is multiplied by -1. In the geometrical approach, the following solution is given to this conundrum. All that you can expect of a symmetry operation is that it conserves observables. Now, the sign of the wave function (or more generally a *phase factor*) for an *isolated* system is not an observable, because the energy in quantum mechanics is a quadratic expression on the wave function. Thus, in the geometrical approach, one deals purely with geometrical symmetry transformations, so that the identity and a rotation by 2π are one and the same thing, but one has then to study separately any phase factors that might appear as a consequence of them. The study of the continuity of the transformations, that is topology, provides a way to fix these phase factors (Chapter **10**). This approach is the one formulated by Weyl (1925) and further discussed in his famous book on group theory, Weyl (1931).

In the physical approach, rotations by 2π are regarded as physically distinct from the identity and any phase factors that appear are taken to be an inherent part of the transformation. Needless to say, the physical approach is very attractive to physicists, and it received considerable impetus in the last twenty years or so in two ways. First, Dirac invented a remarkable machine (see Misner, Thorne, and Wheeler 1970, p. 1148, Biedenharn and Louck 1981, p. 10). I shall not describe the construction and use of this machine in detail: suffice it to say that it consists of a cube, each face of a different colour, joined to the corners of a cubic frame by eight pieces of elastic string that are fixed at each corner of the cube. It is found experimentally that the cube can appear in two *versions*. In version 0, say, nothing has been done, all the strings are nice and straight. In version 1, you have rotated the cube by 2π, the strings are in a mess, and you find that it is impossible, without rotating the cube, to disentangle them again. When the cube is in version 1, if you rotate it by 2π again, the strings still appear entangled but with patience and ingenuity you can now return them to the straight condition (version 0) without rotating the cube. The machine, thus, effects a distinction between a rotation by 0 or 4π and a rotation by 2π. (Within the geometrical school, it will be argued of course that these are rotations of the cube with respect to the frame, and not rotations of an isolated system, the version corresponding to the phase of one subsystem with respect to the other.)

The second piece of support for the physical school was provided by two almost simultaneous and extremely ingenious papers: Aharonov and Susskind (1967) and Bernstein (1967). The first of these, in particular, had a title bound to catch the headlines: 'Observability of the Sign Change of Spinors under 2π Rotations' and I am bound to say that one or two people got a slightly exaggerated idea of what this meant. It meant, in fact, that a change of sign could be observed when you had two *coherent* beams of neutrons and when one of them was rotated *relative to the other* by 2π. Two sets of experiments, appropriately enough in the same year, actually observed this effect: Werner *et al.* (1975) and Rauch *et al.* (1975). Bacry (1977) helped to dispel some misconceptions as to (non-existent) differences between classical and quantum mechanics in the question of rotation of an *isolated* system. Only phase *differences* are measurable. Very useful and clear discussions appear in Anandan (1980), and Page and Wooters (1983), and a review of the whole subject can be found in Biedenharn and Louck (1981).

Of course, the day might come when someone might observe the rotation by 2π of an isolated system. Until such time, the geometrical and the physical approaches are equivalent and they will both be used in this book, although, in order to avoid confusion, the geometrical approach alone will be used in the first twelve chapters. Also, in order to reserve the use of the word '*rotation*' to signify a linear coordinate transformation and nothing else, rotations by 2π understood in the physical sense discussed above will be described as *turns by* 2π. (See §**10**-6.)

Spinor representations

Although I have spent so much time so far on the quaternion story, I have not yet come to one of the major topics of the book, which is the study of the *spinor representations* of the rotation group, i.e. the representations that correspond to half-integral values of the angular momentum j. It is well known that the matrix representative of a given rotation is, in such cases, undetermined as to sign. In order to deal with such representations for a point group G of operations g_i, $i = 1, 2, \ldots, n$, Bethe (1929) used the expedient of artificially introducing a distinction between the null rotation E and the rotation by 2π, $\tilde{E}$, thus defining a new group of $2n$ *operations*, g_i, $i = 1, \ldots, n$, $\tilde{g}_i \equiv g_i\tilde{E}$, $i = 1, \ldots, n$, appropriately called the double group $\tilde{G}$. The advantage of this heuristic procedure (which is the first historic example of the 'physical' approach to 2π rotations described above) is that the ordinary representations of $\tilde{G}$ give precisely the desired spinor representations without the use of any new theoretical ideas.

This heuristic procedure invites the well-known but nevertheless awkward question: why not triple, or for that matter n-fold, groups? This is so because, having introduced the 'rotation by 2π' as a new operation $\tilde{E}$, one is tempted to introduce yet another operation $\tilde{\tilde{E}}$, say, by means of a 'rotation by 4π'. There are two ways in which this conundrum can be solved. One is to use the Dirac machine in order to identify the 'rotation by 4π' with the identity E. The other is to use a topological argument (see §**10**-6) which shows that turns by 2π and by 4π are respectively distinct and indistinguishable from the identity. Thus, $\tilde{\tilde{E}}$ must be the ordinary identity and a turn by 6π must be the same as $\tilde{E}$, and so on.

Bethe's idea was a brilliant one in order to solve the limited range of problems that were handled in the 1930s, but it is my view that, in its original form, it has outlived its time. Its more important drawback is the way in which the method has to be used in practice, which I shall now explain.

The crux of the matter is that although geometry clearly determines the multiplication table of G, the operations of $\tilde{G}$ cannot be multiplied geometrically and no prescriptions were available for doing so. Now, a group without a multiplication table is like a house without foundations and without a roof. This fundamental difficulty was solved by a device due to Opechowski (1940) which is used in practice as follows (a full example is given in Chapter **15**). The rotations g of G are determined by some parameters called the Euler angles (§**3**-1) which in their turn determine two matrices of opposite signs for each g, for a given j (§**4**-7). Normally, these matrices will first be chosen for the lowest possible $j=\frac{1}{2}$. The $2n$ matrices are now assigned to the operations of $\tilde{G}$, the choice of the positive matrix for g and its negative for $\tilde{g}$ being entirely arbitrary (but sensible: one does not want, e.g., the identity to be represented by a negative unit matrix). Once this choice is made, the matrices (two-dimensional for $j=\frac{1}{2}$) are multiplied and their multiplication table is defined as the multiplication table of the group $\tilde{G}$.

If one wants to build now, say, a representation for $j=\frac{7}{2}$ (dimension 8) again the standard expressions for each g in terms of its Euler angles will yield two eight-dimensional matrices. Again, they have to be distributed arbitrarily within each pair g and $\tilde{g}$, but now the multiplication table for the trial set thus obtained has to be worked out and the signs of the matrices adjusted until agreement with the multiplication table for $\tilde{G}$ is reached.

This sounds pretty awful (although with a bit of practice one soon learns the tricks) and I am not sure I have ever seen in print such a frank description of this brute-force approach. Authors (e.g. Altmann 1979, p. 522) tend to clothe this procedure in decent obscurity.

There are two ways in which this problem can be properly handled and in both ways quaternions in the Rodrigues parametrization play a fundamental part. These two ways, as we might expect, depend on what we previously called the geometrical and the physical approaches to rotations by 2π. They will appear to be fundamentally different and yet they are practically the same.

In the geometrical approach, the operations $\tilde{g}$ must be abandoned, since they are not coordinate transformations at all. It is clear that this is all to the good, because we are left with a single group G of order n rather than a double group of order $2n$.

The way in which this is done is basically the following. We are interested in a finite group G, but this is a subgroup of the full rotation group. This is a continuous group (i.e. given an operation g of it, there is another operation g' that differs from g by an infinitesimal quantity) and because it is continuous there are continuity (also called *topological*) conditions to be satisfied. If the product of two operations of G, $g_i g_j = g_k$, is performed while G is embedded in the full rotation group, it is found that the matrix corresponding to g_k appears with a *phase factor* which can uniquely be determined from the topology. This leads to a new type of group representation called a *projective* or *ray representation.* (Chapter **8**.) Let us explain a bit better what goes on. In an ordinary matrix representations, when we multiply the matrices of g_i and g_j, we get the matrix of g_k and nothing else. In the projective representation that appears in our case we get that matrix, but multiplied by the phase factor (± 1) that corresponds to the product $g_i g_j$. Each operation is thus represented by a unique matrix and the 'doubling up' of the matrices which we had before is replaced by a doubling up in the phase factors. This may sound a bit confusing, but the object of the exercise is to keep to (geometrically) well-defined symmetry operations and to (topologically) well-defined phase factors.

This idea is in fact quite old, having been formulated by Weyl (1925) for the full rotation group, although it was not implemented until much later. I myself taught the subject in this way for over ten years since the 1960s at the Charles Coulson Summer School in Theoretical Chemistry. Independently, Brown (1970) made a move in the same direction making a (largely unheeded) plea for the use of projective representations rather than double groups. A great deal of ground had to be covered, however, since certain technical problems with class theorems had to be solved. The difficulty is that although, as is well known, in ordinary representations operations in the same class always have the same characters, this is not necessarily so for the projective representations. Backhouse (1973) made a start on this problem but he used a fairly arbitrary system of phase factors, rather than the systematic way given by Weyl, and the results were cumbersome. The present author (Altmann 1979) gave a method, firmly based on the Weyl programme, that leads to a number of theorems which guarantee the same class structure in the projective as in the ordinary representations. Moreover, if all the rotations are parametrized by Rodrigues's quaternions, then all matrices and all phase factors are uniquely and precisely determined. No trial and error is required and thus the method is ideally suited to calculation by computer.

It will be seen from the above account that some of the properties of the full rotation group as a continuous group will be required in order to determine the phase factors. I shall, however, use an absolute minimum – first, because the theory will appear crisper and clearer in this way and, secondly, because splendid accounts already exist of the rotation group as a continuous group (Gel'fand, Minlos, and Shapiro 1963; Normand 1980).

The 'physical' approach to the spinor representations hardly needs description, since it coincides with the Bethe approach. The question that remains, however, is how to make the distinction between the operation g and the operation $\tilde{g}$, a distinction that is beyond the capability of the traditionally used Euler angles. The answer is so simple that it is astonishing that it was not used right from the beginning as the standard method. If we look at a rotation g parametrized by the Rodrigues quaternion (4), with ϕ from 0 to 2π, it is clear that $\tilde{g}$ is obtained by merely adding 2π to the angle ϕ, whereupon the whole quaternion changes sign. One thus defines a group of quaternions isomorphic to $\tilde{G}$ and, since the quaternion multiplication rules are uniquely defined, so are the multiplication rules of $\tilde{G}$. This is done in Chapter **13**, where a rigorous treatment is provided of the double-group structure. (It must be said that the results obtained by the geometrical method are most useful for this purpose.)

Some applications of the methods developed in this book will be found in Chapter **15**. The first application of the double-group method to space groups was made by Elliott (1954), and Hurley (1966) introduced the study of projective representations in space groups and double space groups. Boyle and Green (1978) dealt extensively with the properties of projective representations of point groups, a subject also further studied by Kim (1981*a*,*b*; 1983*a*,*b*,*c*; 1984*a*,*b*). The calculation of Clebsch–Gordan coefficients within the scheme of the present book is fully discussed by Altmann and Palacio (1979), Altmann and Herzig (1982), and Herzig (1984). One of the fields where very careful and precise definition of the double group is required is in dealing with Dirac Hamiltonians, the symmetry groups of which, for field-free systems, are double groups (see Jansen and Boon 1967). The treatment of such groups has recently been discussed in detail by Altmann and Dirl (1984).

One final reflection on quaternions before we close this introduction. Anyone who has ever used any other parametrization of the rotation group will, within hours of taking up the quaternion parametrization, lament his or her misspent youth: I have no doubt that quaternions will become a major tool in the rotation group specially when applied to point groups. What about Hamilton's hopes – echoes of which I have quoted above – for a deeper rôle for quaternions in physical science? Anyone who attempts to gaze into the crystal ball in order to guess an answer to this question must appreciate the following. Quaternions are inextricably linked to rotations. Rotations, however, are an accident of three-dimensional space. In spaces of any other dimensions, the fundamental operations are reflections (mirrors). The quaternion algebra is, in fact, merely a sub-algebra of the Clifford algebra of order three (see Clifford 1878, and §**12**-10). If the quaternion algebra might be labelled the algebra of rotations, then the Clifford algebra is the algebra of mirrors (see §**12**-10) and it is thus vastly more general than quaternion algebra. It is clear that the Clifford algebra has advanced very rapidly, after the pioneer work of Marcel Riesz (1946, 1958), the book by Hestenes (1966), and the papers by Teitler (1965, 1966). The two recent papers by Greider (1980, 1984) may well signal the coming-of-age of the Clifford algebras in physical applications.

2

ALL YOU NEED TO KNOW ABOUT SYMMETRIES, MATRICES, AND GROUPS

> The Theory of Groups is usually associated with the strictest logical treatment.... Various mathematical tools have been tried for digging down to the basis of physics, and at present this tool seems more powerful than any other.
>
> Sir Arthur Stanley Eddington (1935). *New Pathways in Science. Messenger Lectures* 1934. University Press, Cambridge, p. 257.

This chapter will be a quick revision of some basic material of which the reader will be expected to have some elementary knowledge, as provided e.g. in Chapters 1–3 of Tinkham (1964). Many readers will be able to skip a good deal of the material but a quick glance at some of it is advised in order to grasp the notation that we shall use in the rest of the book. It must be understood that I shall not try to present a condensed account of all the subjects mentioned in the chapter's title; my aim instead is to highlight the critical points.

1. SYMMETRY OPERATORS IN CONFIGURATION SPACE

The reader will be assumed to be familiar with the notion of a geometric symmetry operation and to be aware that they can be performed in either the *active* or the *passive* pictures. In this book, we shall exclusively use the active picture which we shall now describe, for which purpose we must define our *configuration space*, that is the space in which our physical systems, like atoms and molecules, are described. In order to define this space, and its transformations, we take three orthonormal right-handed axes **x**, **y**, **z**, *fixed in space* (never to be transformed). All points of the configuration space will be vectors with components x, y, z along the space-fixed axes, these being triples of real numbers, that is elements of $\mathbb{R}^3$.

We shall be primarily concerned in this book with *point transformations*. As a difference, e.g., with translations, these are transformations that leave one point of the configuration space invariant. This point will be taken to be both the centre of the space-fixed axes **x**, **y**, **z**, and of a *unit ball*, the transformations of which will be a good way to realize the transformations of the whole of the configuration space (of which the ball is just one part and to which it is firmly fixed). It is, in fact, sufficient to consider the surface of the unit ball, that is the *unit sphere.*

In the active picture all symmetry operations are referred to the **x**, **y**, **z** axes; that is, all symmetry planes, axes, etc., are given in fixed positions with respect to the space-fixed axes, and a symmetry operation is one in which all points of the unit sphere are transformed with respect to the given symmetry element. That is, a vector **r** of the configuration space will move with the whole of the configuration space to become a vector $\bar{\mathbf{r}}$ with respect to the space-fixed axes **x**, **y**, **z**. Clearly, a symmetry operation will entail a mapping of $\mathbb{R}^3$ onto itself in which the coordinates x, y, z of each vector **r** change into $\bar{x}$, $\bar{y}$, $\bar{z}$. (In the passive picture of a symmetry operation, this mapping is effected by leaving the configuration space fixed in space and, instead, moving the **x**, **y**, **z** axes.)

Although in the active picture the whole of the configuration space is transformed, i.e. all points of the unit sphere are moved in solidarity, it is often convenient to attach three right-handed orthonormal vectors **i**, **j**, **k** to the unit sphere. These will be transformed like any other vectors of the unit sphere, so that the components of a vector **r** of the configuration space with respect to **i**, **j**, and **k** (which, in practice, should not be used) never change. The advantage of introducing such vectors is that, if they are always taken to coincide with **x**, **y**, **z** when no transformation is made (identity operation E), then their later values $\bar{\mathbf{i}}$, $\bar{\mathbf{j}}$, $\bar{\mathbf{k}}$ after a symmetry operation g will fully specify this symmetry operation. It must be clearly understood that describing symmetry operations by the transformations of **i**, **j**, **k** has nothing to do with going into the passive picture. These axes are firmly attached to the configuration space, which they drag in any transformation, whereas the axes used in the passive picture do not drag the space.

A symmetry operation will change all vectors **r**, **i**, **j**, **k** in configuration space into $\bar{\mathbf{r}}$, $\bar{\mathbf{i}}$, $\bar{\mathbf{j}}$, $\bar{\mathbf{k}}$ respectively and in order to make the relation between an operation and these transforms more explicit, **r** will be written as g**r**, g being the *symmetry operator* associated with the given symmetry operation. Likewise, $\bar{\mathbf{i}}$, $\bar{\mathbf{j}}$, $\bar{\mathbf{k}}$ will be written as g**i**, g**j**, g**k**.

Description of the point-symmetry operations

All point-symmetry operations are fully determined, under certain conditions discussed below, when the transforms of two points $\mathbf{r}_i$ $(i=1,2)$ on the surface of the unit sphere are given. This is illustrated in Fig. 1 for all the symmetry operations that can cover a regular polyhedron, namely: identity, inversion, rotations, reflections, and rotoreflections (a rotation followed by a reflection on a plane normal to the rotation axis). The circle shown in the figure is the horizontal (**x**, **y**) plane of the sphere and the **z**

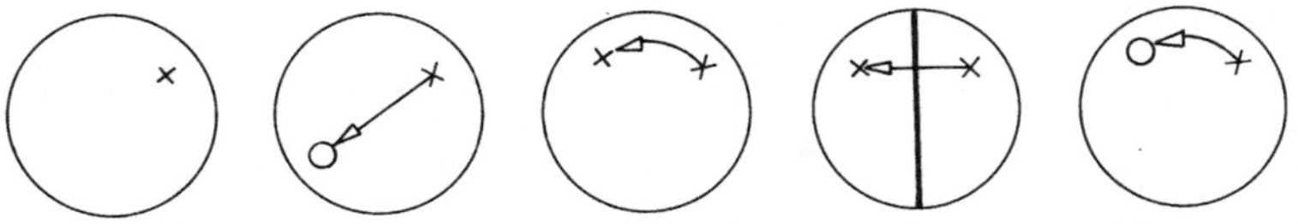

Identity Inversion Rotation Reflection Rotoreflection

$E;e$ $\quad$ $\boldsymbol{i}$ $\quad$ $R\left(\frac{\pi}{2}\mathbf{z}\right) = C_4$ $\quad$ σ $\quad$ S_4

Fig. 2-1.1. Symmetry operations. The last four drawings give the transform of the vector left invariant by the identity in the first drawing. × point on the upper hemisphere; ○ point on the lower hemisphere. The z axis is normal to the drawing.

axis is assumed to be normal to and above the drawing. Points marked with a cross or a circle denote points on the surface of the upper or lower hemisphere respectively. *Rotations* are usually denoted as $R(\phi\mathbf{n})$, $\mathbf{n}$ being the axis of rotation and ϕ the angle of rotation, but in the theory of the symmetry of regular polyhedra (or molecules with such shapes), $R(2\pi m^{-1}\mathbf{n})$ is written as C_m. Thus, a rotation by π, $R(\pi\mathbf{n})$, which is called a *binary rotation*, is denoted with the symbol C_2. The rotation axis $\mathbf{n}$ is in this case left implicit, but it can be added as a further subscript if necessary. The *inversion* will be denoted with the symbol $\boldsymbol{i}$ to make it quite clearly different from the imaginary unit i. *Reflections* will always be denoted as σ with the plane of reflection indicated by a subscript, if necessary. *Rotoreflections* are products of a rotation $R(2\pi m^{-1}\mathbf{n})$ times a reflection σ on a plane normal to $\mathbf{n}$, and they will be denoted with the symbol S_m. Thus S_m is $\sigma R(2\pi m^{-1}\mathbf{n})$ or $\sigma\, C_m$, with $\sigma \perp \mathbf{n}$.

Binary rotations and their corresponding axes, or *binary axes*, will be found to be most important in the study of the rotation group. One particularly significant property which they have is this: if a rotation C_m has a binary axis perpendicular to it, then the two semiaxes of C_m are interchanged by the binary rotation. Rotations C_m thus related to a binary rotation are called for this reason *bilateral rotations*. If the rotation C_m here is itself binary, C_2, then it is called a *bilateral binary rotation*. It is clear in this case that the second C_2 is also bilateral binary; that is, that bilateral binary rotations must always appear as pairs of mutually perpendicular binary axes.

The operations defined above are usually classified in *proper* and *improper* rotations, proper rotations being ordinary rotations and improper rotations being all others. (From Problem 5, they can always be considered as *rotoinversions*; that is, as products of proper rotations times the inversion.)

Specification of the symmetry operations

We must now briefly discuss which is the minimum number of points on the unit sphere, and what conditions they must satisfy, in order uniquely to define a symmetry operation. First, we show that any rotation can uniquely be determined by the transform of two points. Imagine a fixed unit sphere with two points A, B marked on it, covered by a thin Perspex sphere (a shell) that can freely move on the first one, and on which two points A', B' are marked. Imagine also that, at some initial position, A' and B' cover A and B respectively and that a mapping $\mathbf{r} \mapsto \mathbf{r}'$ from each point $\mathbf{r}$ of the fixed sphere onto each point $\mathbf{r}'$ of the moving one is subsequently established by some displacement of the outer sphere. It is intuitively clear that however much the moving sphere is moved, whenever A' and B' cover A and B respectively, then every point $\mathbf{r}'$ maps exactly onto the same point $\mathbf{r}$ as it did before. Thus, the position of the moving sphere is fully determined by the position of two of its points. We must next agree that each arbitrary displacement of the moving sphere is a rotation of the sphere about some axis by some angle (this is Euler's theorem, to be proved below). Thus, every rotation of the sphere around its centre is fully determined by the positions of two of its points, $A'B'$.

Euler's theorem (Euler 1775*a*), which states that every displacement of a sphere with fixed centre is the same as a rotation of the sphere about a diameter (axis of rotation) by some angle is a very beautiful result that we shall prove in two steps. In the first, we use the two spheres described above to show that the points $A'B'$, in any arbitrary position of the moving sphere, can always be returned to register with AB by means of *two* successive rotations. In fact: If O is the centre of the sphere, some rotation around an axis normal to the plane $A'OA$ must return A' to A. Once this is done, a second rotation around OA (coincident with OA') must return B' to B. In the second step we must prove, with Euler, that the product of two arbitrary rotations is still a rotation. This will be done in Chapter **9** by the Rodrigues method (Euler's construction), but it is easy to accept this result at this stage, thus completing the theorem.

Although, as shown, every rotation of the unit sphere is fully determined by the position of any two of its points, some restrictions must be applied on these points when other symmetries are admitted. If the two points are on a great circle normal to a binary axis, this rotation will transform them identically as the inversion at the origin. (See Problem 4.) Likewise, if the two points belong to a symmetry plane, they transform identically as under the identity. Thus, any two points on the surface of the unit sphere that are not on a symmetry plane or on a plane normal to a binary axis are sufficient to denote uniquely all symmetry operations. Although this is so, one often gets a useful picture of what the symmetry operation does by transforming only one point, as we have done in Fig. 1.

On the other hand, as we have said, one often specifies symmetry operations by their effect on the axes **i**, **j**, **k**, which is a somewhat redundant but extremely safe method.

Composition of symmetry operations

If the symmetry operation g_i is performed after the symmetry operation g_j the result is another symmetry operation g_k, say:

$$g_i g_j = g_k. \tag{1}$$

Notice two things: first, that the product on (L1) is read from *right to left*. The other is that $g_i g_j$ and g_k denote identically the same thing. In other words: we cannot think of $g_i g_j$ as being g_k 'performed in a different way', say, because the way in which a transformation is performed is entirely irrelevant. (See Chapter **1**.)

Remember also that $g_i g_j$ is not generally the same as $g_j g_i$. That is, symmetry operations do not necessarily commute, although we shall accept (i) that the identity commutes with everything and (ii) that the *associative property* holds, that is that, if

$$g_h g_i = g_l, \qquad g_i g_j = g_m, \tag{2}$$

then

$$g_l g_j = g_h g_i g_j = g_h g_m. \tag{3}$$

More compactly, in a fairly obvious notation, this is written as

$$(g_h g_i) g_j = g_h (g_i g_j). \tag{4}$$

The *inverse* g^{-1} of a symmetry operation g is defined by the condition $gg^{-1} = E$.

Problems 2-1 (p. 266)

1. Prove that, given that the identity is unique, then g and g^{-1} always commute.
2. Prove that

$$(g_i g_j)^{-1} = g_j^{-1} g_i^{-1}. \tag{5}$$

3. Show by case analysis that the inversion commutes with all symmetry operations. Would this be true if translations were admitted?
4. Show that in two dimensions (a plane alone lacking the third dimension) the inversion at a point and a binary rotation around it are identical operations.
5. Show that a reflection σ is equal to a rotation by π (binary rotation, C_2) around an axis perpendicular to σ followed (or preceded) by the inversion. Thus show that all improper symmetry operations are

given as the product of some rotation with the inversion (*rotoinversion*).

6. Show that the two C_3 rotations around the axes (111) and (11$\bar{1}$) of a cube do not commute. (The numbers here are the vector components with respect to the space-fixed axes **x**, **y**, **z**, parallel to the cube edges.)
7. Show that two reflections commute if and only if their reflection planes are perpendicular.
8. Show that (i) all rotations of the form $R(\phi\mathbf{n})$, $R(\phi'\mathbf{n})$ commute. (Coaxial rotations.) (ii) All rotations $R(\pi\mathbf{m})$ and $R(\pi\mathbf{n})$, $\mathbf{m}\perp\mathbf{n}$ commute. (Either of these two binary rotations swaps the two semiaxes of the other: hence they are bilateral binary rotations; see the section above on the description of operations.)
9. Show that, for orthogonal axes **n**, **p**,

$$R(\pi\mathbf{n})R(\beta\mathbf{p})=R(-\beta\mathbf{p})R(\pi\mathbf{n}). \tag{6}$$

10. Show that if σ_1 and σ_2 are two reflection planes with intersection **n** and at an angle $\frac{1}{2}\phi$, then

$$\sigma_2\sigma_1=R(\phi\mathbf{n}), \tag{7}$$

where ϕ is in the sense $1\rightarrow 2$. (See Fig. 2.)

11. Show that if σ_x and σ_y are reflection plans perpendicular to the orthogonal axes **x** and **y** respectively, then

$$\sigma_y\sigma_x=C_{2z}, \tag{8}$$

where C_{2z} is a binary rotation around the **z** axis.

12. Show that the product of two reflections on parallel planes separated by a vector $\frac{1}{2}\mathbf{t}$ equals a translation by **t**.

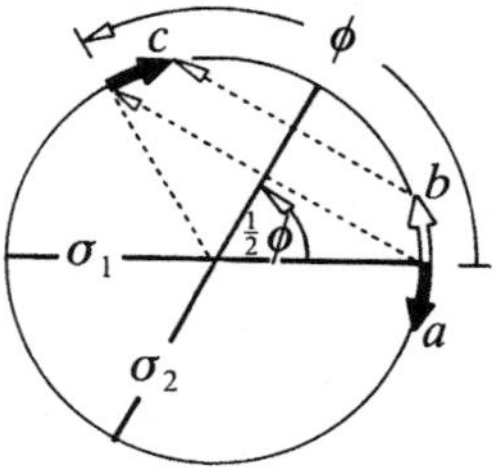

Fig. 2-1.2. The product of two reflections, $\sigma_2\sigma_1$, is a rotation. The first reflection, σ_1, transforms a into b. The second reflection, σ_2, transforms b into c. Thus $\sigma_2\sigma_1$ transforms a into c and it is a rotation by ϕ. The axis **n** is perpendicular to the drawing.

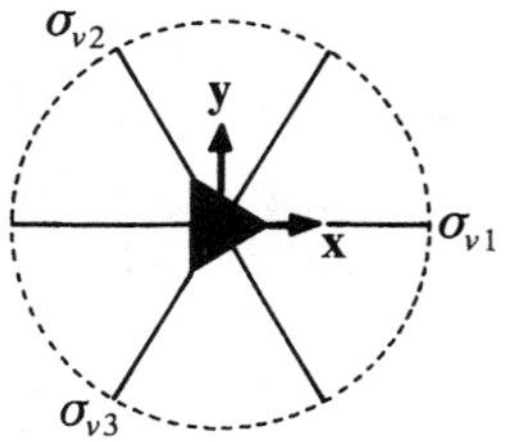

Fig. 2-1.3. The group $\mathbf{C}_{3v}$. The **z** axis is normal to and above the plane **x**, **y**.

13. The set of operations of a triangular pyramid (Fig. 3) is:

$$\mathbf{C}_{3v} : E, C_3^+, C_3^-, \sigma_{v1}, \sigma_{v2}, \sigma_{v3}. \tag{9}$$

(this set is called the group $\mathbf{C}_{3v}$). Form the 36 products of all pairs of operations of this set.

14. Show that the product of two binary rotations is a binary rotation if and only if their axes are perpendicular. (Notice, therefore, that bilateral binary rotations must always form *triples* along orthogonal axes).

2. EIGENVECTORS OF CONFIGURATION SPACE OPERATORS

The geometric verification of the commutation property of symmetry operators discussed in Problems 1.7 and 1.8 is both tedious and unconvincing and it is much better done by considering the *eigenvectors* of the symmetry operators g. We shall illustrate this concept before we define it. A proper rotation $R(\phi\mathbf{n})$ transforms all vectors $\mathbf{r}$ of the configuration space into different vectors $\mathbf{r}'$, except when $\mathbf{r}$ coincides with $\mathbf{n}$, in which case $\mathbf{r}$ is left invariant (multiplied by $+1$). A reflection σ will transform a vector $\mathbf{r}$ into a different vector $\mathbf{r}'$, except when $\mathbf{r}$ is on the plane σ, in which case it is invariant. In the case of reflections, though, a vector $\mathbf{r}$ normal to σ changes into $-\mathbf{r}$, that is, it is multiplied by -1. It will be convenient to regard such vectors also as invariant with respect to the symmetry operation, invariance thus being taken to refer to the behaviour of the line of application of the vector. Vectors which are invariant in this sense with respect to a symmetry operation are called its *eigenvectors*. The factor ± 1 which can affect such eigenvectors on transformation is called the *eigenvalue* of the symmetry operation in question. In general, then, all symmetry operators g must leave invariant some vectors $\mathbf{r}_s$ of the configuration space, in the sense that $g\mathbf{r}_s$ is the same as $\mathbf{r}_s$ multiplied by a scalar a (*eigenvalue*) which can be only ± 1:

$$g\mathbf{r}_s = a\mathbf{r}_s, \qquad a = \pm 1, \tag{1}$$

Such vectors are the *eigenvectors* of g.

If, for a given a, s in (1) can take more than one value, the $\mathbf{r}_s$ taken as eigenvectors are always linearly independent (in usual practice orthonormal). For the identity and the inversion s takes the values 1, 2, 3, and a, $+1$ or -1 respectively. For rotations, s can take only one value (corresponding to $\mathbf{r}_1$ along the axis of rotation) and a equals $+1$. For reflections σ we have either $s = 1$, $a = -1$ (vector $\mathbf{r}_1$ normal to σ) or $s = 1, 2$, $a = 1$ ($\mathbf{r}_1$ and $\mathbf{r}_2$ spanning σ).

The fundamental result as regards commutation can now be enunciated: *Two symmetry operations g_1 and g_2 commute if and only if eigenvectors can be chosen to be common to both operations.* The way in which this choice of eigenvectors can be done is illustrated in Problem 1 below. The way in which the theorem is applied to the study of commutation is shown in Problems 2 to 5. The proof of the theorem is left to §6 after eqn (55).

Problems 2-2 (p. 267)

1. Given two perpendicular reflection planes σ_1 and σ_2 show how three eigenvectors $\mathbf{r}_1$, $\mathbf{r}_2$, and $\mathbf{r}_3$ can be chosen common to both reflections.
2. Prove that the inversion commutes with all symmetry operations.
3. Prove that the only rotation pairs that commute are either coaxial or bilateral binary (see Problem 1.8).
4. Prove that given a rotation $R(\phi\mathbf{n})$ and a reflection σ, they commute if and only if $\sigma \perp \mathbf{n}$ and $\phi = \pi$. (Binary rotation normal to reflection plane.)
5. Prove that two reflections σ_1 and σ_2 commute if and only if they are perpendicular.

3. SYMMETRY OPERATORS IN FUNCTION SPACE

> *Warning.* Equation (1) below requires revision for quantal symmetries. (See §7.)

When an operator g acts on the configuration space changing $\mathbf{r}$ into $g\mathbf{r}$, functional relations implied by a symbol such as $f(\mathbf{r})$ change into new functions that shall be denoted with $\dot{g}f(\mathbf{r})$. (In this expression $\dot{g}f$ must be read as a single symbol that denotes the new function.) That this must be so is illustrated in Fig. 1 for g a translation by t. In the *active* picture which we use, as x is changed into gx (which equals $x + t$), the value of the function originally at x, $f(x)$, is transferred, with the whole of the configuration space, to gx and it is labelled as $\dot{g}f(gx)$. The fundamental

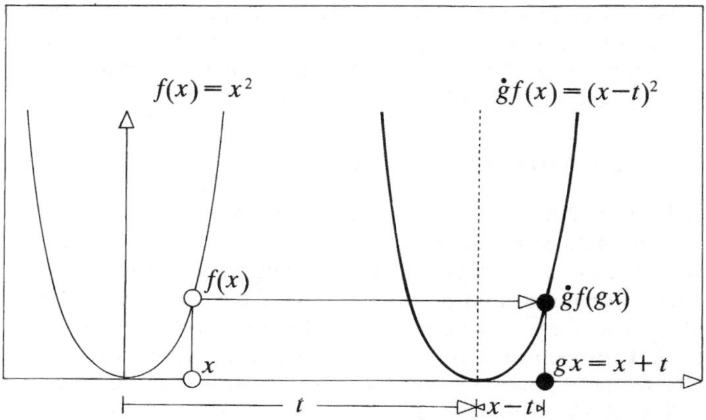

Fig. 2-3.1. Function-space operator.

requirement which we impose is that both values $f(x)$ (old function at old point) and $\dot{g}f(gx)$ (new function at the new point) be equal. The equality must be valid for all x, and in particular for its value $g^{-1}x$, whence we get the fundamental formula for the function space operator,

$$\dot{g}f(x) = f(g^{-1}x). \tag{1}$$

It can readily be seen that this relation gives the correct transformation for the function $f(x) = x^2$ displayed in the figure, as it is shown there. The product $\dot{g}_i\dot{g}_j$ must be understood, when acting on f, as $\dot{g}_i$ acting on the function $\dot{g}_j f$ (i.e. not as $\dot{g}_i$ acting on both sides of eqn 1 written for $\dot{g}_j$: this is not correct because eqn 1 is not a relation between functions, on which the operators $\dot{g}$ operate, but a relation between the *values* of functions at given points: in a functional relation, such as the equality of $\sin 2x$ to $2 \sin x \cos x$, the variables in all the functions involved must be the same throughout. See Wigner 1959, p. 106.) Thus,

,1

$$\dot{g}_i\dot{g}_j f(x) = \dot{g}_i\{(\dot{g}_j f)(x)\} = \dot{g}_j f(g_i^{-1}x) = f(g_j^{-1}g_i^{-1}x) \tag{2}$$

1.5 | $R2$

$$= f\{(g_i g_j)^{-1}x\}. \tag{3}$$

Equation (3) agrees with (1) applied for the operator $(g_i g_j)^{\dot{}}$ so that we obtain

$$(g_i g_j)^{\dot{}} = \dot{g}_i\dot{g}_j. \tag{4}$$

Problem 2-3 (p. 267)

1. Consider the well known atomic orbitals

$$p_x = \sin\theta\ \cos\varphi, \qquad p_y = \sin\theta\ \sin\varphi, \tag{5}$$

represented graphically as in, e.g. McWeeny (1979), p. 31. Transform them under a rotation g equal to $R(\frac{1}{2}\pi\ \mathbf{z})$. Compare the result with the active rotation of your graph. Would the agreement be maintained if you had used g instead of g^{-1} on (R1)?

4. MATRICES AND OPERATORS

The reader is certainly familiar with the use of matrices: the object of this section is to emphasize the need for extreme care with the notation. We show in Fig. 1 an active rotation of the configuration space by α around the **z** axis, as demonstrated by the rotation of the unit ball with axes **i**, **j**, **k**, in which a vector **r** is identified. After the rotation these vectors change into $\bar{\mathbf{i}}$, $\bar{\mathbf{j}}$, $\bar{\mathbf{k}}$, $\bar{\mathbf{r}}$ respectively. In the figure, for simplicity, we take **r** in the **xy** plane. Thus, there are two ways in which the transformation of Fig. 1 can be manifested by a matrix. One is through the transformation of the components of **r** (always with respect to **x**, **y**: components with respect to **i**, **j**, **k** are never used because they never change), x, y into $\bar{x}$, $\bar{y}$ and the other is through the transformation of the unit vectors **i**, **j** into $\bar{\mathbf{i}}$, $\bar{\mathbf{j}}$. The problem is that, as we shall see, these two descriptions do not lead to the same matrix. It is, in fact, very easy to perform the necessary trigonometry in Fig. 1 and to obtain the following two equations:

$$\begin{bmatrix}\bar{x}\\ \bar{y}\end{bmatrix}=\begin{bmatrix}\cos\alpha & -\sin\alpha\\ \sin\alpha & \cos\alpha\end{bmatrix}\begin{bmatrix}x\\ y\end{bmatrix}; \qquad \begin{bmatrix}\bar{\mathbf{i}}\\ \bar{\mathbf{j}}\end{bmatrix}=\begin{bmatrix}\cos\alpha & \sin\alpha\\ -\sin\alpha & \cos\alpha\end{bmatrix}\begin{bmatrix}\mathbf{i}\\ \mathbf{j}\end{bmatrix}. \tag{1}$$

In both cases, of course, the matrix represents the same operation, $R(\alpha\mathbf{z})$,

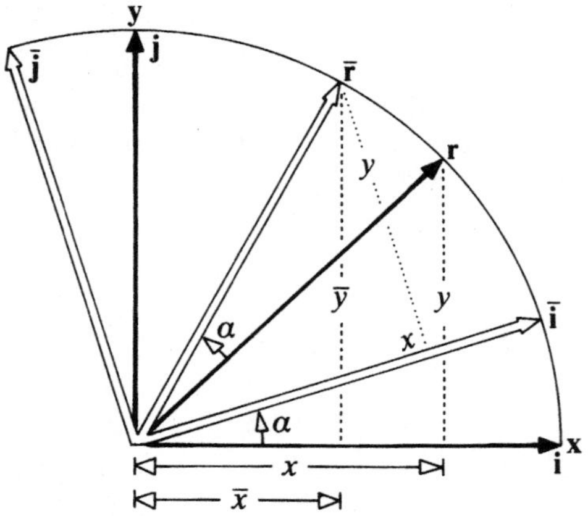

Fig. 2-4.1. Active rotation of the configuration space. The **z** axis is normal to and above the plane **x**, **y**.

and it is, of course, highly inconvenient to have two different matrices representing in this manner the same operation. This difficulty is easily corrected by writing the second equation in (1) as follows

$$\begin{bmatrix}\bar{\mathbf{i}} & \bar{\mathbf{j}}\end{bmatrix} = \begin{bmatrix}\mathbf{i} & \mathbf{j}\end{bmatrix} \begin{bmatrix}\cos\alpha & -\sin\alpha \\ \sin\alpha & \cos\alpha\end{bmatrix}. \tag{2}$$

We shall generalize this result but, before we do so, it is convenient to improve the notation. First, because we must often use row as well as column vectors we shall agree that row and column vectors of elements $a_1, a_2, \ldots, a_n$ (whatever these elements be: vector components, unit vectors or functions) shall be written as $\langle a_1 a_2 \ldots a_n|$ for the row vector and $|a_1 a_2 \ldots a_n\rangle$ for the column vector. Secondly, we shall use the symmetry operators of §1 in order to make it explicit the origin of transformations such as (1) and (2). If, for generality, we call g the operation, $R(\alpha\mathbf{z})$ in our case, that effects the transformation, then we shall write the column vector $|\bar{x}\bar{y}\rangle$ as $|gx, gy\rangle$ or, in short, $g\,|x\,y\rangle$. Correspondingly, the matrix in the first of the equations in (1) will be denoted with the symbol $\hat{g}$. Thus, we can write this equation, as well as eqn (2) as follows:

$$|gx, gy\rangle \equiv g\,|xy\rangle = \hat{g}\,|xy\rangle, \tag{3}$$

$$\langle g\mathbf{i}, g\mathbf{j}| \equiv g\,\langle\mathbf{ij}| = \langle\mathbf{ij}|\,\hat{g}. \tag{4}$$

The column or row vectors that allow us to define the transformation matrix as in (L1) and (2) or (3) and (4) are called *bases*. Any transformation can be represented by two different types of bases, one in which the elements are the n components $r_1, r_2, \ldots, r_n$ (always measured with respect to the space-fixed axes) of any one vector $\mathbf{r}$ of the configuration space, and the other having n unit vectors of the configuration space $\mathbf{e}_1, \mathbf{e}_2, \ldots, \mathbf{e}_n$, as elements. We agree to define the matrix representative $\hat{g}$ of an operator g as the matrix that transforms the *column vector* $|r_1 r_2 \ldots r_n\rangle$ and then this matrix will also transform the *row-vector* basis $\langle\mathbf{e}_1\mathbf{e}_2 \ldots \mathbf{e}_n|$. (See Problem 3.)

Several points here are worthy of comment. First, the convention whereby we give priority to the matrix defined by the components written as a column vector is precisely the one we have used in going from (1) to (2), instead of preserving the second matrix in (1). Secondly, for every transformation g two bases can be defined, row and column type respectively, that transform by identically the same matrix and that will be said to be one the *dual* of the other. (This term, though has not exactly the same meaning as that used in pure mathematics.) Finally, it must be emphatically restated that the distinction made between the transformation laws by row or column vectors is done entirely within the active picture and has nothing whatever to do with the distinction between the active and passive pictures. Any reader going through Problem 3 will see, in fact, that the components of the unit vectors of the configuration space transform exactly like the components of any vector in that space.

As regards the matrix multiplication rules, we shall assume it to be known that they are so chosen that the matrix multiplication rules are preserved when going from the operators themselves to their matrices:

$$g_i g_j = g_k \qquad \Rightarrow \qquad \hat{g}_i \hat{g}_j = \hat{g}_k. \tag{5}$$

It is, however, essential for this purpose to position the matrices carefully with respect to the bases pre- or post-multiplying them as the case might be. This, however, is not the end of the story because abominable mistakes can still be made, as we shall now show. In polar coordinates, the Cartesian coordinates x, y of a point on the unit sphere are

$$x = \sin\theta \cos\varphi, \tag{6}$$

$$y = \sin\theta \sin\varphi. \tag{7}$$

Just as in Fig. 1 or in (L1), let us operate on $|xy\rangle$ with $g = R(\alpha\mathbf{z})$. Since a rotation by α changes φ into $\varphi + \alpha$, on using (3.1), we have

$$\dot{g}x = \sin\theta\ \cos(\varphi - \alpha) = x\cos\alpha + y\sin\alpha, \tag{8}$$

$$\dot{g}y = \sin\theta\ \sin(\varphi - \alpha) = -x\sin\alpha + y\cos\alpha, \tag{9}$$

which leads to

$$\dot{g}\,|xy\rangle = \begin{bmatrix} \cos\alpha & \sin\alpha \\ -\sin\alpha & \cos\alpha \end{bmatrix} |xy\rangle, \quad \text{(wrong)}, \tag{10}$$

clearly with the wrong matrix with respect to (L1). The origin of the confusion should be clear, though: x, y in (1) are the independent variables whereas the same symbols in (6) and (7) stand for the functions $x(\theta\varphi)$, $y(\theta\varphi)$, which can cause endless trouble. Let us use the symbols x, y to denote the functions $x(\theta\varphi)$ and $y(\theta\varphi)$ respectively. Then, in order to stick to the same matrix as in (L1), we must rewrite (10) in row-vector notation:

$$\dot{g}\,\langle\mathsf{xy}| = \langle\mathsf{xy}| \begin{bmatrix} \cos\alpha & -\sin\alpha \\ \sin\alpha & \cos\alpha \end{bmatrix}. \tag{11}$$

We thus realize, on comparing with (2), that functions of vectors in the configuration space transform contrary to them and like the unit vectors of the space. That is, bases constructed from functions must be used as row and not as column vectors. In practice, most of the bases used in this book will be made up of functions and thus will be row vectors, but column vectors must still be used for the independent variables in configuration space, which means that column vectors will only appear in our case in two- and three-dimensional bases.

It should be noticed that the symmetry behaviour of x and y described in (11) does not depend, of course, on them being regarded as functions of θ and φ. If they are taken as functions (monomials of degree one) then one has to act on the corresponding variables with the function space operator, whence the inverse of (L1) has to be used (α changed by $-\alpha$) and (R8) and thus (11) follows. The typographical distinction which we have made above between the independent variables x, y, z and the dependent variables x, y, z, say, is not one that can always be made explicit and it is necessary to learn to read it from the context. Thus, if x appears in a sequence x, x^2, $x^3 \ldots$ it is clear that we are dealing with functions and that the transform of x must be effected by a function space operator, as in (8) and not be obtained from (L1).

Problems 2-4 (p. 268)

1. Show by working in configuration space that the matrix $\hat{\sigma}_\alpha$ corresponding to a reflection line on the **xy** plane at an angle α from **x** (measured in the positive direction from **x** to **y**) is

$$\hat{\sigma}_\alpha = \begin{bmatrix} \cos 2\alpha & \sin 2\alpha \\ \sin 2\alpha & -\cos 2\alpha \end{bmatrix}. \tag{12}$$

2. Show that the action of σ, as defined in Problem 1, on the polar coordinate φ on the plane is given by

$$\sigma_\alpha \varphi = 2\alpha - \varphi. \tag{13}$$

   Apply the operator $\hat{\sigma}_\alpha$ on the functions x and y defined in (6) and (7) and obtain the matrix of the transformation.

3. A vector space spanned by fixed unit vectors $\mathbf{E}_i$ $(i=1, \ldots, n)$, has elements $\mathbf{r}=\sum_i r_i \mathbf{E}_i$, written as $|\mathbf{r}\rangle = |r_1 \ldots r_n\rangle$. An active transformation of this space leaves all $\mathbf{E}_i$ invariant and effects the transformation $\mathbf{r} \mapsto \bar{\mathbf{r}}$ such that $|\bar{\mathbf{r}}\rangle = A\,|\mathbf{r}\rangle$, for some matrix A. Moving unit vectors $\mathbf{e}_1, \ldots, \mathbf{e}_n$ are defined which, under the same transformation, are mapped as follows: $\mathbf{e}_i \mapsto \bar{\mathbf{e}}_i$. These unit vectors satisfy the condition that, before the transformation, $\mathbf{e}_i \equiv \mathbf{E}_i$, $\forall i$. Prove that the basis $\langle \mathbf{e}| = \langle \mathbf{e}_1 \ldots \mathbf{e}_n|$ transforms under the given linear transformation into the basis $\langle \bar{\mathbf{e}}| = \langle \bar{\mathbf{e}}_1 \ldots \bar{\mathbf{e}}_n|$, such that $\langle \bar{\mathbf{e}}| = \langle \mathbf{e}|\, A$.

4. Consider a cube with right-handed axes **x**, **y**, **z**, at its centre and parallel to its edges. The cube-fixed axes **i**, **j**, **k**, coincide with **x**, **y**, **z**, respectively, for the identity E. Find the transform of the components $|x\, y\, z\rangle$ of a vector **r** under the following rotations: $R(\pi\mathbf{z})$, $R(\pi(011))$,

$R(\frac{2}{3}\pi(111))$, $R(-\frac{2}{3}\pi(1\bar{1}1))$. (A rotation-axis vector such as $(1\bar{1}1)$ indicates a vector with components 1, −1, 1, respectively, along **x**, **y**, and **z**.) Compare your results with the tables of Onodera and Okazaki (1966).

5. GROUPS

Of the four group properties – existence of identity, existence of inverse, associative property of the product, and closure – it is only the latter that requires any attention when dealing with finite groups of symmetry operators. The associative property (1.4) of the product of symmetry operations is always satisfied, and in any set of symmetry operators the identity and inverses are most easily included.

We shall use the following notation for a group. The group G will have elements g labelled g_i, $i = 1, \ldots, |G|$, the number of elements $|G|$ of the group being the order. Thus, to establish *closure*, we need

$$G = \{g\}, \text{ such that: } g_i \in G,\ g_j \in G \quad \Rightarrow \quad g_i g_j \in G. \tag{1}$$

The *conjugate* g_i^g of an element g_i by g is defined by

$$g_i^g = g g_i g^{-1}. \tag{2}$$

Conjugation is an equivalence relation which permits partitioning a group in disjoint sets such that any two elements of each set are conjugate and any two elements of different sets are never conjugate. Such sets are called *classes*. A class can be labelled by any one of its elements g as $C(g)$:

$$C(g) = \sum\nolimits_{\forall g \in G} g_i^g. \tag{3}$$

Here the sum is taken in the sense of the collection of elements of the stated form and, in practice, repeated elements must be removed from it. (Whenever g commutes with g_i, g_i would otherwise be repeated in eqn 3.)

Two groups are homomorphic when there is a mapping between them which preserves the multiplication rules. More precisely, G is homomorphic onto G' if every element g_i of G is mapped by one and only one element g_i' of G' and every element of G' is the image of some element of G, in such a way that $g_r' g_s'$ is the image of $g_r g_s$. It is clear from our work so far that $\{g\}$, $\{\dot{g}\}$, and $\{\hat{g}\}$ are all homomorphic and we shall refer indistinctly to them with the same symbol G, since the context will make it clear which set we mean. (Remember, though, that unless care is taken at every stage to ensure the preservation of the multiplication rules, great confusion may arise.) In practice, we shall be mostly interested in the set $\{\dot{g}\}$ and we shall write the operators of this set simply as g, since whenever we see this operator acting on a function we know that we mean $\dot{g}$.

A basis $\{\varphi\}$ for a group G is a set of functions that forms an invariant space under G, by which we mean that the transform $g\varphi$ of each function of the set is a linear combination of all the functions of the set. This means that the transformed row-vector $g\,\langle\varphi_1 \ldots \varphi_n|$ can be written as the original vector times a matrix. (Row-vector notation is here used in keeping with the note after eqn 4.11, but see also the remark after eqn 10 below.) We use the following notation:

$$g\,\langle\varphi| = \langle\varphi|\,\hat{G}(g), \tag{4}$$

where $\langle\varphi|$ stands for the whole row-vector $\langle\varphi_1 \ldots \varphi_n|$. Notice also that whereas before we denoted the matrix on (R4) simply as $\hat{g}$ we now use a fuller notation.

If a group G is homomorphic onto a group of matrices $\hat{G}$, so that $\hat{G}(g_i)$ is the image of g_i in $\hat{G}$, then $\hat{G}$ is a *representation* of G. The matrices $\hat{G}(g)$ are called *representatives*. We verify that (4) entails a representation. Consider two operations of G:

$$g_j\,\langle\varphi| = \langle\varphi|\,\hat{G}(g_j), \tag{5}$$

$$g_i\,\langle\varphi| = \langle\varphi|\,\hat{G}(g_i). \tag{6}$$

Then

5 $$g_i g_j \langle\varphi| = g_i\,\langle\varphi|\,\hat{G}(g_j) \tag{7}$$

6 | 7 $$= \langle\varphi|\,\hat{G}(g_i)\hat{G}(g_j), \tag{8}$$

4 $$g_i g_j\,\langle\varphi| = \langle\varphi|\,\hat{G}(g_i g_j). \tag{9}$$

8,9 $$\hat{G}(g_i)\hat{G}(g_j) = \hat{G}(g_i g_j), \tag{10}$$

thus preserving the multiplication rules. Notice that if the same work were repeated with column-vector notation, it would fail. On the other hand, bases can also be defined in the configuration space, in which case they must be used as column vectors, as done in elementary matrix work, thus preserving the group multiplication rules. Fallacies in the work leading to eqn (8) are possible, as illustrated in Problem 16.

It is clear that if we choose $\hat{G}(g) = 1$ for all g in G these matrices satisfy (10) and therefore they form a representation. Appropriately, this is called the *trivial representation.* Notice also that the matrices $\hat{G}(g)$ that form a representation need not be all distinct (the trivial representation is a dramatic example). If, however, they are all distinct, then the representation is called a *faithful representation.*

The representation of a group depends on the basis. Thus, for two different bases like in (4) we write

$$g\,\langle\varphi^1| = \langle\varphi^1|\,{}^1\hat{G}(g), \tag{11}$$

$$g\,\langle\varphi^2| = \langle\varphi^2|\,{}^2\hat{G}(g). \tag{12}$$

Such bases, and the corresponding matrices, can be entirely unrelated. One important case, however, arises when one is a linear combination of the other with a non-singular matrix L (i.e. such that L^{-1} exists),

$$\langle\varphi^2| = \langle\varphi^1| L. \tag{13}$$

Then:

13|12 $$g\langle\varphi^1| L = \langle\varphi^1| L\, {}^2\hat{G}(g), \tag{14}$$

14 $$g\langle\varphi^1| = \langle\varphi^1| L\, {}^2\hat{G}(g) L^{-1}, \tag{15}$$

11,15 $${}^1\hat{G}(g) = L\, {}^2\hat{G}(g) L^{-1}. \tag{16}$$

Such a transformation is called a *similarity*.

Because notation often varies here (L^{-1} is often substituted for L on R16), notice the following relation, from (16) and (13)

$$\text{rep: } L\,\hat{G}\,L^{-1}; \qquad \text{basis: } \langle\varphi| L^{-1}, \tag{17}$$

where $\langle\varphi|$ is the basis for $\hat{G}$. (See 12.)

It can be seen that similarity preserves the multiplication rules (Problem 8). Thus, if $\hat{G}$ is a representation, then $L\,\hat{G}\,L^{-1}$ is automatically a representation. Such representations are called *equivalent*.

An invariant of a representation under similarity (or equivalence) is the *trace* of the matrix representatives (the trace, $\operatorname{Tr} A$, of a matrix A is the sum of its diagonal elements). Thus, in (17) the trace of $\hat{G}(g)$ before equivalence, equals the trace of the new representative $L\,\hat{G}\,(g)\,L^{-1}$. (See Problem 9.)

Because the trace of the matrix representative of an operator g in the representation $\hat{G}$ is invariant under equivalence, it is a useful quantity that is called the *character* $\chi(g \mid \hat{G})$ of g in $\hat{G}$:

$$\chi(g \mid \hat{G}) = \operatorname{Tr} \hat{G}(g). \tag{18}$$

We shall now need the definition of a new matrix operation, called the *direct sum of matrices.* Given the square matrices (not necessarily of the same dimension) $A_1, A_2, \ldots, A_n$, their direct sum A is the block-diagonal matrix that contains those matrices in that order as diagonal elements and zeros elsewhere. We shall use the following notations:

$$A = A_1 \oplus A_2 \oplus \ldots \oplus A_n = \textstyle\sum_i A_i. \tag{19}$$

The *structure* of a block-diagonal matrix so defined is given by the ordered set of the dimensions of the matrices along the diagonal.

If, under equivalence of a representation $\hat{G}(g)$ with a matrix L all the matrices of the new representation, $L\hat{G}(g)L^{-1}$ are block-diagonal matrices with the same structure, we say that the representation has been *reduced.* It can immediately be seen that the set of matrices corresponding to each block of the reduced matrices forms itself a representation of G and that, at the same time, the reduced basis $\langle\varphi| L^{-1}$ (see 17) splits up into disjoint sub-bases with dimensions corresponding to the structure of $L\hat{G}L^{-1}$, in such a way that each new representation formed as just explained corresponds to one of these sub-bases. This splitting of the basis is often the most important feature of the reduction process from the point of view of the user: the basis is made up of the very functions one uses in calculations, whereas the matrices are mere tools used in order to simplify these functions.

If a representation (or a basis) cannot be further reduced the representation (or the basis) is said to be *irreducible.* Functions belonging to different irreducible bases cannot be transformed one into another by any operation of the group (this follows from the definition of reduction), and if the different irreducible bases span different representations, then they can be shown to be orthogonal, a very desirable property in calculations.

The reader is assumed to be familiar with the fact that the number of irreducible representations of a group is equal to the number of its classes. Also with the test that reveals that a given representation is irreducible (the sum of the square of the moduli of the characters over the group must be equal to the order of the group).

There is one further technicality that we must revise, which is the idea of a *direct product* (also called *tensor* or *Kronecker product*) of bases and representations. We shall consider two representations of G as in (4) but changing the notation somewhat for simplicity:

$$g\,\langle u_1 \dots u_m| = \langle u_1 \dots u_m|\,\hat{U}, \tag{20}$$

$$g\,\langle v_1 \dots v_n| = \langle v_1 \dots v_n|\,\hat{V}. \tag{21}$$

We define the *direct product* of the bases $\langle u|$ and $\langle v|$ as the array of all products $u_i v_j$ in *dictionary* order (for each i, j to vary from 1 to n, and the i's to be arranged in the order 1 to m):

$$\langle u| \otimes \langle v| = \langle u_1 v_1 \dots u_1 v_n, u_2 v_1 \dots u_2 v_n, \dots, u_m v_1 \dots u_m v_n|. \tag{22}$$

This operation is useful because the array on (R22), which we shall denote with $\langle w|$, is itself a basis for a representation of G, as we shall now prove.

It simplifies results very much to label this array, rather than $\langle w_1 \dots w_{m\times n}|$, by a double suffix in dictionary order with $u_r v_s$, say, written as w_{rs}. We can now proceed. Re-write:

20
$$g u_i = \sum\nolimits_r u_r \hat{U}_{ri}, \tag{23}$$

21
$$g v_j = \sum\nolimits_s v_s \hat{V}_{sj}. \tag{24}$$

In order to test (22) we must transform $u_i v_j$, say, with g, and we must remember that $g(u_i v_j)$ equals $g u_i g v_j$. (Of course, since g acts on the whole of the function space.) Then:

23,24 $$g u_i v_j = \sum_{rs} u_r v_s \hat{U}_{ri} \hat{V}_{sj}. \tag{25}$$

25 $$g w_{ij} = \sum_{rs} w_{rs} \hat{U}_{ri} \hat{V}_{sj}. \tag{26}$$

Remember now that *ij* is the single suffix for one column (element) of the basis and that the rows of the corresponding matrix must also be labelled by a double suffix. Thus, as in (23),

$$g w_{ij} = \sum_{rs} w_{rs} \hat{W}_{rs,ij}, \tag{27}$$

which leads to

26,27 $$\hat{W}_{rs,ij} = \hat{U}_{ri} \hat{V}_{sj}. \tag{28}$$

The matrix $\hat{W}$ is called the *direct product* (or *tensor* or *Kronecker product*) of $\hat{U}$ and $\hat{V}$:

$$\hat{W} = \hat{U} \otimes \hat{V}. \tag{29}$$

In (6.54) we shall give a very quick way of forming this matrix.

It will be useful to revise briefly the definitions of subgroups and various related concepts. If a group G of order $|G|$ has a set H of elements $h_1, h_2, \ldots, h_{|H|}$ which satisfies the group postulates, then H is a *subgroup* of G, $H \subset G$. The *index* of H is the order of G divided by the order of H, $|G|/|H|$.

If $H \subset G$ and g is some element of G (which as a particular case could belong to H), then the set $gh_1, gh_2, \ldots, gh_{|H|}$ is a *left coset* of H by g. Similarly, *right cosets* are defined. It is easy to prove that two cosets of H by two elements of G are either identical or disjoint (have no element in common). (See Problems 18, 19.) It is this property that makes cosets important, since it permits G to be partitioned as a sum of the disjoint cosets $g_1 H, \ldots, g_n H$, for some choice of the elements $g_1, \ldots, g_n$, which are called the *coset representatives*. (Problem 20.) They equal in number the index of the subgroup, but their choice is not unique.

Given $H \subset G$, the set gHg^{-1} means the set of all elements of H conjugated by g, that is, the set $gh_1 g^{-1}, \ldots, gh_{|H|} g^{-1}$. If gHg^{-1} equals H for all $g \in G$, then H is invariant under conjugation and it is called an *invariant subgroup* of G, $H \lhd G$. (See Problems 22, 23, 24.)

One final remark about groups: although most of the work of this book will be concerned with finite groups, we shall always consider them as subgroups of some group with an infinite number of elements. Let us consider an example. The point group $\mathbf{C}_3$ of operations E, C_3^+, C_3^-, around the $\mathbf{z}$ axis is a subgroup of the group of all rotations around the $\mathbf{z}$ axis, $\mathbf{C}_\infty$, which, as the symbol implies, is of infinite order. Moreover, the operations of this group form a continuous set, as follows from the fact that each operation of $\mathbf{C}_\infty$ is fully defined by a parameter, the angle of rotation ϕ, which varies continuously from 0 to 2π. The group properties for this set are, in fact, easy to verify, so that we have a *continuous group.* The elements of this group can be considered as functions of a continuous parameter and Sophus Lie discovered that if some simple mathematical behaviour is required of these functions (namely, that they be analytic) then such groups, called *one-parameter Lie groups,* have a number of important properties (see e.g. §7). More generally, *n-parameter Lie groups* can also be defined and they are a particularly important type of continuous group. (See Hamermesh 1962, p. 283.) The most important example of such groups for us is the rotation group itself. Many of the results of this book, in fact, are possible because the rotation group is a Lie group, as very clearly discussed by Normand (1980), but we shall be prepared to obtain such results on an *ad hoc* basis, rather than from the general Lie theory.

Problems 2-5 (p. 269)

1. Prove that the set (1.9) is a group. (This is the group $\mathbf{C}_{3v}$.)
2. Form all the classes of $\mathbf{C}_{3v}$.
3. Form a representation of this group on the basis $|xy\rangle$, where x and y are the coordinates of a vector $\mathbf{r}$ on the axes shown in Fig. 1.3.
4. Form a representation of $\mathbf{C}_{3v}$ on the basis $\langle d_{x^2-y^2}, d_{xy}|$, where these functions (which are the atomic d orbitals) are given by

$$d_{x^2-y^2} = \sin^2\theta \cos 2\varphi, \qquad d_{xy} = \sin^2\theta \sin 2\varphi. \tag{30}$$

   Is this representation equivalent to that obtained in Problem 3? Is it irreducible?
5. Form a representation of $\mathbf{C}_{3v}$ on the basis $\langle \mathsf{z}\, \mathsf{x}\, \mathsf{y}|$, where x and y are the functions defined in (4.6) and (4.7) and $\mathsf{z} = \cos\theta$. Show that it is already reduced into two representations. Would this be the case if the z axis in Fig. 1.3 were tipped off the vertical by an arbitrary angle?
6. Show that the two representations obtained in Problem 5 are irreducible. Are these all the possible irreducible representations?
7. Verify that the representations shown with conventional labels on the first column of Table 1 are all irreducible and that they are all the non-equivalent irreducible representations in this group.

Table **2**-5.1. Characters of the irreducible representations of $\mathbf{C}_{3v}$.

	E	$2C_3$	$3\sigma_v$
A_1	1	1	1
A_2	1	1	-1
E	2	-1	0

8. Prove that, if $\hat{G}(g_i)\hat{G}(g_j) = \hat{G}(g_ig_j)$, then the same rule obtains for the matrices $\hat{G}'(g)$ obtained by the similarity $L\hat{G}(g)L^{-1}$.
9. Prove that Tr (AB) equals Tr (BA) and hence that the trace of a representative is invariant under an equivalence or similarity.
10. Prove that characters are class functions (i.e. that elements of the same class have the same characters).
11. Prove that the set of the two elements E and $\boldsymbol{i}$ is a group (this is the group $\mathbf{C}_i$). From an arbitrary function $f(\mathbf{r})$ form the function $g(\mathbf{r})$ equal to $\boldsymbol{i}f(\mathbf{r})$, show that $\langle f, g|$ is a basis for $\mathbf{C}_i$, and find and reduce the corresponding representation.
12. Show that the following operations

$$E, C_{2x}, C_{2y}, C_{2z}, (\mathbf{D}_2), \tag{31}$$

where the C_2 are binary rotations about the $\mathbf{x}$, $\mathbf{y}$, $\mathbf{z}$ axes as shown in Fig. 1, form a group. (This is called $\mathbf{D}_2$.) In doing this, verify the multiplication table of the group, shown in Table 2. Why is it that all operations of $\mathbf{D}_2$ commute?
13. Verify that the four sets of four one-dimensional matrices shown in Table 3 form each a representation of $\mathbf{D}_2$ and show that they are all irreducible and that they are all the irreducible representations possible in this group. The labels of the representations on the left column are conventional symbols.

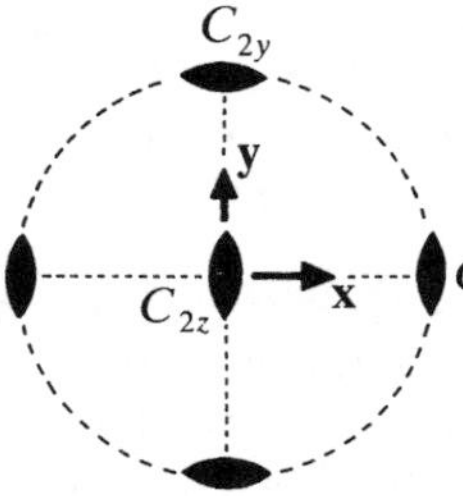

Fig. 2-5.1. The group $\mathbf{D}_2$. The $\mathbf{z}$ axis is normal to and above the plane $\mathbf{x}, \mathbf{y}$.

Table **2**-5.2. $\mathbf{D}_2$. Table of products $g_i g_j$. The product $g_i g_j$ appears in the intersection of the row g_i with the column g_j.

	E	C_{2x}	C_{2y}	C_{2z}
E	E	C_{2x}	C_{2y}	C_{2z}
C_{2x}	C_{2x}	E	C_{2z}	C_{2y}
C_{2y}	C_{2y}	C_{2z}	E	C_{2x}
C_{2z}	C_{2z}	C_{2y}	C_{2x}	E

14. Form a representation of $\mathbf{D}_2$ (Fig. 1) on the basis $\langle \mathbf{i}\,\mathbf{j}\,\mathbf{k}|$. Show that this representation is reducible and reduce it.
15. Prove that the eigenfunctions of the equation $\nabla^2\psi(\mathbf{r}) = k\,\psi(\mathbf{r})$ must be either gerade or ungerade. (Warning: this result is not general in quantum mechanics. See Problem **2**-7.4.)
16. Given eqns (5) and (6), it is tempting to write $g_i g_j \langle\varphi|$ as $g_i\{g_j \langle\varphi|\}$ and thus as $\{g_j \langle\varphi|\}\hat{G}(g_i)$, which equals $\langle\varphi|\,\hat{G}(g_j)\hat{G}(g_i)$, in apparent contradiction with (8). What is wrong here?
17. Show that, given the group $\mathbf{C}_{3v}$ defined in Problem 1, then the set E, C_3^+, C_3^- is a subgroup of it. (This subgroup is the group $\mathbf{C}_3$.)
18. Prove that, if $H \subset G$ and $g \in H$, then the coset gH is identical with H.
19. Prove that, given $H \subset G$, the cosets g_iH, g_jH are either identical or disjoint.
20. Prove that, if $H \subset G$, then G can be written as $g_1H, \ldots, g_nH$, with $n = |G|/|H|$, the index of H.
21. Prove that $\mathbf{C}_{3v}$ can be written as the sum of the two cosets $E\,\mathbf{C}_3$ and $\sigma_{v1}\mathbf{C}_3$.
22. Prove that $\mathbf{C}_3$ is an invariant subgroup of $\mathbf{C}_{3v}$.
23. Prove that, if $H \lhd G$, then, for all g, $gH = Hg$.
24. Prove that a subgroup of index two must be invariant. Apply this result to $\mathbf{C}_3 \subset \mathbf{C}_{3v}$.

Table **2**-5.3. $\mathbf{D}_2$. Irreducible representations.

	E	C_{2x}	C_{2y}	C_{2z}
A_1	1	1	1	1
B_1	1	−1	−1	1
B_2	1	−1	1	−1
B_3	1	1	−1	−1

6. ALL ABOUT MATRIX PROPERTIES

We shall collect in this section a number of matrix definitions and properties which will be useful in this book; for generality, all our matrix and vector elements, unless statements to the contrary, will be taken to be complex. We shall start with the discussion of the two invariants of a matrix A under the similarity LAL^{-1} (see eqn 5.17). The first, we already know, is the trace:

$$\mathrm{Tr}\, A = \sum_i A_{ii}, \tag{1}$$

$$\mathrm{Tr}\,(AB) = \mathrm{Tr}\,(BA). \tag{2}$$

Equation (2) is proved in Problem 5.9, where invariance under similarity is also established. The second invariant is det A, the determinant of A. I shall assume known the result

$$\det(\mathrm{AB}) = \det \mathrm{A} \det \mathrm{B}, \tag{3}$$

from which, since $\mathrm{A}^{-1}\mathrm{A} = \mathbf{1}$, and $\det \mathbf{1} = 1$, it follows that

$$\det(\mathrm{A}^{-1}\mathrm{A}) = \det A^{-1} \det \mathrm{A} = 1. \tag{4}$$

Therefore,

4

$$\det A^{-1} = (\det A)^{-1}, \tag{5}$$

which means that, if $\det \mathrm{A} = 0$, then $\det \mathrm{A}^{-1}$ is infinite, so that A^{-1} cannot exist. Such matrices are called *singular*. From here the proof of the invariance is immediate (Problem 1).

We shall also assume as known the relation between Tr A, det A, and the eigenvalues of A (Problem 2):

$$\mathrm{Tr}\, A = \text{sum of the eigenvalues of } A. \tag{6}$$

$$\det A = \text{product of the eigenvalues of } A. \tag{7}$$

We must next concern ourselves with the *scalar* or *inner product* of vectors. In order to write them in matrix notation we agree that vectors, like **r**, **s**, etc., are always understood as column vectors, and that, for complex or real components $r_1, \ldots, r_n$, corresponding row vectors are defined as follows:

$$\mathbf{r}\text{: column vector of components } r_1 \ldots r_n. \tag{8}$$

$$\mathbf{r}^{\mathrm{T}}\text{: row vector of components } r_1 \ldots r_n. \tag{9}$$

$\mathbf{r}^{\mathrm{T}}$ is called the *transpose* of **r**. We can now write inner products in matrix notation and we must remember that two different such products are used:

$$\text{inner product:} \qquad \mathbf{r}^{\mathsf{T}}\mathbf{s}, \tag{10}$$

$$\text{Hermitian inner product:} \qquad (\mathbf{r}^*)^{\mathsf{T}}\mathbf{s}. \tag{11}$$

In order to simplify the notation the *adjoint* $\mathbf{r}^\dagger$ of the vector $\mathbf{r}$ in (8) is defined:

$$\mathbf{r}^\dagger = (\mathbf{r}^*)^{\mathsf{T}}: \text{ row vector of components } r_1^* \dots r_n^*. \tag{12}$$

Thus (11) can be rewritten:

$$\text{Hermitian inner product:} \qquad \mathbf{r}^\dagger\mathbf{s}. \tag{13}$$

The virtue of the Hermitian product is that

$$\mathbf{r}^\dagger\mathbf{r} = 0 \qquad \Rightarrow \qquad \mathbf{r} \equiv \mathbf{0}. \tag{14}$$

In other words, since we can always exclude null vectors from the work, we can always assume

$$\mathbf{r}^\dagger\mathbf{r} \neq 0. \tag{15}$$

Warning: (15) is not true for $\mathbf{r}^{\mathsf{T}}\mathbf{r}$ (see Problem 3) and serious fallacies often arise from the use of this false result.

We shall now consider the definitions and properties of matrices which for convenience are listed in Table 1. We shall assume the following

Table **2**-6.1. Matrix properties.
$\mathbf{1}_n = n \times n$ unit matrix, abbreviated $\mathbf{1}$. $a\mathbf{1}$ = constant matrix. Tr $A = \sum_i A_{ii}$.

Matrix	Notation	Definition	
Transpose	A^{T}	$(A^{\mathsf{T}})_{ij} = A_{ji}$	
Conjugate	A^*	$(A^*)_{ij} = (A_{ij})^*$	
Adjoint	$A\dagger$	$A\dagger = (A^{\mathsf{T}})^*$	a
Reciprocal	A^{-1}	$A^{-1}A = \mathbf{1}$	b
Real (Re)		$A^* = A$	
Orthogonal		$A^{\mathsf{T}}A = \mathbf{1}$	b
Unitary		$A\dagger A = \mathbf{1}$	b
Symmetric		$A^{\mathsf{T}} = A$	
Skew-symmetric		$A^{\mathsf{T}} = -A$	
Hermitian		$A\dagger = A$	
Skew-Hermitian		$A\dagger = -A$	
Real orthogonal		$A^{\mathsf{T}}A = \mathbf{1}$ and $A^* = A \Rightarrow A$ unitary	c
Real symmetric		$A^{\mathsf{T}} = A$ and $A^* = A \Rightarrow A$ Hermitian	c
Real skew-symmetric		$A^{\mathsf{T}} = -A$ and $A^* = A \Rightarrow A$ skew-Hermitian	c
Non singular		$A^{-1}\exists$ (det $A \neq 0$)	
Traceless		Tr $A = 0$	

[a] The symbols ^{T} and * can be interchanged.
[b] The matrices on the left can be interchanged if A is non-singular.
[c] The converse entailment is not valid.

results (see Problem 4):

$$(AB)^{\mathrm{T}} = B^{\mathrm{T}}A^{\mathrm{T}}, \qquad (AB)^{\dagger} = B^{\dagger}A^{\dagger}, \qquad (AB)^{-1} = B^{-1}A^{-1}. \qquad (16)$$

Orthogonal matrices

All geometric symmetry operations must leave invariant the lengths of all vectors and the angles between two vectors, i.e. the inner products between any two vectors of the ordinary three-dimensional space ($\mathbb{R}^3$).

,16 $$\mathbf{r}^{\mathrm{T}}\mathbf{s} = (A\mathbf{r})^{\mathrm{T}}A\mathbf{s} = \mathbf{r}^{\mathrm{T}}A^{\mathrm{T}}A\mathbf{s} \qquad \Rightarrow A^{\mathrm{T}}A = \mathbf{1}. \qquad (17)$$

(A and A^{T} can be commuted on R17, see Problem 5.) A matrix with this property is an *orthogonal matrix*. All rotations, proper or improper, must be represented in $\mathbb{R}^3$ by real orthogonal matrices. They can be distinguished by their determinant:

17,3 $$\det(A^{\mathrm{T}}A) = 1 \qquad \Rightarrow \qquad \det A^{\mathrm{T}} \det A = 1. \qquad (18)$$

18 $$(\det A)^2 = 1 \qquad \Rightarrow \qquad \det A = \pm 1. \quad (A \text{ real}). \qquad (19)$$

Proper rotations never change the handedness of a triple $\mathbf{r}$, $\mathbf{s}$, $\mathbf{t}$, whereas improper rotations acting on the mixed triple product of those vectors, must multiply it by -1. Therefore, in the transformation $\mathbf{r}, \mathbf{s}, \mathbf{t} \mapsto \bar{\mathbf{r}}, \bar{\mathbf{s}}, \bar{\mathbf{t}}$ that is,

$$A\begin{bmatrix} r_1 & s_1 & t_1 \\ r_2 & s_2 & t_2 \\ r_3 & s_3 & t_3 \end{bmatrix} = \begin{bmatrix} \bar{r}_1 & \bar{s}_1 & \bar{t}_1 \\ \bar{r}_2 & \bar{s}_2 & \bar{t}_2 \\ \bar{r}_3 & \bar{s}_2 & \bar{t}_3 \end{bmatrix}, \qquad (20)$$

the transformed volume $\bar{v}$ on (R20) must be $(\pm 1)v$, the latter being the volume spanned by $\mathbf{r}$, $\mathbf{s}$, $\mathbf{t}$. From (3) on (L20),

19 $$(\det A)v = \bar{v} \qquad \Rightarrow \qquad \pm v = \bar{v}. \qquad (21)$$

Real 3×3 orthogonal matrices with determinant $+1$ are called *Special Orthogonal* matrices (SO) and represent proper rotations in $\mathbb{R}^3$. Real 3×3 orthogonal matrices with determinant -1 represent improper rotations in $\mathbb{R}^3$. The product of two orthogonal matrices is orthogonal (Problem 6). Thus the set of all 3×3 real orthogonal matrices forms a group called the *Orthogonal group* O(3). Clearly, the set of all SO matrices of O(3) forms a subgroup of it called the *Special Orthogonal group* SO(3). (The set of all improper orthogonal matrices, with determinant -1 is, obviously, not a group.)

The columns and rows of orthogonal matrices are orthonormal (Problem 7):

$$\sum_r A_{ri}A_{rj} = \sum_r A_{ir}A_{jr} = \delta_{ij}. \qquad (22)$$

We now consider the properties of the eigenvalues and eigenvectors of orthogonal matrices. We write the eigenvalue equation as follows, (**r** eigenvector, a eigenvalue),

$$A\mathbf{r} = a\mathbf{r}, \tag{23}$$

16 $$\mathbf{r}^{\mathrm{T}}A^{\mathrm{T}} = a\mathbf{r}^{\mathrm{T}}, \tag{24}$$

23,24|10 $$\mathbf{r}^{\mathrm{T}}A^{\mathrm{T}}A\mathbf{r} = a^2\mathbf{r}^{\mathrm{T}}\mathbf{r}, \tag{25}$$

17|25 $$\mathbf{r}^{\mathrm{T}}\mathbf{r} = a^2\mathbf{r}^{\mathrm{T}}\mathbf{r}. \tag{26}$$

The temptation to cancel $\mathbf{r}^{\mathrm{T}}\mathbf{r}$ here indiscriminately must be resisted (see Warning after eqn 15). It is only if **r** is real that $\mathbf{r}^{\mathrm{T}}\mathbf{r}$ must be different from zero and can be cancelled. Thus, all that we can say is that *the real eigenvectors of an orthogonal matrix must correspond to eigenvalues a such that a^2 equals unity*. More can be said about *real orthogonal* matrices as a special case of interest (see also below, eqn 47). In this case, since $A = A^*$,

23 $$A^*\mathbf{r}^* = a^*\mathbf{r}^* \qquad \Rightarrow \qquad A\mathbf{r}^* = a^*\mathbf{r}^*. \tag{27}$$

Equations (23) and (27) show that *for real orthogonal matrices eigenvalues and eigenvectors come in complex conjugate pairs*. Furthermore, if the eigenvector **r** is real,

27 $$A\mathbf{r} = a^*\mathbf{r}, \tag{28}$$

so that, on equating (R23) and (R28), it follows that the corresponding eigenvalue must also be real. Therefore, on applying the result italicized after (26), we find that *the real eigenvectors of a real orthogonal matrix must correspond to eigenvalues* ± 1.

We shall apply these results for the real orthogonal matrices of O(3). Because their eigenvectors must come in conjugate pairs, it follows that only two cases are possible:

(a) 1 real, 2 complex eigenvectors (conjugate), (29)

(b) 3 real eigenvectors. (30)

From the italicized results after (27) and (28) we know that the real eigenvectors must correspond to eigenvalues ± 1, and that the complex eigenvectors must correspond to conjugate eigenvalues. We can therefore classify in more detail the two cases above, now with respect to the eigenvalues. We start with (a), for which the first choice of eigenvalues is:

$$(a1)\ \ a_1 = 1,\ a_2 = \omega,\ a_3 = \omega^* \qquad \Rightarrow \qquad \det A = \omega\omega^*. \tag{31}$$

Since $\omega\omega^*$ is positive, $\det A$ (which equals $a_1\omega\omega^*$) cannot be negative, so that $\det A$ must be $+1$. This is a *proper rotation*, the eigenvector corresponding to a_1 being along the rotation axis. (Notice that it follows that $\omega\omega^*$ is unity. See the remark below eqn 47.) The second alternative for (29) is:

$$(a2)\ a_1 = -1,\ a_2 = \omega,\ a_3 = \omega^*. \tag{32}$$

In the same manner as above, det A equals -1 and this must be an *improper rotation*, in fact a *rotoreflection*, the axis of rotation now corresponding to the eigenvalue -1, since the corresponding eigenvector is reflected by this operation.

As regards (b) in (30), we have clearly only four possible cases:

$$(b1)\ a_1 = 1,\ a_2 = 1,\ a_3 = 1 \quad \Rightarrow \quad \det A = 1, \quad \text{identity}, \tag{33}$$

$$(b2)\ a_1 = 1,\ a_2 = 1,\ a_3 = -1 \quad \Rightarrow \quad \det A = -1, \quad \text{reflection}, \tag{34}$$

$$(b3)\ a_1 = 1,\ a_2 = -1,\ a_3 = -1 \quad \Rightarrow \quad \det A = 1, \quad \text{binary rotation}, \tag{35}$$

$$(b4)\ a_1 = -1,\ a_2 = -1,\ a_3 = -1 \quad \Rightarrow \quad \det A = -1, \quad \text{inversion}. \tag{36}$$

We have now obtained all symmetry operations. It is interesting that binary rotations, which are proper rotations, do not appear as a special case of (a1). This is significant, because we shall see that binary rotations play a special rôle in SO(3). To make a distinction between the identity and binary rotations, on one hand, and the proper rotations that satisfy (31) on the other, we shall refer to the latter as *general rotations*. (They have only one real eigenvector.)

The final point that we want to discuss is the nature of the matrix elements of a real orthogonal matrix. If n is its dimension, there are n^2 matrix elements, n normalization conditions (one for each column) and $\frac{1}{2}n(n-1)$ orthogonality conditions between the columns. This leaves altogether $\frac{1}{2}n(n-1)$ free matrix elements, which means that all the matrix elements of a 2×2 and 3×3 real orthogonal matrix can be expressed in terms of 1 or 3 parameters respectively. We shall first consider the case $n = 2$. Since the first column of the matrix must be normalized, it is obvious to take it, in terms of the single parameter, as $|\cos\lambda, \sin\lambda\rangle$. (The opposite choice, although possible, is more inconvenient.) The second column must be chosen by orthogonality and two cases arise:

$$A = \begin{bmatrix} \cos\lambda & -\sin\lambda \\ \sin\lambda & \cos\lambda \end{bmatrix}, \qquad B = \begin{bmatrix} \cos\lambda & \sin\lambda \\ \sin\lambda & -\cos\lambda \end{bmatrix}. \tag{37}$$

From (L4.1) A is a rotation in the xy plane around the z axis. From (4.12), B is a reflection matrix with λ equal to twice the angle α therein defined. This interpretation, of course, correctly agrees with the determinants of A and B.

Similar work for n equals 3 is much harder and will be done in §§**3**-2, **3**-3.

Unitary matrices

A *unitary matrix* is one that leaves invariant the Hermitian inner product. (In analogy to orthogonal matrices, for the reasons explained before eqn 17, they can be regarded as symmetry operations in a complex vector space. This will be important in quantum mechanics, as will be seen in Chapter **3**.) Thus we must have,

,16 $$\mathbf{r}^{\dagger}\mathbf{s} = (A\mathbf{r})^{\dagger}A\mathbf{s} = \mathbf{r}^{\dagger}A^{\dagger}A\mathbf{s} \quad \Rightarrow \quad A^{\dagger}A = \mathbf{1}. \tag{38}$$

(A and $A^{\dagger}$ can be commuted on R38; see Problem 8. It can also be seen that the product of any number of unitary matrices is unitary; see Problem 10.) It follows that the complex number $\det A$ is unimodular:

38 $$\det A^{\dagger} \det A = 1 \quad \Rightarrow \quad \det (A^*)^{\mathrm{T}} \det A = (\det A)^* \det A = 1, \tag{39}$$

from which it follows that

$$|\det A| = 1. \tag{40}$$

The columns and rows of a unitary matrix are orthonormal in the Hermitian sense, i.e. with respect to the Hermitian inner product, eqn (11), (see Problem 9):

$$\sum_r A^*_{ri}A_{rj} = \delta_{ij}; \qquad \sum_r A_{ir}A^*_{jr} = \delta_{ij}. \tag{41}$$

We shall now prove that the eigenvalues of a unitary matrix are unimodular. Consider:

$$A\mathbf{r} = a\mathbf{r}. \tag{42}$$

Form the adjoint of (42), remembering that the adjoint of a is a^*:

16 $$\mathbf{r}^{\dagger}A^{\dagger} = a^*\mathbf{r}^{\dagger}. \tag{43}$$

42,43 $$\mathbf{r}^{\dagger}A^{\dagger}A\mathbf{r} = aa^*\mathbf{r}^{\dagger}\mathbf{r}. \tag{44}$$

38|44 $$\mathbf{r}^{\dagger}\mathbf{r} = aa^*\mathbf{r}^{\dagger}\mathbf{r} \quad \Rightarrow \quad aa^* = 1; \quad |a| = 1, \tag{45}$$

since $\mathbf{r}^{\dagger}\mathbf{r}$ is positive definite.

The form of the most general 2×2 unitary matrix can easily be obtained (see Problem 11):

$$A = \begin{bmatrix} a & b \\ -b^*\Delta & a^*\Delta \end{bmatrix}, \tag{46}$$

where Δ is an arbitrary complex number of unit modulus (notice that it is $\det A$), and aa^*+bb^* must be unity.

We shall now prove that a *real orthogonal* matrix is a particular case of a unitary matrix. It must satisfy, of course, $A=A^*$ and

$$A^{\mathsf{T}}A=\mathbf{1} \qquad \Rightarrow \qquad (A^{\mathsf{T}})^*A^*=\mathbf{1} \qquad \Rightarrow \qquad A^{\dagger}A=\mathbf{1}. \tag{47}$$

Thus the eigenvalues of a real orthogonal matrix must satisfy (45), i.e. be unimodular, as we found above, after eqn (31).

Hermitian and skew-Hermitian matrices

Hermitian and skew-Hermitian matrices satisfy respectively the conditions

$$A=A^{\dagger}, \qquad A=-A^{\dagger}. \tag{48}$$

Notice that the product of two Hermitian matrices is not Hermitian unless the matrices commute (Problem 12). Hermitian and skew-Hermitian matrices have real and pure-imaginary eigenvalues respectively. (Problem 13.) Eigenvectors belonging to different eigenvalues are orthogonal in both cases. (Problem 14.)

Real symmetric and *real skew-symmetric* matrices are particular cases of Hermitian and skew-Hermitian matrices respectively:

$$A=A^*, \qquad A=\pm A^{\mathsf{T}} \qquad \Rightarrow \qquad A=\pm A^{\dagger}. \tag{49}$$

Thus real symmetric matrices have real eigenvalues, real skew-symmetric matrices have imaginary eigenvalues and both have orthogonal eigenvectors.

The remainder of this section will be concerned with a few useful matrix operations.

Supermatrices and the direct product

Any matrix A of matrix elements A_{ij} can be regarded as a *supermatrix*, the matrix elements of which are themselves matrices. To denote that this is so, such supermatrix matrix elements will be denoted with suffices surrounded by square brackets, such as $A_{[ij]}$. Thus a 3×3 matrix A can be denoted as a 2×2 supermatrix with $A_{[11]}$ equal to A_{11}, $A_{[12]}$ the row vector $\langle A_{12}A_{13}|$, $A_{[21]}$ the column vector $|A_{21}A_{31}\rangle$ and $A_{[22]}$ the square matrix with A_{22}, A_{23} and A_{32}, A_{33} for the first and second rows respectively.

A matrix element of a supermatrix product will itself be a sum of matrix products, and for this to make sense one must ensure that the matrices that appear in such products are *conformable*. By this we mean the following. Consider a matrix A of m rows by n columns and write these values as A^{mn}. Then for any matrix product $A^{mn}B^{pq}$ to be possible, it is required that n and p be equal, in which case the matrices are said to be conformable. We shall now assume our supermatrices to be always square and of dimension ν as a supermatrix. If we now write a $\nu \times \nu$ matrix in which we put the superscript mn of $A_{[ij]}^{mn}$ in the ij position, we must get an array, which we shall call the *structure* of the supermatrix (since it gives the dimensions of its elements), that must look like this:

$$\begin{matrix} (mm) & (mn) & (mp) \\ (nm) & (nn) & (np). \\ (pm) & (pn) & (pp) \end{matrix}$$

In fact: all matrices on the same row (column) of the supermatrix must have the same number of rows (columns), and we further require (for reasons which will presently be clear) that the element in the ij position of this array be obtained by transposition of the element in the ji position of the array. It follows at once that as long as we multiply supermatrices with the same structures, then the matrices that appear in the product

$$\sum_r A_{[ir]}B_{[rj]} = C_{[ij]} \tag{50}$$

will be conformable, the number of columns of the first factor $A_{[ir]}$ being m, n, p . . . and the number of rows of the second factor being precisely this sequence also.

Supermatrix notation simplifies profoundly the direct-product expressions. We recall that the direct-product matrix $\hat{W}_{rs,ij}$ (see eqns 5.28, 5.22) has its rows and columns labelled by the double suffices

$$11, 12, \ldots, 1n, 21, 22, \ldots, 2n, 31, 32, \ldots, 3n, \ldots . \tag{51}$$

Thus all elements of the form, say $\hat{W}_{3s,2j}$ $(s = 1, \ldots, n; j = 1, \ldots, n)$ will form an $n \times n$ matrix that can conveniently be labelled as $\hat{W}_{[32]}$, and so on. Hence the multiplication rule (5.28)

$$\hat{W}_{rs,ij} = \hat{U}_{ri}\hat{V}_{sj} \tag{52}$$

can be written as

$$\hat{W}_{[ri]} = \hat{U}_{ri}\hat{V}, \tag{53}$$

since, as it follows from the above, $\hat{V}_{sj}$ for $s = 1, \ldots, n$, $j = 1, \ldots, n$ gives the matrix $\hat{V}$.

In general, for any two matrices A and B,

$$(A \otimes B)_{[ri]} = A_{ri}B. \tag{54}$$

That is, each supermatrix element is the product of the corresponding matrix element of the first matrix times the whole of the second matrix.

(Remember that in matrix work a constant a multiplying a $n \times n$ matrix is always understood to mean the constant matrix $a\,\mathbf{1}_n$ and thus it multiplies each element of the post-factor.)

The following important direct-product equality is very easy to prove in supermatrix notation (see Problem 16):

$$(A \otimes B)(C \otimes D) = AC \otimes BD. \tag{55}$$

Commutation of matrices

We have seen in §2 that the question of commutation of symmetry operations can best be handled by that of their matrix representatives, since the latter follow the multiplication rules of the operators. We shall now briefly discuss, therefore, the theorems that establish the commutativity of two matrices.

We consider first the simpler case of diagonal matrices. It is pretty obvious that two diagonal matrices always commute. It does not follow, however, that a matrix B which commutes with a diagonal matrix A must itself be diagonal (a result sometimes erroneously assumed). This bogus result happens to be true if all the matrix elements of the diagonal matrix A are different, but not if some of them are equal. Without loss of generality, we shall always assume that all equal elements are gathered in blocks along the diagonal, that is that the diagonal matrix A *is a diagonal supermatrix the diagonal elements of which are constant matrices.* (Since the diagonal elements are the eigenvalues of the matrix, each element of the supermatrix will correspond to one degenerate eigenvalue and we can safely assume that no two supermatrix elements are equal.) Under these conditions, *any matrix B that commutes with A must be a diagonal supermatrix of the same structure as A.* (Clearly, B is not diagonal, but rather block-diagonal.) The proof is as follows. From the commutation of A and B and the supermatrix multiplication rule, we have, on choosing the structure of B as that determined by that of A,

$$\sum_r A_{[ir]}B_{[rj]} = \sum_s B_{[is]}A_{[sj]}. \tag{56}$$

Because A is diagonal, it follows that

$$A_{[ii]}B_{[ij]} = B_{[ij]}A_{[jj]}. \tag{57}$$

57 $$(A_{[ii]} - A_{[jj]})B_{[ij]} = \mathbf{0} \quad \Rightarrow \quad B_{[ij]} = \mathbf{0} \quad \text{for} \quad i \neq j. \tag{58}$$

It is not too difficult to prove now that *A and B commute if and only if common eigenvectors to both matrices can be found.* (See Problem 20.)

Matrix functions

The reader is expected not to be worried by expressions such as the sine of a matrix A. The general definition of a function $f(A)$ of a matrix is very simple: you first expand $f(x)$ in power series. In the latter, the terms x^n are still meaningful when x is replaced by A, giving A^n. Thus:

$$f(x)=\sum_n c_n x^n \qquad \Rightarrow \qquad f(A)=\sum_n c_n A^n. \tag{59}$$

Clearly, problems of convergence will arise, but we shall not be concerned with them.

It should be noticed from (59) that

$$\{f(A)\}^{\mathrm{T}}=f(A^{\mathrm{T}}), \tag{60}$$

and, similarly, as long as the power series is real, that

$$\{f(A)\}^{\dagger}=f(A^{\dagger}). \tag{61}$$

A particularly useful matrix function is the exponential. We shall prove with it a very important relation:

If A is hermitian exp (iA) *is unitary.* Proof:

$$\begin{aligned}\{\exp(\mathrm{i}A)\}^{\dagger}\exp(\mathrm{i}A)&=\exp(-\mathrm{i}A^{\dagger})\exp(\mathrm{i}A)\\&=\exp \mathrm{i}(A-A^{\dagger})=\exp \mathbf{0}=\mathbf{1}.\end{aligned} \tag{62}$$

(Notice that $\mathbf{0}$ is the zero matrix and that $\exp \mathbf{0}$ is thus $\mathbf{1}_n$, n being the dimension of A.) The converse of the italicized statement is in fact true. (See Fano 1971, p. 91.) Further examples appear in Problems 21–23.

Warning: We have assumed in the above proof a result which is not general, namely that $\exp U \exp V=\exp(U+V)$. This is true if and only if U and V commute, as can be seen by multiplying the corresponding power series. This is valid in (62) because for Hermitian matrices A and $A^{\dagger}$ (i.e. A and A) always commute. The general form of the expression $\exp U \exp V$ is given by the Baker–Campbell–Hausdorff theorem. (Normand 1980, p. 437.)

Problems 2-6 (p. 272)

1. Prove that $\det A=\det(L\,A\,L^{-1})$.
2. Prove that the trace and determinant of a matrix are respectively the sum and the product of the eigenvalues.
3. Find $\mathbf{r}^{\mathrm{T}}\mathbf{r}$ and $\mathbf{r}^{\dagger}\mathbf{r}$ for $\mathbf{r}=|1, i\rangle$.
4. Find $(AB)^{\mathrm{T}}$; $(AB)^{\dagger}$, $(AB)^{-1}$.
5. Prove that, subject to a condition, $A^{\mathrm{T}}A=\mathbf{1} \Rightarrow AA^{\mathrm{T}}=\mathbf{1}$. Under what condition is this true?
6. Prove that if A and B are orthogonal, so is AB.
7. Prove that the columns and rows of an orthogonal matrix are orthonormal.
8. Prove that $A^{\dagger}A=\mathbf{1} \quad \Rightarrow \quad AA^{\dagger}=\mathbf{1}$.

9. Prove that the columns and rows of a unitary matrix are orthonormal.
10. Prove that if A and B are unitary, so is AB.
11. Find the most general form of a 2×2 unitary matrix.
12. Prove that if A, B are Hermitian, AB is Hermitian if and only if $[A, B]=\mathbf{0}$.
13. Prove that Hermitian (skew-Hermitian) matrices have real (imaginary) eigenvalues.
14. Prove that eigenvectors $\mathbf{r}$, $\mathbf{s}$, belonging to two different eigenvalues of a Hermitian or skew-Hermitian matrix, $A\mathbf{r}=a\mathbf{r}$, $A\mathbf{s}=b\mathbf{s}$, $a\neq b$, are orthogonal.
15. Form $A\otimes B$ for $A=\begin{bmatrix}1 & \mathrm{i}\\ \mathrm{i} & -1\end{bmatrix}$, $B=\begin{bmatrix}2 & 1\\ -1 & 2\end{bmatrix}$.
16. Prove: $(A\otimes B)(C\otimes D)=AC\otimes BD$.
17. Prove that $\operatorname{Tr}(A\otimes B)=\operatorname{Tr}A\ \operatorname{Tr}B$ and hence find the character of g_ig_j in the representation given by ${}^r\hat{G}(g_i)\otimes{}^s\hat{G}(g_j)$.
18. Prove that $(A\otimes B)^\dagger=A^\dagger\otimes B^\dagger$. (63)
19. Prove that if A and B are unitary, so is $A\otimes B$.
20. Prove that the necessary and sufficient condition for $[A, B]=\mathbf{0}$ is that common eigenvectors for A and B can be chosen.
21. Prove that A skew-symmetric entails $\exp A$ orthogonal.
22. Prove that if A is real skew-symmetric then $\exp A$ is unitary, but that, if $\exp A$ is unitary, then A is skew-Hermitian.
23. Prove that if A is $\mathrm{i}\mathbf{1}$, then $\exp(\phi A)$ is $\mathbf{1}\cos\phi+A\sin\phi$.
24. Prove that a real symmetric matrix cannot be a general rotation matrix.

7. QUANTAL SYMMETRY. OBSERVABLES AND INFINITESIMAL GENERATORS

When we defined the function-space operators in §3 we required that the transformed function at the transformed point kept precisely the same value as the original function at the original point. This condition is too strong in quantum mechanics since it is well known that the wave functions $\psi(x)$ and $\omega\psi(x)$, for any constant ω, denote identical physical states. Thus, the fundamental relation (3.1) must be somewhat modified:

$$\dot{g}f(x)=\omega f(g^{-1}x). \tag{1}$$

We shall see that ω must be a phase factor, that is a unimodular complex number. As before, we shall drop the dot on $\dot{g}$, since it will be clear that the operators in this section are function-space operators.

A quantal symmetry operator must conserve three things: (i) all transition probabilities, (ii) mean values of all observables, (iii) eigenvalues of all observables. Let us consider the first condition. The transition probability P from the initial state φ to the final state ψ is given in terms of the inner product in Dirac's notation as $\langle\psi\mid\varphi\rangle^*\langle\psi\mid\varphi\rangle$ where $\langle\psi\mid\varphi\rangle$ is $\int \psi^*\varphi\, d\tau$. Therefore, for any symmetry operator g we must have

$$\langle g\psi \mid g\varphi\rangle^*\langle g\psi \mid g\varphi\rangle = \langle\psi\mid\varphi\rangle^*\langle\psi\mid\varphi\rangle, \tag{2}$$

which requires g to satisfy one of the following two relations:

$$\langle g\psi \mid g\varphi\rangle = \langle\psi\mid\varphi\rangle, \tag{3}$$

$$\langle g\psi \mid g\varphi\rangle = \langle\psi\mid\varphi\rangle^*. \tag{4}$$

In the first case, remembering that any operator can always be taken to the other side of a bracket as its adjoint, we have

$$\langle g^\dagger g\psi \mid \varphi\rangle = \langle\psi\mid\varphi\rangle, \tag{5}$$

which entails that

$$g^\dagger g = 1, \tag{6}$$

(we write here 1 for the identity operator). This, of course, defines a *unitary operator*. (Compare with the definition in eqn 6.38 of a unitary matrix.) In the second case, eqn (4), the operator is called *antiunitary*. It is known that all operators related to space transformations are unitary, whereas the time reversal operator is antiunitary. We shall be concerned only with the first type of operator in this book, so that we shall assume that all our operators are unitary.

It is clear from eqn (3) that $g\varphi$ and $g\psi$ can both be multiplied by the same factor ω, as long as $\omega^*\omega = 1$. (Remember that because $g\psi$ is on the left of the bracket, it is taken as complex conjugate.) Thus, each g can be associated with one such phase factor in the manner of eqn (1). This phase factor, though, must be the same for all operands of g if (3) is to be maintained, and a lot of technical problems may arise if phase factors, which otherwise do not affect single states, are not consistently used.

We shall now consider the second conservation condition, that of the mean values, and we shall show that the necessary and sufficient condition for the conservation of the mean value $\langle A\rangle$ of an operator A under the action of an operator g is

$$[g, A] = 0. \tag{7}$$

The proof is as follows. The mean value of the operator A is

$$\langle A\rangle = \langle\psi|\, A\, |\psi\rangle, \tag{8}$$

but when ψ is changed into $g\psi$, A changes into A':

$$\langle A'\rangle = \langle g\psi|\, A'\, |g\psi\rangle \tag{9}$$

$$= \langle\psi|\, g^\dagger A' g\, |\psi\rangle. \tag{10}$$

From (8) and (10), it is necessary and sufficient for $\langle A\rangle$ to equal $\langle A'\rangle$ that

$$g^{\dagger}A'g = A \qquad \Leftrightarrow \qquad A' = gAg^{\dagger}. \tag{11}$$

On the other hand, the unitary property of g in (9) ensures the conservation of the inner product:

$$\langle A'\rangle = \langle\psi|\, A'\, |\psi\rangle \tag{12}$$

11 $$= \langle\psi|\, gAg^{\dagger}\, |\psi\rangle. \tag{13}$$

On comparing (8) and (13), it follows that it is necessary and sufficient for the conservation of the mean value that the operators commute as in (7). All symmetry operators, therefore, must commute with the Hamiltonian H.

The third conservation condition, that of the eigenvalues, is a consequence of the above. Consider the eigenvalue equation

$$A\psi = a\psi, \tag{14}$$

and change ψ into $\psi' = g\psi$, whereupon A and a change also into primed quantities

$$A'\psi' = a'\psi'. \tag{15}$$

11|15 $$gAg^{\dagger}g\psi = a'g\psi. \tag{16}$$

6|16 $$gA\psi = a'g\psi. \tag{17}$$

Now, because g is linear, on acting with it on (14), it gives

$$gA\psi = ag\psi. \tag{18}$$

17,18 $$a = a'.$$

As a corollary, ψ and $g\psi$ are degenerate, because, commuting gA on (L18), we have

$$Ag\psi = ag\psi, \tag{19}$$

which, on comparison with (14) establishes the degeneracy.

Symmetries and observables

It is well known that, subject to certain conditions, the mean value of an observable A is time independent if

$$[A, H] = 0, \tag{20}$$

where H is the Hamiltonian operator (see Problem 1). Such an observable is called a *constant of the motion*. It is a remarkable result that a symmetry operation g can be associated with every constant of the motion, by the relation

$$g = \exp(iA) = \sum_n (iA)^n/n!. \tag{21}$$

That g is a symmetry operator follows from the fact that it satisfies the following four properties:

(i) g is linear (because R21 is linear).

(ii) g is unitary because $g^\dagger g$ is $\exp(-iA)\exp(iA)$, clearly unity.

(iii) $[g, A]=0$, as follows from (R21).

(iv) $[g, H]=0$, as follows from (R21) and (20).

Infinitesimal operators and observables

It is important to realize that although, as we have just seen, every constant of the motion leads to a symmetry operation, the converse result is not true in general. (Time reversal, e.g., is not associated with a constant of the motion.) For the unitary operators in $\mathbb{R}^3$ which we are considering, however, it is always possible to find an observable associated with every symmetry operation that forms a one-parameter continuous group. (More precisely, a one-parameter Lie group, see the end of §5.) We shall illustrate this for one-dimensional function-space translation operators $\dot{t}$. These operators form a one-parameter continuous group because their corresponding displacements t in configuration space vary continuously, t being the group parameter that denotes each group element. Our proof will demonstrate a remarkable property which obtains for such groups, as long as they satisfy certain analytical conditions. This property is that all their operations can be generated by means of infinitesimal operations which, infinitely repeated (i.e. integrated) produce any group operation. It will be easy to demonstrate, in its turn, that the infinitesimal operator (unitary) is given in terms of an operator which is Hermitian and thus an observable. This observable thus related to the infinitesimal operator is called an *infinitesimal generator* of the group in question. We shall now see how all this works.

Consider an infinitesimal displacement δ along the x axis, which we assume to be a symmetry operation, so that it commutes with H. Its corresponding translation operator $\dot{\delta}$ is given by

$$\dot{\delta}f(x)=f(x-\delta)=f(x)-\delta\frac{\mathrm{d}}{\mathrm{d}x}f(x)=\left(1-\delta\frac{\mathrm{d}}{\mathrm{d}x}\right)f(x). \tag{22}$$

Here we have expressed $f(x-\delta)$ from the definition of the derivative $\mathrm{d}f(x)/\mathrm{d}x$, which equals $\delta^{-1}\{f(x)-f(x-\delta)\}$, for $\delta\to 0$. From (22):

$$\dot{\delta}=1-\delta\,\mathrm{d}/\mathrm{d}x. \tag{23}$$

Call

$$\mathrm{d}/\mathrm{d}x =_{\mathrm{def}} \mathrm{i}p. \tag{24}$$

Then

$$\dot{\delta}=1-\mathrm{i}p\delta, \tag{25}$$

and it follows that, to the first order (as kept in eqn 22),

$$\dot{\delta}\dot{\delta}^{\dagger}=(1-\mathrm{i}p\delta)(1+\mathrm{i}p^{\dagger}\delta)=1+\mathrm{i}(p^{\dagger}-p)\delta+O\delta^{2}. \tag{26}$$

Since $\dot{\delta}$ is unitary, whence (L26) is unity, then from (R26) $p^{\dagger}-p$ must vanish, so that p is Hermitian. Thus (25) shows how the infinitesimal translation $\dot{\delta}$ leads to an observable p. Moreover, since $\dot{\delta}$ must commute with H (as δ does) it follows at once from (R25) that p also commutes with H and it is thus a constant of the motion. (This is, of course, the *linear momentum.*) We now have to show how $\dot{\delta}$ can be integrated into a finite translation $\dot{t}$, which can be expressed as follows:

$$\dot{t}f(x)=f(x-t),\ t=\lim_{n\to\infty} n\delta \quad \Rightarrow \quad \dot{t}=\lim_{n\to\infty}(\dot{\delta})^{n}. \tag{27}$$

(The last expression can be obtained by repeated application of $\dot{\delta}$ on L22.) Now

25|27
$$\dot{t}=\lim_{n\to\infty}\left(1-\mathrm{i}p\frac{t}{n}\right)^{n}=\exp(-\mathrm{i}pt). \tag{28}$$

We see in (28) that the infinitesimal generator p (constant of the motion) determines in fact each (finite) group operation.

The reader should notice that all we have done in the above is to re-derive the Taylor series in a very compact form. (See Problem 2.)

Problems 2-7 (p. 276)

1. Show that, given an operator A that does not explicitly depend on the time, i.e. $\partial A/\partial t=0$, then the necessary and sufficient condition for $\langle A\rangle$ to be time-independent is $[A, H]=0$.
2. Given an infinitesimal operator $\dot{\tau}$ such that $\dot{\tau}f(x)=f(x+\delta)$, derive $f(x+t)$ for finite t. Show that $f(x+t)=\exp(t\mathrm{d}/\mathrm{d}x)f(x)$ and that this coincides with the Taylor series. Compare with (28).
3. Prove the commutator identities

$$[u\,v, w]=u[v, w]+[u, w]v. \tag{29}$$

$$[u, vw]=[u, v]w+v[u, w]. \tag{30}$$

4. The following argument is sometimes used for the inversion: $\boldsymbol{i}\psi(x)=\psi(-x)$, whence $\boldsymbol{i}^{2}\psi(x)=\psi(x)$, so that $\boldsymbol{i}^{2}=E$. Thus, the inversion classifies all functions under the symmetry group $\mathbf{C}_i$ (Problem 5.11) whence all functions must be gerade or ungerade. Show that this argument is not necessarily valid in quantum mechanics.

3

A PRIMER ON ROTATIONS AND ROTATION MATRICES

> For example two spherical triangles on opposite hemispheres which have an arc of the equator as their common base can be completely equal, in respect of sides as well as angles, so that nothing is found in either, when it is described alone and completely, which does not appear in the description of the other, and yet one cannot be put in the place of the other (on the opposite hemisphere).
>
> I. Kant (1783), *Prolegomena to any future metaphysics that will be able to present itself as a science.* Trans. P. G. Lucas. University Press, Manchester (1953), §13.

We shall be concerned in this chapter with a fairly dull problem which is, first, to identify the proper rotations in $\mathbb{R}^3$ (alluded to in the epigraph) and secondly, to obtain their corresponding 3×3 matrices. We shall thus work entirely in configuration space.

It is clear that we shall be concerned with the special orthogonal matrices (determinant $+1$), of SO(3) as defined in §**2**-6. Because these matrices represent proper rotations in $\mathbb{R}^3$, the symbol SO(3) is also used to designate the proper rotation group, whereas O(3) (see §**2**-6) denotes the full rotation group, that is the group of all proper and improper rotations around a point. We have seen in §**2**-6 (after eqn 36), that 3×3 orthogonal matrices are fully determined by three parameters. They will be chosen in this chapter in two different ways. The first, discussed in §1, uses three angles, called the Euler angles. The second (§§2,3) uses the angle of rotation ϕ and the axis of rotation $\mathbf{n}$, i.e. the already-used $R(\phi\mathbf{n})$ symbol (see §**2**-1). Since the three components of the unit vector $\mathbf{n}$ are related by one normalization condition, this slightly redundant parametrization also reduces to three free parameters. Later on, we shall discuss the parametrization of SO(3) in more detail (§6) since we shall see that some delicate points have to be carefully considered. Also later (§**9**-4) we shall introduce another set of parameters, the Euler–Rodrigues or quaternion parameters, which, as we shall see, present notable advantages.

Before starting the work properly, we must discuss the range of the angles used in this parametrization. As one would expect, this range often (but not always) is, for some angle ω,

$$0 \leqslant \omega < 2\pi. \tag{1}$$

Although this choice is very common, it is also very injudicious. It is easy to guess that a zero angle will be associated with the identity of SO(3). Now, the identity of any group is an important thing and the identity of a continuous group is a precious thing because, as we have seen in §**2**-7, the elements in its neighbourhood (i.e. the infinitesimal generators) generate, under certain conditions, the whole group. We must therefore define the range in such a way that the whole of the neighbourhood of the identity – that is, all small positive as well as all small negative angles – be contained in the range, which is not the case in (1). It is necessary for this reason to keep the identity safe and sound in the middle of the range, as follows,

$$-\pi < \omega \leqslant \pi, \tag{2}$$

and whenever necessary we shall stick to this choice.

1. EULER ANGLES

Euler angles have been so much used that, like all strong currencies, they have suffered much counterfeiting. For this reason (also because they will not be much traded in this book) we shall not discuss the very many different ways in which they can be presented. We have chosen our definition so that it agrees with the one that appears now to be most popular. It coincides exactly with the notation used by Biedenharn and Louck (1981), Brink and Satchler (1968), Butler (1981), Rose (1957), and, on replacing their ψ, θ, φ by α, β, γ respectively, with Fano and Racah (1959). Also, although Messiah (1964) presents his definitions in a different way, they entirely agree with the ones used here.

In order to get our conventions right, we restate that we always use the active convention in configuration space (transform the vectors of the space). Correspondingly, the functions in this space are transformed actively. Moreover, the matrices of transformation are always defined so as to transform the basis functions by post-multiplying the row basis which they form. Since such a basis transforms precisely in the same manner as the unit vectors **i**, **j**, **k** in configuration space (compare eqns **2**-4.11 and **2**-4.2), we define the rotation matrices, and thus the rotations themselves, by their action on **i**, **j**, **k**. This does not mean that we have gone passive: that we are in the active convention will be expressed through the fact that we shall transform **i**, **j**, **k** in every case *with respect to fixed axes* **x**, **y**, **z** *in space* and that **i**, **j**, **k** carry solidly with them the whole of the configuration space.

A rotation will change the set **i**, **j**, **k** into a new set and the definition of the three Euler angles entails three successive rotations around the fixed axes which transform the old **i**, **j**, **k** into the new ones. These three rotations are defined as follows.

$$\text{First rotation by } \gamma \text{ around } \mathbf{z},\ R(\gamma\mathbf{z}). \quad (1)$$

$$\text{Second rotation by } \beta \text{ around } \mathbf{y},\ R(\beta\mathbf{y}). \quad (2)$$

$$\text{Third rotation by } \alpha \text{ around } \mathbf{z},\ R(\alpha\mathbf{z}). \quad (3)$$

Thus the total rotation, designated $R(\alpha\beta\gamma)$, is given by:

$$R(\alpha\beta\gamma) = R(\alpha\mathbf{z})R(\beta\mathbf{y})R(\gamma\mathbf{z}). \quad (4)$$

Notice: (i) that α is not the first but the third angle; (ii) that older definitions in which the second rotation is taken around **x** do not agree with current angular momentum conventions (see §**4**-8 and Problem **6**-7.1).

It should be noticed that the above definition is not sufficient to determine unequivocally the angles α, β, γ, as can best be seen by an example. We illustrate in Fig. 1 a rotation by $-\frac{2}{3}\pi$ around the (111) axis of a cube. On taking the fixed **x**, **y**, **z** directions shown, we can go from Fig. 1*a* to Fig. 1*b* in two ways, always following (4), as shown in Fig. 2. Because the final **i** is along **z** it cannot be altered by the third rotation and must thus be set in position by the second one. In the first rotation the **i**, **k** plane must be set normal to **y**, so as to allow **i** to be rotated to its final position by the second rotation. But there are two ways in which this can be achieved, with $\mathbf{i}\times\mathbf{k}$ along positive or negative **y**, as shown in the bottom and top of Fig. 2 respectively. The top and bottom circuits would lead respectively to

$$\gamma = 0, \qquad \beta = -\tfrac{1}{2}\pi, \qquad \alpha = -\tfrac{1}{2}\pi, \quad (5)$$

$$\gamma = \pi, \qquad \beta = \tfrac{1}{2}\pi, \qquad \alpha = \tfrac{1}{2}\pi. \quad (6)$$

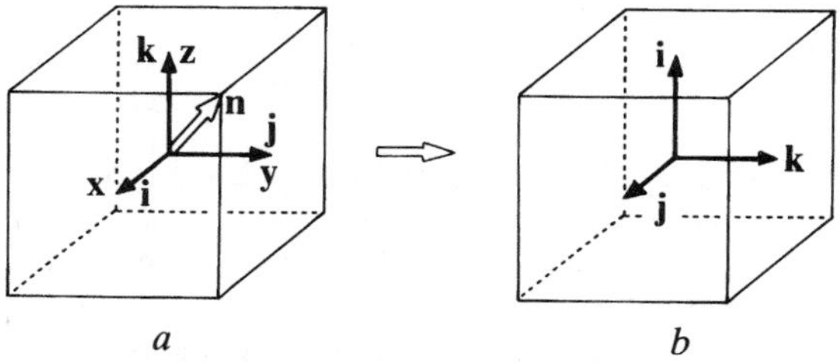

Fig. 3-1.1. Rotation by $-\frac{2}{3}\pi$ around the axis **n**, which is (111).

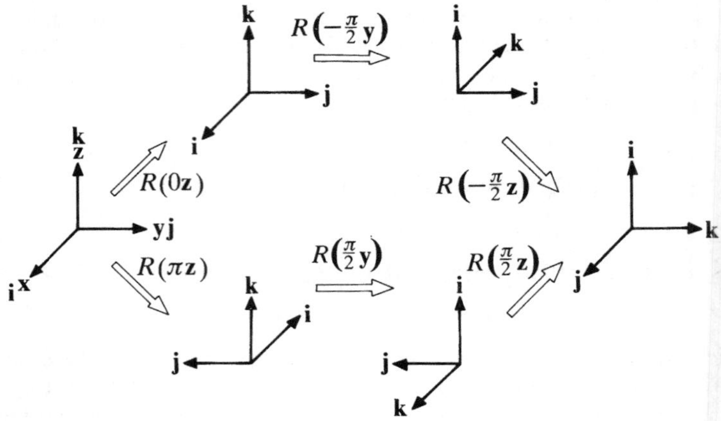

Fig. 3-1.2. Choice of Euler angles. The axes **x**, **y**, **z** are space-fixed axes.

The device used for changing the sign of β here should be noted. It is based on (**2**-1.6) from which,

$$R(\pi\mathbf{z})R(\beta\mathbf{y})=R(-\beta\mathbf{y})R(\pi\mathbf{z}). \tag{7}$$

Therefore, for a general rotation $R(\alpha,-\beta,\gamma)$,

4 $$R(\alpha,-\beta,\gamma)=R(\alpha\mathbf{z})R(-\beta\mathbf{y})R(\pi\mathbf{z})R(\pi\mathbf{z})R(\gamma\mathbf{z}) \tag{8}$$

2-1.6 $$=R(\alpha\mathbf{z})R(\pi\mathbf{z})R(\beta\mathbf{y})R(\pi\mathbf{z})R(\gamma\mathbf{z}) \tag{9}$$

$$=R(\alpha\pm\pi,\mathbf{z})R(\beta\mathbf{y})R(\gamma\pm\pi,\mathbf{z})=R(\alpha\pm\pi,\beta,\gamma\pm\pi). \tag{10}$$

We have exploited here the fact that $R(\pi\mathbf{z})$ can also be interpreted as $R(-\pi\mathbf{z})$, which must sometimes be done in order to get an angle such as $\alpha\pm\pi$ in a desired range.

In order to avoid the ambiguity shown by (5) and (6) or, more generally, by (10), we agree to *take the second Euler angle* β *always as positive*. The ranges for the angles, taking into account the convention stated in eqn (2) of the last section, are:

$$-\pi<\gamma\leqslant\pi, \qquad 0\leqslant\beta\leqslant\pi, \qquad -\pi<\alpha\leqslant\pi. \tag{11}$$

Since all rotations are around the fixed **x**, **y**, **z** axes, a positive rotation is defined as that which takes the **x** into the **y** axis when viewed from the positive branch of the **z** axis, with cyclic permutations of **x**, **y**, **z**.

Some unavoidable ambiguities remain, however: when β is zero, the only two non-vanishing rotations are about the **z** axis, so that their sum $\gamma+\alpha$ is determined, but not the individual values of γ and α. Also, if

$\beta = \pi$, then (from Problem 1), $R(\alpha\pi\gamma)$ and $R(\alpha+\omega, \pi, \gamma+\omega)$ entail the same rotation for any arbitrary angle ω.

Given $R(\alpha\beta\gamma)$, its inverse, clearly, is obtained from (4) by taking the inverses in the opposite order:

$$\{R(\alpha\beta\gamma)\}^{-1} = R(-\gamma, -\beta, -\alpha) \tag{12}$$

10|12
$$= R(-\gamma \pm \pi, \beta, -\alpha \pm \pi). \tag{13}$$

(Because of the special relation for $\beta = \pi$ just discussed, the π's here can be ignored when $\beta = \pi$.) Equation (12) is, however, the one most commonly, but incorrectly, quoted in the literature. (Remember that β must be always positive.)

It should be evident that the determination of the Euler angles for an arbitrary rotation is a delicate art. In crystallographic point-groups, however, one of the **i**, **j**, **k** axis, say **u**, must always be taken to the **z** direction (positive or negative) by any rotation in the group. By the same principles used in Fig. 2, γ must be chosen so as to orient the **uz** plane normal to **y** in such a way that a positive rotation around **y** should take **u** to its final setting. This determines β, of course, and α follows at once. Formulae will later on be given to determine α, β, γ from the parameters $\phi\mathbf{n}$ of a rotation.

Rotation matrix in terms of the Euler angles

This is extremely easy to obtain, which accounts partly for the popularity of the Euler angles. All we need are the matrices $\hat{R}(\alpha\mathbf{z})$, $\hat{R}(\beta\mathbf{y})$, $\hat{R}(\gamma\mathbf{z})$ in (4). These are most easily obtained from the rotation matrices in §**2**-4. Admittedly, in (**2**-4.2), we only consider the basis $\langle\mathbf{i}, \mathbf{j}|$ but, since **k** is invariant under rotation around **z**, the change is trivial. For $R(\beta\mathbf{y})$ a permutation of the variables leads easily to the matrix (Problem 3). The matrices are shown in (14) and their product (4) is given in (15):

$$\overset{\hat{R}(\alpha\mathbf{z})}{\begin{bmatrix} \cos\alpha & -\sin\alpha & 0 \\ \sin\alpha & \cos\alpha & 0 \\ 0 & 0 & 1 \end{bmatrix}}, \quad \overset{\hat{R}(\beta\mathbf{y})}{\begin{bmatrix} \cos\beta & 0 & \sin\beta \\ 0 & 1 & 0 \\ -\sin\beta & 0 & \cos\beta \end{bmatrix}}, \quad \overset{\hat{R}(\gamma\mathbf{z})}{\begin{bmatrix} \cos\gamma & -\sin\gamma & 0 \\ \sin\gamma & \cos\gamma & 0 \\ 0 & 0 & 1 \end{bmatrix}}. \tag{14}$$

$$\hat{R}^1_r(\alpha\beta\gamma) = $$
$$= \begin{bmatrix} \cos\alpha\cos\beta\cos\gamma - \sin\alpha\sin\gamma & -\cos\alpha\cos\beta\sin\gamma - \sin\alpha\cos\gamma & \cos\alpha\sin\beta \\ \sin\alpha\cos\beta\cos\gamma + \cos\alpha\sin\gamma & -\sin\alpha\cos\beta\sin\gamma + \cos\alpha\cos\gamma & \sin\alpha\sin\beta \\ -\sin\beta\cos\gamma & \sin\beta\sin\gamma & \cos\beta \end{bmatrix}. \tag{15}$$

The reader must realize that indiscriminate use of any matrix quoted in group theory can lead into mortal trouble: one must know precisely on what basis and in what form the matrix is supposed to operate. Matrices

(14) and (15) transform: the components $x\,y\,z$ of a vector $\mathbf{r}$, $|xyz\rangle$, (matrix pre-multiplying, of course); the basis $\langle\mathbf{ijk}|$; or the functions $\langle$xyz| of §**2**-4, in the latter two cases the matrix post-multiplying, of course. The matrix symbol used in (L15) indicates, in fact, that it is a 3×3 matrix (3 given as $2j+1$, j being the superscript of $\hat{R}$) and that the basis is $\mathbf{r}$.

The reader should verify that (13) gives the correct results in (15) (Problem 4) and that the latter describes correctly the transformation in Fig. 1. (Problem 5.)

Problems 3-1 (p. 277)

1. Show that $R(\alpha\pi\gamma)=R(\alpha+\omega,\,\pi,\,\gamma+\omega)$, for an arbitrary angle ω.
2. Given the rotation $R(\pi\mathbf{n})$ for $\mathbf{n}=(\frac{\sqrt{2}}{2},\frac{\sqrt{2}}{2},0)$, show geometrically that $R(-\frac{1}{2}\pi,\,\pi,\,0)$ and $R(-\frac{1}{4}\pi,\,\pi,\,\frac{1}{4}\pi)$ are both correct parametrizations and compare this result with Problem 1. Show that the inverse of the first set, as given by (13), returns the transformed axes $\mathbf{i},\mathbf{j},\mathbf{k}$ to their original positions. Compare this inverse with $R(\pi\mathbf{n})$.
3. Obtain $\hat{R}(\beta\mathbf{y})$ in (14).
4. Verify that the parameters (13) for the inverse of $R(\alpha\beta\gamma)$, when substituted in the rotation matrix $\hat{R}^1_{\mathbf{r}}(\alpha\beta\gamma)$ in (15), give the inverse of $\hat{R}^1_{\mathbf{r}}(\alpha\beta\gamma)$.
5. Obtain by inspection the matrix of the rotation displayed in Fig. 1 and show that it agrees with that derived from (15) with the Euler angles in (6).
6. Find the Euler angles and transformation matrix for $R(\phi\mathbf{n})$, $\phi=\frac{2}{3}\pi$, $\mathbf{n}=3^{-1/2}(111)$. Compare with (6).
7. Obtain the Euler angles for all operations of $\mathbf{C}_{3v}$ in Fig. **2**-1.3.

2. ANGLE AND AXIS OF AN ORTHOGONAL MATRIX

In this and the following section we shall be concerned with the parametrization of a rotation in the form $R(\phi\mathbf{n})$. This parametrization is, of course, obvious. There are three easy points to bear in mind. First, ϕ must be in the accepted range from $-\pi$ to π in (2) at the beginning of this chapter. Secondly, $\mathbf{n}$ must be taken to be of unit length. Thirdly,

$$R(\phi\mathbf{n})\equiv R(-\phi,-\mathbf{n}). \qquad (1)$$

This last relation is most important and the reader who is in any doubt should perform the following experiment, for which a pencil and a friend are needed. The pencil will be used to represent the vector $\mathbf{n}$, the tip of the pencil being the head of $\mathbf{n}$. The reader and the friend should view the head and tail of the pencil, respectively: when the reader rotates the pencil by ϕ, i.e. performs $R(\phi\mathbf{n})$, the friend will view (from $-\mathbf{n}$) the rotation of the pencil by $-\phi$, i.e. he or she will view $R(-\phi,-\mathbf{n})$.

Whereas, given the Euler parameters, it is easy to obtain the corresponding rotation matrix, this is not so for the present parameters. We commence, though with a simpler problem. We know that every real orthogonal 3×3 matrix must correspond to a rotation, because it leaves lengths and angles invariant. Also that if their determinant is $+1$ (special orthogonal matrix) then they correspond to a proper rotation. (§**2**-6.) Thus, the problem we want to solve is, given the special, real, orthogonal matrix

$$A=\begin{bmatrix} A_{11} & A_{12} & A_{13} \\ A_{21} & A_{22} & A_{23} \\ A_{31} & A_{32} & A_{33} \end{bmatrix}, \tag{2}$$

to determine ϕ, $\mathbf{n}$.

Determining the angle of rotation is easy. Under some similarity transformation (which leaves the trace invariant) A must transform into a matrix like $\hat{R}(\alpha\mathbf{z})$ in (1.14). This means that under the similarity the $\mathbf{z}$ axis has been chosen to coincide with the rotation axis: the resultant matrix must have the same form as $\hat{R}(\alpha\mathbf{z})$ in (1.14) with ϕ substituted for α. Since the trace of this matrix, $1+2\cos\phi$, must equal that of A, it follows that

$$\cos\phi=\tfrac{1}{2}(\operatorname{Tr} A-1). \tag{3}$$

The axis of rotation $\mathbf{n}$ is the eigenvector corresponding to the eigenvalue $+1$. (The other two eigenvalues are complex conjugate, see eqn **2**-6.31.) The problem therefore, is simple in principle, since it consists in finding the solution of the characteristic equation for A corresponding to the root $+1$. This, however, is tedious and we shall adopt a short cut based on the following observation. A real symmetric matrix (even orthogonal) can never be a general rotation matrix (Problem **2**-6.24). On the other hand, every matrix can be written as a sum of a symmetric and a skew-symmetric matrix:

$$A=\tfrac{1}{2}\{(A+A^{\mathsf{T}})+(A-A^{\mathsf{T}})\}. \tag{4}$$

We shall call S the skew-symmetric 'component' $A-A^{\mathsf{T}}$ of A and we must expect that it will play an important rôle in establishing the properties of rotation matrices, as will indeed be the case. We shall write it as follows:

$$S=A-A^{\mathsf{T}}=\begin{bmatrix} 0 & A_{12}-A_{21} & A_{13}-A_{31} \\ A_{21}-A_{12} & 0 & A_{23}-A_{32} \\ A_{31}-A_{13} & A_{32}-A_{23} & 0 \end{bmatrix}$$

$$=_{\text{def}}\begin{bmatrix} 0 & c & b \\ -c & 0 & a \\ -b & -a & 0 \end{bmatrix}. \tag{5}$$

We shall bring S into the eigenvector equation of A for its eigenvalue equal to unity:

$$A\mathbf{n} = \mathbf{n}. \tag{6}$$

From the orthogonality condition $AA^{\mathsf{T}} = \mathbf{1}$, we obtain

6 $$\mathbf{n} = A^{\mathsf{T}}\mathbf{n}. \tag{7}$$

6|L7 $$A\mathbf{n} = A^{\mathsf{T}}\mathbf{n}. \tag{8}$$

8 $$(A - A^{\mathsf{T}})\mathbf{n} = \mathbf{0} \qquad \Rightarrow \qquad S\mathbf{n} = \mathbf{0}, \tag{9}$$

an equation evidently easier to solve than (6). We write it as follows:

$$S\mathbf{n} = \begin{bmatrix} 0 & c & b \\ -c & 0 & a \\ -b & -a & 0 \end{bmatrix} \begin{bmatrix} n_x \\ n_y \\ n_z \end{bmatrix} = 0. \tag{10}$$

From the first two equations we obtain n_x, n_y in terms of n_z:

$$n_x = an_z/c, \qquad n_y = -bn_z/c, \tag{11}$$

from which expressions, and the normalization condition of $\mathbf{n}$, n_z follows:

$$n_z = \pm c(a^2 + b^2 + c^2)^{-1/2}. \tag{12}$$

In order to obtain $a^2 + b^2 + c^2$, it is useful to notice that it appears on the trace of SS^{T}. In fact, when these matrices are multiplied, one obtains

$$a^2 + b^2 + c^2 = \tfrac{1}{2}\,\mathrm{Tr}\,(SS^{\mathsf{T}}). \tag{13}$$

On the other hand,

$$SS^{\mathsf{T}} = (A - A^{\mathsf{T}})(A^{\mathsf{T}} - A) = 2\mathbf{1} - A^2 - (A^{\mathsf{T}})^2. \tag{14}$$

A and A^2, on diagonalization, will have the form (see eqn **2**-6.31):

$$A' = \begin{bmatrix} 1 & & \\ & e^{i\phi} & \\ & & e^{-i\phi} \end{bmatrix}, \qquad (A')^2 = \begin{bmatrix} 1 & & \\ & e^{i2\phi} & \\ & & e^{-i2\phi} \end{bmatrix}, \tag{15}$$

so that the trace of A^2 must be $1 + 2\cos 2\phi$. The trace of $(A^{\mathsf{T}})^2$ must be the same since A^{T} being the reciprocal of A, its rotation angle is equal and opposite to that of A. Thus, in (14), $\mathrm{Tr}\,(SS^{\mathsf{T}})$ is $8\sin^2\phi$, giving

13 $$a^2 + b^2 + c^2 = 4\sin^2\phi. \tag{16}$$

From (12) n_z follows, whence, from (11),

$$n_x = \pm a(2\sin\phi)^{-1}, \qquad n_y = \mp b(2\sin\phi)^{-1}, \qquad n_z = \pm c(2\sin\phi)^{-1}. \tag{17}$$

The signs are decided by comparing with $R(\phi\mathbf{z})$ from (1.14), for which

$\mathbf{n}=(001)$, whereas c, which from (5) is $A_{12}-A_{21}$, is $-2\sin\phi$ thus requiring the lower sign in (17) in order to get the right result for $\mathbf{n}$. (It is also prudent to check n_x, n_y similarly. See Problem 1.) We thus obtain, on identifying explicitly a, b, c from (5), in (3) and (17):

$$\cos\phi=\tfrac{1}{2}(A_{11}+A_{22}+A_{33}-1), \tag{18}$$

$$n_x=\frac{A_{32}-A_{23}}{2\sin\phi}, \qquad n_y=\frac{A_{13}-A_{31}}{2\sin\phi}, \qquad n_z=\frac{A_{21}-A_{12}}{2\sin\phi}. \tag{19}$$

Notice that (18) leads to two signs of $\sin\phi$. When they are used, $R(\phi\mathbf{n})$ and $R(-\phi, -\mathbf{n})$ are respectively obtained but, from (1), they are the same rotation. Also, when $\phi=0$, $\mathbf{n}$ is undetermined, which is reasonable, since the rotation axis loses sense in this case. An application of (18) and (19) is given in Problem 2.

Problems 3-2 (p. 278)

1. On considering the matrices $R(\phi\mathbf{x})$, $R(\phi\mathbf{y})$, obtained as from eqn (1.14), check the signs of n_x and n_y in (19).
2. Find the axis and angle of rotation implied by the matrix

$$P=\begin{bmatrix} & & 1 \\ 1 & & \\ & 1 & \end{bmatrix}. \tag{20}$$

Compare with Problem 1.6.

3. Prove that conjugate rotations (see eqn **2**-5.2) have the same rotation angle.

3. THE MATRIX OF A ROTATION $R(\phi\mathbf{n})$

We deal in this section with the converse of the problem treated in §2. That is, given the rotation angle ϕ and the rotation axis $\mathbf{n}=(n_xn_yn_z)$, we want to write down the corresponding orthogonal matrix.

The method of solution is very simple. In the matrix A of (2.2) we have nine unknowns $A_{11}, A_{22}, \ldots, A_{33}$. Equation (2.19) provides three equations and the orthonormality conditions of A, six more. In principle then, the nine unknowns can be found. (In practice, the condition $\det A=1$ will also have to be used in order to adjust one arbitrary parameter in the solution.) The work to solve the simultaneous quadratic equations required is, however, ghastly, so that I shall use an ingenious method based on the already proven assumption that the skew-symmetric component of A, the matrix S in (2.5) must be useful in determining the properties of A. We conjecture that, in fact, S will determine A. We first observe that given any skew-symmetric matrix S, an orthogonal matrix A can instantly be constructed by the formula

$$\mathsf{A} = \exp \mathsf{S}. \tag{1}$$

In fact, because S^{T} is $-\mathsf{S}$,

$$\mathsf{A}\mathsf{A}^{\mathsf{T}} = \exp \mathsf{S} \exp(-\mathsf{S}) = \mathbf{1}. \tag{2}$$

We then go on to prove that, if S is chosen from (2.5), the axis of rotation of A coincides with that of A. This axis of rotation, in fact, satisfies $S\mathbf{n} = \mathbf{0}$ (see eqn 2.9) whence $S^m\mathbf{n} = \mathbf{0}, \forall m$. Now, with S in (1)

$$\mathsf{A} = \exp S = \mathbf{1} + S + \frac{1}{2!} S^2 + \dots. \tag{3}$$

$$\mathsf{A}\mathbf{n} = \mathbf{n} + S\mathbf{n} + \frac{1}{2!} S^2\mathbf{n} + \dots = \mathbf{n}, \tag{4}$$

as enunciated above. All that remains to prove is that A, now given by (3), has the same rotation angle as A (i.e. the same trace). In order to do this, we re-define a little the skew-symmetric matrix S so as to introduce in it the rotation axis. For this, we use S as given in (2–10) and replace a, b, c by their values from (2.17), with the lower signs but ignoring for convenience the $2 \sin \phi$ factor, to be absorbed by any scalar factor multiplying the matrix. We thus conveniently construct the skew-symmetric matrix

$$Z = \begin{bmatrix} 0 & -n_z & n_y \\ n_z & 0 & -n_x \\ -n_y & n_x & 0 \end{bmatrix}. \tag{5}$$

Clearly, all that we have proved above for A, that is, for $\exp S$, must still be true for

$$\mathbf{A} = \exp(\phi Z), \tag{6}$$

and when this form is taken for $\mathbf{A}$, it is possible to prove in an elementary way that $\mathbf{A}$ is given in terms of ϕ and Z by the relation

$$\mathbf{A} = \mathbf{1} + (\sin \phi) Z + (1 - \cos \phi) Z^2. \tag{7}$$

(See Problem 1 or, for a more sophisticated proof, Murnaghan 1938, p. 237.) So far, we know that $\mathbf{A}$ is orthogonal and that it has the same rotation axis as A. We now prove that it has the same trace (so that $\mathbf{A}$ is the end of our quest). On forming Z^2 from (5) and remembering the normalization condition for $\mathbf{n}$, it follows that

$$\operatorname{Tr} Z^2 = -2. \tag{8}$$

Thus, in (7), since $\operatorname{Tr} Z = 0$,

$$\operatorname{Tr} \mathbf{A} = 3 - 2(1 - \cos \phi) = 1 + 2 \cos \phi, \tag{9}$$

which shows that $\mathbf{A}$ coincides with the desired matrix A, which can now be identified with the symbol $\hat{R}^1_{\mathbf{r}}(\phi\mathbf{n})$, as the matrix of the rotation $R(\phi\mathbf{n})$ on the basis $|xyz\rangle$. (Notice that we have verified our conjecture that this matrix is fully determined by its skew-symmetric component.)

It is convenient to rewrite (7):

$$A = \hat{R}^1_{\mathbf{r}}(\phi\mathbf{n}) = \mathbf{1} + (\sin\phi)Z + 2(\sin^2 \tfrac{1}{2}\phi)Z^2, \tag{10}$$

and on forming Z^2 from (5), one immediately obtains

$$A = \hat{R}^1_{\mathbf{r}}(\phi\mathbf{n}) =$$
$$\begin{bmatrix} 1-2(n_y^2+n_z^2)\sin^2\frac{1}{2}\phi & -n_z\sin\phi+2n_xn_y\sin^2\frac{1}{2}\phi & n_y\sin\phi+2n_zn_x\sin^2\frac{1}{2}\phi \\ n_z\sin\phi+2n_xn_y\sin^2\frac{1}{2}\phi & 1-2(n_z^2+n_x^2)\sin^2\frac{1}{2}\phi & -n_x\sin\phi+2n_yn_z\sin^2\frac{1}{2}\phi \\ -n_y\sin\phi+2n_zn_x\sin^2\frac{1}{2}\phi & n_x\sin\phi+2n_yn_z\sin^2\frac{1}{2}\phi & 1-2(n_x^2+n_y^2)\sin^2\frac{1}{2}\phi \end{bmatrix}. \tag{11}$$

This matrix must be used in the same manner as (1.15). Notice that, as expected, $R(\phi\mathbf{n})$ and $R(-\phi, -\mathbf{n})$ lead to the same matrix in (11). Equation (10) is very useful to find the transform $R(\phi\mathbf{n})\mathbf{r}$ of a vector $\mathbf{r}$ under a rotation. This, of course, from the definition of the matrix representative, is the same as $\hat{R}^1_{\mathbf{r}}(\phi\mathbf{n})\mathbf{r}$, for $\mathbf{r}$ equal to $|xyz\rangle$. Thus, the matrix product $Z\mathbf{r}$ that we shall need is easily obtained:

5 $$Z\mathbf{r} = \mathbf{n}\times\mathbf{r}. \tag{12}$$

Therefore,

10 $$R(\phi\mathbf{n})\mathbf{r} = \mathbf{r} + (\sin\phi)Z\mathbf{r} + 2(\sin^2\tfrac{1}{2}\phi)Z^2\mathbf{r} \tag{13}$$

12|13 $$= \mathbf{r} + (\sin\phi)(\mathbf{n}\times\mathbf{r}) + 2(\sin^2\tfrac{1}{2}\phi)Z(\mathbf{n}\times\mathbf{r}) \tag{14}$$

12|14 $$= \mathbf{r} + (\sin\phi)(\mathbf{n}\times\mathbf{r}) + 2(\sin^2\tfrac{1}{2}\phi)\mathbf{n}\times(\mathbf{n}\times\mathbf{r}). \tag{15}$$

This transformation is called the *conical transformation* of $\mathbf{r}$ because under it this vector moves on a cone around the rotation axis $\mathbf{n}$. (See §**9**-5.)

Problems 3-3 (p. 278)

1. Prove that, with Z given by (5),

$$\exp \phi Z = \mathbf{1} + (\sin\phi)Z + (1-\cos\phi)Z^2. \tag{16}$$

(This is the proof of eqn 7).

2. Prove that eqn (15) verifies the transformation matrix (11).
3. Given the rotation $R(\frac{2}{3}\pi\mathbf{n})$ with $\mathbf{n} = 3^{-1/2}(111)$, find, on using (15), the transform of $\mathbf{r} = \mathbf{i}$ and compare with Problem 1.6.

4. EULER ANGLES IN TERMS OF THE ANGLE AND AXIS OF ROTATION

If we write the Euler matrix $\hat{R}^1_{\mathbf{r}}(\alpha\beta\gamma)$ in (1–15) as A, with matrix elements A_{ij}, then the Euler angles are fully determined by the following relations:

$$\cos\beta = A_{33} \tag{1}$$

$$\tan\alpha = A_{23}/A_{13}, \qquad \sin\alpha = A_{23}/\sin\beta, \qquad \cos\alpha = A_{13}/\sin\beta. \tag{2}$$

$$\tan\gamma = -A_{32}/A_{31}, \qquad \sin\gamma = A_{32}/\sin\beta, \qquad \cos\gamma = -A_{31}/\sin\beta. \tag{3}$$

The angle β, of course, is fully determined by its cosine, since β is always smaller than or equal to π.

Of course, the Euler matrix $\hat{R}^1_{\mathbf{r}}(\alpha\beta\gamma)$ (1.15) must equal the matrix for the same rotation given in the form $\hat{R}^1_{\mathbf{r}}(\phi\mathbf{n})$ in (3.11). Thus, introducing this matrix into eqns (1) to (3), we obtain:

$$\cos\beta = 1-2(n_x^2+n_y^2)\sin^2\tfrac{1}{2}\phi. \tag{4}$$

$$\tan\alpha = (-n_x\sin\phi+2n_yn_z\sin^2\tfrac{1}{2}\phi)(n_y\sin\phi+2n_zn_x\sin^2\tfrac{1}{2}\phi)^{-1}. \tag{5}$$

$$\sin\alpha = (-n_x\sin\phi+2n_yn_z\sin^2\tfrac{1}{2}\phi)(\sin\beta)^{-1}. \tag{6}$$

$$\cos\alpha = (n_y\sin\phi+2n_zn_x\sin^2\tfrac{1}{2}\phi)(\sin\beta)^{-1}. \tag{7}$$

$$\tan\gamma = (n_x\sin\phi+2n_yn_z\sin^2\tfrac{1}{2}\phi)(n_y\sin\phi-2n_zn_x\sin^2\tfrac{1}{2}\phi)^{-1}. \tag{8}$$

$$\sin\gamma = (n_x\sin\phi+2n_yn_z\sin^2\tfrac{1}{2}\phi)(\sin\beta)^{-1}. \tag{9}$$

$$\cos\gamma = (n_y\sin\phi-2n_zn_x\sin^2\tfrac{1}{2}\phi)(\sin\beta)^{-1}. \tag{10}$$

These formulae, admittedly, are not pretty, but they are the only workable alternative to doing pictorial geometry as in Fig. 1.2. Notice that $R(\phi\mathbf{n})$ and $R(-\phi, -\mathbf{n})$ lead to precisely the same Euler angles. From (2.1) these rotations, of course, are identical but it will be seen later on, nevertheless, that it is important in some cases to be able to distinguish between them (see §**9**-1) for which purpose the Euler angles are of no use.

Problem 3-4 (p. 278)

1. Use eqns (4) to (10) to find the Euler angles for $R(-\tfrac{2}{3}\pi\mathbf{n})$ with $\mathbf{n}=3^{-1/2}(111)$. Compare your result with (1.6).

5. A ROTATION IN TERMS OF ROTATIONS ABOUT ORTHOGONAL AXES

We might expect that a rotation may be expressed in terms of rotations around the orthogonal axes $\mathbf{x}, \mathbf{y}, \mathbf{z}$ and we shall find here that this is indeed the case. In a sense, however, this section will be somewhat abortive, since when this problem is solved, we shall see that the solution is not unique. Thus, we cannot speak of a rotation $R(\phi\mathbf{n})$ as having components along $\mathbf{x}, \mathbf{y}, \mathbf{z}$, certainly not in the same sense in which a vector has components along these axes. That this must be so should be reasonably obvious, because general rotations do not commute, so that the order in which the presumptive components of a rotation along **x**, **y**, **z** are effected matters, whereas the order in which vector components are taken is immaterial. It is a fact, however, that *in some sense* that cannot be the one obtaining for a vector, rotations are expressed in terms of orthogonal rotations, as we shall see at the end of this section, and it is for this reason that it is worthwhile investigating this problem, since confusion can easily arise.

Let us call $X(\xi)$, $Y(\eta)$, $Z(\zeta)$ three rotations around **x**, **y**, **z** respectively, and W the product $Y(\eta)Z(\zeta)$. Their corresponding matrices can be obtained from (1.14):

$$\overset{\displaystyle\hat{X}(\xi)}{\begin{bmatrix} 1 & 0 & 0 \\ 0 & \cos\xi & -\sin\xi \\ 0 & \sin\xi & \cos\xi \end{bmatrix}}, \quad \overset{\displaystyle\hat{Y}(\eta)}{\begin{bmatrix} \cos\eta & 0 & \sin\eta \\ 0 & 1 & 0 \\ -\sin\eta & 0 & \cos\eta \end{bmatrix}}, \quad \overset{\displaystyle\hat{Z}(\zeta)}{\begin{bmatrix} \cos\zeta & -\sin\zeta & 0 \\ \sin\zeta & \cos\zeta & 0 \\ 0 & 0 & 1 \end{bmatrix}}. \tag{1}$$

$$\hat{\mathsf{W}} = \hat{Y}(\eta)\hat{Z}(\zeta) = \begin{bmatrix} \cos\eta\cos\zeta & -\cos\eta\sin\zeta & \sin\eta \\ \sin\zeta & \cos\zeta & 0 \\ -\sin\eta\cos\zeta & \sin\eta\sin\zeta & \cos\eta \end{bmatrix}. \tag{2}$$

We want to write $\hat{R}(\phi\mathbf{n})$ in the form $\hat{X}(\xi)\hat{Y}(\eta)\hat{Z}(\zeta)$, that is as $\hat{X}(\xi)\hat{\mathsf{W}}$, where $\hat{\mathsf{W}}$ is some matrix that can be put into the form (2), for which it is sufficient to ensure that $\hat{\mathsf{W}}_{23}$ vanishes. (Notice that the rest of $\hat{\mathsf{W}}$ follows from orthonormality!) On writing $\{\hat{X}(\xi)\}^{-1}\hat{R}$, that is $\hat{X}(-\xi)\hat{R}$, equal to $\hat{\mathsf{W}}$, we have the condition

$$\cos\xi\,\hat{R}_{23}+\sin\xi\,\hat{R}_{33}=0 \qquad \Rightarrow \qquad \tan\xi=-\hat{R}_{23}\hat{R}_{33}^{-1}. \tag{3}$$

On the other hand, from (2), $\hat{Y}(-\eta)\hat{\mathsf{W}}$ must be $\hat{Z}(\zeta)$ for some ζ. It is sufficient for this purpose that this product gives two zeros in the last column, as shown for $\hat{Z}(\zeta)$ in (1), since the rest of the matrix form follows from orthonormality. In this product, however, $(\hat{Y}\hat{\mathsf{W}})_{23}$, from (1) and (2), will necessarily be zero. It is thus sufficient to take $(\hat{Y}\hat{\mathsf{W}})_{13}$ to vanish for $\hat{Y}=\hat{Y}(-\eta)$, which leads to the condition

$$\cos\eta\,\hat{\mathsf{W}}_{13}-\sin\eta\,\hat{\mathsf{W}}_{33}=0 \qquad \Rightarrow \qquad \tan\eta=\hat{\mathsf{W}}_{13}\hat{\mathsf{W}}_{33}^{-1}. \tag{4}$$

Since $\hat{\mathsf{W}}=\hat{X}(-\xi)\hat{R}$, we have,

$$\hat{\mathsf{W}}_{13}=\hat{R}_{13}, \qquad \hat{\mathsf{W}}_{33}=-\sin\xi\,\hat{R}_{23}+\cos\xi\,\hat{R}_{33}. \tag{5}$$

5|4 $$\tan\eta=\hat{R}_{13}(-\sin\xi\,\hat{R}_{23}+\cos\xi\,\hat{R}_{33})^{-1}. \tag{6}$$

Finally, from (2), $\cos\zeta$ and $\sin\zeta$ are respectively $\hat{\mathsf{W}}_{22}$ and $\hat{\mathsf{W}}_{21}$, which are obtained as in (5):

$$\cos\zeta=\cos\xi\,\hat{R}_{22}+\sin\xi\,\hat{R}_{32}, \tag{7}$$

$$\sin\zeta=\cos\xi\,\hat{R}_{21}+\sin\xi\,\hat{R}_{31}. \tag{8}$$

The solutions of (3) and (6) are, at best, two-valued and they can be entirely undetermined when the matrix elements on their right vanish. The expansion of $\hat{R}(\phi\mathbf{n})$ in the form $\hat{X}(\xi)\hat{Y}(\eta)\hat{Z}(\zeta)$ is thus not unique even though a specified order of the non-commuting rotations in it is given. (An example is given in Problem 1.) Given this strong limitation, it is indeed remarkable that every rotation matrix can be uniquely determined in terms of three rotation matrices corresponding to the orthogonal axes, a fact that we shall prove by a bit of experimental work. From (3.6), with the notation of (3.10), it follows that

$$\hat{R}^1_{\mathbf{r}}(\phi\mathbf{n}) = \exp(\phi Z). \tag{9}$$

We now rewrite Z from (3.5), in a strange but nevertheless obvious manner:

$$Z = \begin{bmatrix} 0 & -n_z & n_y \\ n_z & 0 & -n_x \\ -n_y & n_x & 0 \end{bmatrix}$$
$$= n_x\begin{bmatrix} 0 & 0 & 0 \\ 0 & 0 & -1 \\ 0 & 1 & 0 \end{bmatrix} + n_y\begin{bmatrix} 0 & 0 & 1 \\ 0 & 0 & 0 \\ -1 & 0 & 0 \end{bmatrix} + n_z\begin{bmatrix} 0 & -1 & 0 \\ 1 & 0 & 0 \\ 0 & 0 & 0 \end{bmatrix}. \tag{10}$$

This expression can be much simplified on defining a 'vector' $\hat{\mathbf{J}}$ with components $\hat{J}_x$, $\hat{J}_y$, $\hat{J}_z$ that are the matrices in (10) in the order given therein. Then Z is $\mathbf{n}\,.\,\hat{\mathbf{J}}$ and (9) takes the form

$$\hat{R}^1_{\mathbf{r}}(\phi\mathbf{n}) = \exp(\phi\mathbf{n}\,.\,\hat{\mathbf{J}}). \tag{11}$$

Expression (11) is undoubtedly true, but the meaning of the $\hat{J}$ matrices in it is, of course, rather obscure at this stage. It can be accepted, though, that $\hat{J}_x$, say, must pertain to a rotation around the x axis, since the matrix elements in $\hat{J}_x$ that correspond to n_y and n_z in Z all vanish. Likewise for $\hat{J}_y$ and $\hat{J}_z$. Their meaning will be further interpreted in §**4**-3, where the importance of these matrices will be demonstrated.

Another very important result of this section is that infinitesimal rotations can indeed be uniquely expanded in terms of orthogonal (infinitesimal) rotations. (Problem 3.)

Warning. Equation (11) must not be written as a product of three exponentials because the matrices $n_x\hat{J}_x$, $n_y\hat{J}_y$, $n_z\hat{J}_z$ do not commute, as can easily be verified. (See Warning after eqn **2**-6.62.)

Problems 3-5 (p. 279)

1. Prove and verify that the rotation $R(\frac{2}{3}\eta\mathbf{n})$ with $\mathbf{n} = 3^{-1/2}(111)$, can be written in the form $X(\xi)Y(\eta)Z(\zeta)$ with $\xi = 0$, $\eta = \frac{1}{2}\pi$, $\zeta = \frac{1}{2}\pi$ or $\xi = \frac{1}{2}\pi$, $\eta = \frac{1}{2}\pi$, $\zeta = 0$.

2. Find the angle and axis of rotation and expressions for them in terms of orthogonal rotations, of the matrices

$$A=\begin{bmatrix} \frac{1}{2} & \frac{1}{2} & -\frac{\sqrt{2}}{2} \\ \frac{1}{2} & \frac{1}{2} & \frac{\sqrt{2}}{2} \\ \frac{\sqrt{2}}{2} & -\frac{\sqrt{2}}{2} & 0 \end{bmatrix}, \qquad B=\begin{bmatrix} \frac{2}{3} & \frac{2}{3} & -\frac{1}{3} \\ -\frac{1}{3} & \frac{2}{3} & \frac{2}{3} \\ \frac{2}{3} & -\frac{1}{3} & \frac{2}{3} \end{bmatrix}.$$

3. Write down the matrix (11) of $R(\phi\mathbf{n})$ for an infinitesimal angle ϕ and show that this infinitesimal rotation can always be written in the form $\hat{X}(\xi)\hat{Y}(\eta)\hat{Z}(\zeta)$ for infinitesimal angles ξ, η, ζ, which are respectively ϕn_x, ϕn_y, ϕn_z.

6. COMMENTS ON THE PARAMETRIZATION OF ROTATIONS

We have discussed a number of parametric forms for a rotation in this chapter but we must now refine our conditions on the parameters used. As we have already said, in a continuous group the neighbourhood of the identity is a most important region from which, under certain conditions which in fact obtain for the rotation group, the whole group can be constructed, as will indeed be verified in Chapter **4**. For this reason, it is an important requirement of any parametrization that it should describe the infinitesimal elements of the group in such a way that they go smoothly, i.e. without singularities, into the identity. The parametrization $R(\alpha\beta\gamma)$ of rotations by the Euler angles is not without fault in this respect because, for vanishing β, α and γ are undetermined, only their sum being significant. Thus the identity has in its neighbourhood elements with large finite values of α and γ, as long as their sum be infinitesimal.

Now consider the parametrization $R(\phi\mathbf{n})$. The identity corresponds to $\phi=0$ and its neighbourhood would thus contain all sets with infinitesimal ϕ and arbitrary finite rotation vectors $\mathbf{n}$. This would not do at all. In order to eliminate this difficulty, we must understand the symbol $\phi\mathbf{n}$ as a *single vector parameter*, namely a vector parallel to the rotation axis with modulus equal to the rotation angle. In other words, we have three parameters ϕn_x, ϕn_y, ϕn_z, *all* of which vanish for the identity $\phi=0$ and *all* of which are infinitesimal in its neighbourhood.

Once this is understood, it must be realized that, although the parametrization in the form $X(\xi)Y(\eta)Z(\zeta)$ is not satisfactory, because it is undetermined for finite $R(\phi\mathbf{n})$, it works excellently in the neighbourhood of the identity, where the vector $\phi\mathbf{n}$ can be decomposed in its three components in such a way that ξ, η, and ζ are uniquely ϕn_x, ϕn_y, ϕn_z. (See Problem 5.3 and also §**4**-1.)

4

ROTATIONS AND ANGULAR MOMENTUM

No sé si volveremos en un ciclo segundo
Como vuelven las cifras de una fracción periódica;
Pero sé que una oscura rotación pitagórica
Noche a noche me deja en un lugar del mundo

Jorge Luis Borges (1940). *La Noche Cíclica. Poemas* 1922–1943, (1943). Losada, Buenos Aires, p. 165.

I do not know if we shall return in a second cycle,
like the figures in a periodic fraction;
but I know that a dark Pythagorean rotation
night after night takes me to some place in the world

This chapter will start with a discussion of infinitesimal rotations that will lead naturally to the concept of the angular momentum as the infinitesimal generator of the rotation group. After that, we shall revise some well-known concepts about angular momentum operators that will lead to the form of the irreducible representations of SO(3).

1. INFINITESIMAL ROTATIONS

We know that, because two general rotations $R(\phi\mathbf{n})$, $R(\psi\mathbf{m})$ do not commute, their rotation vector parameters $\phi\mathbf{n}$, $\psi\mathbf{m}$ cannot be added. That is, if the product $R(\phi\mathbf{n})R(\psi\mathbf{m})$ is $R(\chi\mathbf{p})$ we cannot expect $\chi\mathbf{p}=\phi\mathbf{n}+\psi\mathbf{m}$ since this expression is commutative, while the rotation product is not. Fortunately, this is completely changed for infinitesimal rotations, because they always commute as we shall soon see.

Equation (**3**-3.15) gives the transformation of a vector $\mathbf{r}$ under a rotation (conical transformation):

$$R(\phi\mathbf{n})\mathbf{r}=\mathbf{r}+\sin\phi(\mathbf{n}\times\mathbf{r})+2\sin^2\tfrac{1}{2}\phi\ \ \mathbf{n}\times(\mathbf{n}\times\mathbf{r}). \tag{1}$$

Therefore, for infinitesimal ϕ, to the first order,

$$R(\phi\mathbf{n})\mathbf{r}=\mathbf{r}+\phi(\mathbf{n}\times\mathbf{r}), \tag{2}$$

(see Fig. 1). For a second rotation $R(\psi\mathbf{m})$, similarly,

$$R(\psi\mathbf{m})\mathbf{r}=\mathbf{r}+\psi(\mathbf{m}\times\mathbf{r}), \tag{3}$$

(see the 'white' vector $\mathbf{b}$ in Fig. 1).

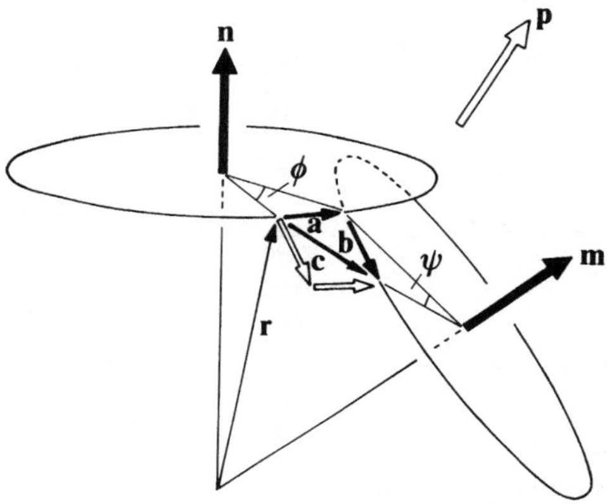

Fig. 4-1.1. Infinitesimal rotations and their commutation. The rotation by ϕ around the unit vector $\mathbf{n}$ takes $\mathbf{r}$ into $\mathbf{r}+\mathbf{a}$, where, from the figure, $\mathbf{a}=\phi\mathbf{n}\times\mathbf{r}$. $R(\psi\mathbf{m})$ adds $\mathbf{b}=\psi\mathbf{m}\times(\mathbf{r}+\mathbf{a})$ to the transformed $\mathbf{r}$. The result is that $\mathbf{r}$ is transformed into $\mathbf{r}+\mathbf{c}$, which corresponds to some rotation around an axis $\mathbf{p}$. (The circle perpendicular to $\mathbf{p}$ and tangential to $\mathbf{c}$ is not drawn.) The same result is achieved through the 'white' circuit, in which $R(\psi\mathbf{m})$ is performed first and it is followed by $R(\phi\mathbf{n})$. The angles ϕ and ψ are infinitesimal and second-order terms are neglected.

On applying $R(\psi\mathbf{m})$ to (2), keeping first order terms only, we have,

$$R(\psi\mathbf{m})R(\phi\mathbf{n})\mathbf{r}=\mathbf{r}+\phi(\mathbf{n}\times\mathbf{r})+\psi[\mathbf{m}\times\{\mathbf{r}+\phi(\mathbf{n}\times\mathbf{r})\}]$$
$$=\mathbf{r}+\phi(\mathbf{n}\times\mathbf{r})+\psi(\mathbf{m}\times\mathbf{r}). \quad (4)$$

On applying $R(\phi\mathbf{n})$ to (3), we have instead,

$$R(\phi\mathbf{n})R(\psi\mathbf{m})\mathbf{r}=\mathbf{r}+\psi(\mathbf{m}\times\mathbf{r})+\phi(\mathbf{n}\times\mathbf{r}). \quad (5)$$

From (4) and (5) the two rotations commute. Indeed, on comparison with (2), the vector parameter $\chi\mathbf{p}$ of the resultant transformation is given in either case by

$$\chi\mathbf{p}=\phi\mathbf{n}+\psi\mathbf{m}. \quad (6)$$

That is, the vector parameters (see §**3**-6) of the infinitesimal rotations add up vectorially. In fact, we have seen (§**3**-6 and Problem **3**-5.3) that an infinitesimal rotation $\delta\phi\ \mathbf{n}=\delta\phi\ \mathbf{n}_x+\delta\phi\ \mathbf{n}_y+\delta\phi\ \mathbf{n}_z$ can indeed be decomposed uniquely as a product of three rotations $X(\delta\phi\ \mathbf{n}_x)\ Y(\delta\phi\ \mathbf{n}_y)\ Z(\delta\phi\ \mathbf{n}_z)$. Thus, for infinitesimal $\delta\phi$, and observing that the symbol $X(\delta\phi\ \mathbf{n}_x)$ already entails in the X the axis of rotation, we can write

$$R(\delta\phi\ \mathbf{n})=X(\xi)\ Y(\eta)\ Z(\zeta), \quad (7)$$

where

$$\xi = \delta\phi\ |\mathbf{n}_x|, \qquad \eta = \delta\phi\ |\mathbf{n}_y|, \qquad \zeta = \delta\phi\ |\mathbf{n}_y|, \tag{8}$$

are the angles of the infinitesimal rotations around **x**, **y**, and **z** respectively.

2. THE INFINITESIMAL GENERATOR: ANGULAR MOMENTUM

We shall now write the infinitesimal rotation in (1.2) in terms of an infinitesimal angle $\delta\phi$:

$$R(\delta\phi\ \mathbf{n})\mathbf{r} = \mathbf{r} + \delta\phi(\mathbf{n} \times \mathbf{r}). \tag{1}$$

Thus, on acting on a function $f(\mathbf{r})$ (with the corresponding function space operator, clearly), we have,

$$R(\delta\phi\ \mathbf{n})f(\mathbf{r}) = f(R^{-1}\mathbf{r}) = f(\mathbf{r} - \delta\phi\ \mathbf{n} \times \mathbf{r}). \tag{2}$$

Since, from the definition of the gradient, to first order in a small vector **u**,

$$f(\mathbf{r} + \mathbf{u}) = f(\mathbf{r}) + \mathbf{u} \cdot \nabla f(\mathbf{r}), \tag{3}$$

we can rewrite (2) as follows,

$$R(\delta\phi\ \mathbf{n})f(\mathbf{r}) = f(\mathbf{r}) - \delta\phi\ \mathbf{n} \times \mathbf{r} \cdot \nabla f(\mathbf{r}), \tag{4}$$

$$= f(\mathbf{r}) - \delta\phi\ \mathbf{n} \cdot \mathbf{r} \times \nabla f(\mathbf{r}). \tag{5}$$

We shall write

$$\mathbf{r} \times \nabla = \boldsymbol{\nabla}, \tag{6}$$

since this operator is itself a sort of gradient operator as the following consideration shows. The z component of $\mathbf{r} \times \nabla$ is $x\ \partial/\partial y - y\ \partial/\partial x$ which, as can easily be verified, is $\partial/\partial\phi$ in polar coordinates. Now ϕ, in the notation used in (1.8) is identical with ζ, the angle of rotation around the **z** axis, and by symmetry it follows at once that

$$\boldsymbol{\nabla} = (\partial/\partial\xi, \partial/\partial\eta, \partial/\partial\zeta), \tag{7}$$

ξ, η, ζ being the angles of rotation around the **x**, **y**, and **z** axes respectively. On introducing this notation in (5), and on taking operators (E the identity) we have,

$$R(\delta\phi\ \mathbf{n}) = E - \delta\phi\ \mathbf{n} \cdot \boldsymbol{\nabla}. \tag{8}$$

The operator $\boldsymbol{\nabla}$ is skew-Hermitian, as follows from the unitary property of R: from (8), to first-order, we must have

$$R^{\dagger}R = (E - \delta\phi\ \mathbf{n} \cdot \boldsymbol{\nabla}^{\dagger})(E - \delta\phi\ \mathbf{n} \cdot \boldsymbol{\nabla}) \tag{9}$$

$$= E - \delta\phi\ \mathbf{n} \cdot (\boldsymbol{\nabla}^{\dagger} + \boldsymbol{\nabla}) = E, \tag{10}$$

which requires

$$\boldsymbol{\nabla}^{\dagger} = -\boldsymbol{\nabla}, \tag{11}$$

as stated. For this reason, it is convenient to define a Hermitian operator related to $\boldsymbol{\nabla}$

$$\mathbf{I} = -\mathrm{i}\boldsymbol{\nabla}, \qquad \boldsymbol{\nabla} = \mathrm{i}\mathbf{I}, \tag{12}$$

whence, in (8),

$$R(\delta\phi\ \mathbf{n}) = E - \mathrm{i}\ \delta\phi\ \mathbf{n}\,.\,\mathbf{I}. \tag{13}$$

As in §**2**-7, we find that associated with the symmetry (unitary) operator R there is a Hermitian operator (observable), which is called the infinitesimal generator of the group of R, in this case SO(3). This observable $\mathbf{I}$, defined in (12), is the *angular momentum*. It should be noticed that the opposite sign in (12) would have done equally well from the point of view of the Hermitian property. The sign taken was chosen so that, from (6), $\mathbf{I}$ equals $\mathbf{r}\times(-\mathrm{i}\nabla) = \mathbf{r}\times\mathbf{p}$, the operator $\mathbf{p}$ being given in the form used in quantum mechanics for the momentum operator, with $\hbar = 1$. Thus, the identification of $\mathbf{I}$ with the angular momentum is obvious.

In (13), $\mathbf{I}$ is called the infinitesimal generator because this relation can now be integrated to give any finite rotation $R(\phi\mathbf{n})$ of the group, by a procedure already used in §**2**-7. Let us take ϕ as $\lim m\,\delta\phi$ (for $m \to \infty$). Thus

$$R(\phi\mathbf{n}) = \lim\,\{R(\delta\phi\ \mathbf{n})\}^m \tag{14}$$

13

$$= \lim\left(E - \mathrm{i}\frac{\phi}{m}\mathbf{n}\,.\,\mathbf{I}\right)^m = \exp\,(-\mathrm{i}\phi\mathbf{n}\,.\,\mathbf{I}). \tag{15}$$

(In eqn **2**-7.28 we wrote 1 for the identity, which makes it a little easier to read eqn 15 as an exponential, but it must be appreciated that this is the exponential of an operator, as understood by a power series in operators.)

In principle, (15) can be used to get the matrix representative $\hat{R}(\phi\mathbf{n})$ of $R(\phi\mathbf{n})$ by just replacing $\mathbf{I}$ by $\hat{\mathbf{I}}$ on its right-hand side. We do not have access, at present, to the matrices of the angular momentum operator, so that we shall again have to perform a similar piece of work for the matrices. This will also throw light on the meaning of the infinitesimal operator.

Problems 4-2 (p. 279)

1. Obtain the following forms of the angular momentum operators in polars

$$I_x = -\mathrm{i}(-\sin\varphi\ \partial/\partial\theta - \cot\theta\ \cos\varphi\ \partial/\partial\varphi). \tag{16}$$

$$I_y = -\mathrm{i}(\cos\varphi\ \partial/\partial\theta - \cot\theta\ \sin\varphi\ \partial/\partial\varphi). \tag{17}$$

$$I_z = -\mathrm{i}\partial/\partial\varphi. \tag{18}$$

$$\mathbf{I}^2 = -\ (\partial^2/\partial\theta^2 + \cot\theta\ \partial/\partial\theta + (\sin\theta)^{-2}\ \partial^2/\partial\varphi^2). \tag{19}$$

2. On comparing (19) with expressions for ∇^2 in polars from standard tables, verify that $-\mathbf{I}^2$ (which is $\boldsymbol{\nabla}^2$ from eqn 12) equals ∇^2 for $\mathbf{r}=1$, an operator that is called the Legendrian Λ. On using (7), give an argument to explain the equality of $\boldsymbol{\nabla}^2$ with Λ.

3. ROTATION MATRICES

In order to obtain the rotation matrix $\hat{R}(\delta\phi\ \mathbf{n})$ we relate it to the matrix for the identity, $\hat{R}(\mathbf{0})$, by eqn (2.3) with $\mathbf{0}$ substituted for $\mathbf{r}$:

$$\hat{R}(\delta\phi\ \mathbf{n})=\hat{R}(\mathbf{0})+\delta\phi\ \mathbf{n}\,.\,\boldsymbol{\nabla}\hat{R}(\mathbf{0}). \tag{1}$$

Notice that we have also changed ∇ by $\boldsymbol{\nabla}$, for the following reason. The differentiations involved in the ∇ in (2.3) must be taken with respect to the variables that define the components of $\mathbf{u}$, in our case the components of $\delta\phi\ \mathbf{n}$, which are, from (1.8) ξ, η, and ζ. Thus the gradient has to be taken with respect to these variables and it has to be $\boldsymbol{\nabla}$, as given in (2.7). We must also understand clearly the meaning of $\boldsymbol{\nabla}\hat{R}(\mathbf{0})$, which is a shorthand for 'differentiate $\hat{R}(\xi\eta\zeta)$ with respect to ξ and then take the limit for ξ, η, ζ going to zero', etc. What do we mean by $\hat{R}(\xi\eta\zeta)$ here? The idea is simple: because we are going to take the limit to zero, the rotation matrices that matter are those in the neighbourhood of the identity, for which the expansion (1.7) in orthogonal rotations is valid. Thus, if we call $\hat{\mathbf{J}}$ the matrix 'vector' $\boldsymbol{\nabla}\hat{R}(\mathbf{0})$ with matrix components $\hat{J}_x, \hat{J}_y, \hat{J}_z$, then

$$\hat{J}_x=\lim_{\xi,\eta,\zeta\to 0}\frac{\partial}{\partial\xi}\hat{X}(\xi)\ \hat{Y}(\eta)\ \hat{Z}(\zeta), \tag{2}$$

and similar expressions for $\hat{J}_y$ and $\hat{J}_z$.

In (2), $\hat{Y}$ and $\hat{Z}$ are kept constant on differentiation and in the limit they must go to the unit matrix (as is in any case obvious when their form in eqn **3**-5.1 is inspected). Also, from (**3**-5.1)

$$\hat{J}_x=\lim_{\xi\to 0}\begin{bmatrix}0 & 0 & 0\\ 0 & -\sin\xi & -\cos\xi\\ 0 & \cos\xi & -\sin\xi\end{bmatrix}=\begin{bmatrix}0 & 0 & 0\\ 0 & 0 & -1\\ 0 & 1 & 0\end{bmatrix}. \tag{3}$$

It is thus clear that, proceeding in the same way for $\hat{J}_y$ and $\hat{J}_z$,

$$\mathbf{n}\,.\,\boldsymbol{\nabla}\hat{R}(\mathbf{0})=\mathbf{n}\,.\,\hat{\mathbf{J}}, \tag{4}$$

with $\mathbf{n}\,.\,\hat{\mathbf{J}}$ defined precisely as in (**3**-5.10) and (**3**-5.11). Thus, in (1),

$$\hat{R}(\delta\phi\ \mathbf{n})=\mathbf{1}+\delta\phi\ \mathbf{n}\,.\,\hat{\mathbf{J}}, \tag{5}$$

and, exactly as in (2.14) and (2.15),

$$\hat{R}(\phi\mathbf{n})=\exp\,(\phi\mathbf{n}\,.\,\hat{\mathbf{J}}), \tag{6}$$

which coincides precisely with our previous 'experimental' results in (**3**-5.11). The symbol for the matrix in (L6) should properly be written as $\hat{R}^1_{\mathbf{r}}(\phi\mathbf{n})$ since it is a 3×3 matrix which acts on $|xyz\rangle$, as follows from (3) and similar equations. It follows at once from the orthogonality of $\hat{R}(\phi\mathbf{n})$ that $\hat{\mathbf{J}}$ must be skew-symmetric, as is indeed the case in (**3**-5.10). It is important to notice that if we were to replace $\boldsymbol{\nabla}$ in (4) by $\mathrm{i}\hat{\mathbf{I}}$ (as in eqn 2.12, making allowance for the change from operator to matrix), we would get in (6)

$$\hat{R}(\phi\mathbf{n}) = \exp(+\mathrm{i}\phi\mathbf{n}\cdot\hat{\mathbf{I}}) \tag{7}$$

with opposite sign to (2.15). There is no contradiction here, however. The main point is that the matrices in this section are defined as operating on the vectors $\mathbf{r}$ of the configuration space, whereas in §2 we worked with operators acting on the function space. Thus, strictly speaking, the matrix $\hat{\mathbf{I}}$ to be inserted in (2.15) should be given a somewhat different symbol from that of the matrix $\hat{\mathbf{I}}$ in (7). We shall see in §9, in fact, that when the nature of the operations to be effected is carefully taken into account, (2.15) entirely agrees with (7). Some further comments are appropriate. First, we must remember that the sign in (2.12) is conventional, and consequently also the sign in (2.15), whereas the sign in (6) is essential and it is clear that without it the correct agreement with (**3**-5.11) would not arise. Secondly, whereas $\hat{\mathbf{J}}$ and thus $\hat{\mathbf{I}}$ in (7) is fully fixed by defining the $\mathbf{x}, \mathbf{y}, \mathbf{z}$ axes, the matrix $\hat{\mathbf{I}}$ to be taken in (2.15), like all representatives, depends within a similarity on the basis. What we must do, of course, is to choose the basis so that (2.15) coincides with (7) and we shall see in §9 that our conventions allow for this to be done.

4. COMMUTATION

The angular momentum operator $\mathbf{I}$ is related to an infinitesimal rotation $R(\delta\phi\ \mathbf{n})$, to be abbreviated as $R(\phi\mathbf{n})$ in this section (in which all angles ϕ, ψ will be taken as infinitesimal), by the equation

2.13
$$R(\phi\mathbf{n}) = E - \mathrm{i}\phi\mathbf{n}\cdot\mathbf{I}. \tag{1}$$

The commutation relations of the components I_x, I_y, I_z of $\mathbf{I}$ will clearly depend on those of infinitesimal rotations like $R(\phi\mathbf{n})$. We know that infinitesimal operators commute to first-order but it should be clear from (1) that if we drop second-order terms in the angles, then we shall drop all the terms that contain products of angular momentum components. Thus, in order to obtain commutators such as $[I_x, I_y]$ we have to work out the commutation relations of infinitesimal rotations afresh, now to second-order. From (1.2) two infinitesimal rotations about the unit vectors $\mathbf{x}$ and $\mathbf{y}$ are

$$R(\phi\mathbf{x})\mathbf{r} = \mathbf{r} + \phi(\mathbf{x}\times\mathbf{r}), \qquad R(\psi\mathbf{y})\mathbf{r} = \mathbf{r} + \psi(\mathbf{y}\times\mathbf{r}). \tag{2}$$

Therefore,

$$R(\phi\mathbf{x})R(\psi\mathbf{y})\mathbf{r} = R(\phi\mathbf{x})\{\mathbf{r}+\psi(\mathbf{y}\times\mathbf{r})\}$$

$$= \mathbf{r}+\psi(\mathbf{y}\times\mathbf{r})+\phi\mathbf{x}\times\{\mathbf{r}+\psi(\mathbf{y}\times\mathbf{r})\} \quad (3)$$

$$= \mathbf{r}+\psi(\mathbf{y}\times\mathbf{r})+\phi(\mathbf{x}\times\mathbf{r})+\phi\psi\mathbf{x}\times(\mathbf{y}\times\mathbf{r}) \quad (4)$$

$$= \mathbf{r}+\psi(\mathbf{y}\times\mathbf{r})+\phi(\mathbf{x}\times\mathbf{r})+\phi\psi(\mathbf{x}\,.\,\mathbf{r})\mathbf{y} \quad (5)$$

$$= \mathbf{r}+\psi(\mathbf{y}\times\mathbf{r})+\phi(\mathbf{x}\times\mathbf{r})+\phi\psi x\mathbf{y}. \quad (6)$$

Similarly,

$$R(\psi\mathbf{y})R(\phi\mathbf{x})\mathbf{r} = \mathbf{r}+\phi(\mathbf{x}\times\mathbf{r})+\psi(\mathbf{y}\times\mathbf{r})+\psi\phi y\mathbf{x}. \quad (7)$$

6,7 $$[R(\phi\mathbf{x}), R(\psi\mathbf{y})]\mathbf{r} = \phi\psi(x\mathbf{y}-y\mathbf{x}) = \phi\psi(\mathbf{z}\times\mathbf{r}). \quad (8)$$

On the other hand, as in (2),

$$R(\phi\psi\mathbf{z})\mathbf{r} = \mathbf{r}+\phi\psi(\mathbf{z}\times\mathbf{r}). \quad (9)$$

9|8 $$[R(\phi\mathbf{x}), R(\psi\mathbf{y})] = R(\phi\psi\mathbf{z})-E. \quad (10)$$

We now introduce, on both sides of this equation, eqn (1) for **n** equal to **x**, **y**, and **z** and with the appropriate angles:

$$(E-\mathrm{i}\phi I_x)(E-\mathrm{i}\psi I_y)-(E-\mathrm{i}\psi I_y)(E-\mathrm{i}\phi I_x) = -\mathrm{i}\phi\psi I_z. \quad (11)$$

11 $$-\phi\psi I_x I_y+\psi\phi I_y I_x = -\mathrm{i}\phi\psi I_z. \quad (12)$$

12 $$[I_x, I_y] = \mathrm{i}I_z. \quad (13)$$

The other commutation relations follow from (13) by cyclic permutation.

The major job we now want to do is to construct irreducible bases for SO(3), for which purpose the shift operators to be studied in the next section have proved to be a fundamental tool.

Problem 4-4 (p. 280)

1. Define the total angular momentum operator

$$\mathbf{I}^2 = I_x^2+I_y^2+I_z^2. \quad (14)$$

and show that

$$[\mathbf{I}^2, I_x] = [\mathbf{I}^2, I_y] = [\mathbf{I}^2, I_z] = 0. \quad (15)$$

5. SHIFT OPERATORS

The angular momentum operators I_x, I_y, I_z are conveniently replaced by the set

$$I_\pm = I_x \pm \mathrm{i}I_y, \qquad I_z = I_z, \quad (1)$$

$I_\pm$ being called the *shift operators*.

The commutation relations (4.13) and (4.15) are now replaced by the relations (see Problem 1),

$$[I_z, I_\pm] = \pm I_\pm, \tag{2}$$

$$[I_+, I_-] = 2I_z, \tag{3}$$

$$[\mathbf{I}^2, I_\pm] = [\mathbf{I}^2, I_z] = 0, \tag{4}$$

and $\mathbf{I}^2$ now takes the form

$$\mathbf{I}^2 = I_+ I_- - I_z + I_z^2. \tag{5}$$

The property that makes $I_\pm$ a fundamental tool is the result of the *shift theorem* (which gives these operators their names). Assume that we know an eigenfunction of I_z, u_m say, labelled by the corresponding eigenvalue m of I_z:

$$I_z u_m = m u_m. \tag{6}$$

Then the theorem asserts that $I_\pm u_m$ is another eigenfunction of I_z corresponding to the eigenvalue $m \pm 1$.

Proof:

2 $$I_z(I_\pm u_m) = (\pm I_\pm + I_\pm I_z) u_m \tag{7}$$

6|R7 $$= (\pm I_\pm + I_\pm m) u_m \tag{8}$$

8 $$= (m \pm 1)(I_\pm u_m). \tag{9}$$

It is important to remember that the shift operators are not Hermitian, as it is clear from the following relation (Problem 2):

$$I_\pm^\dagger = I_\mp. \tag{10}$$

Problems 4-5 (p. 280)

1. Prove eqns (2) to (5).
2. Obtain the adjoints of $I_\pm$.
3. Prove that

$$[I_+ I_-, I_z] = [I_- I_+, I_z] = 0. \tag{11}$$

4. Prove that, in polar coordinates, the shift operators are:

$$I_\pm = \exp(\pm \mathrm{i}\varphi)(\pm \partial/\partial\theta + \mathrm{i} \cot\theta \, \partial/\partial\varphi). \tag{12}$$

6. THE EIGENFUNCTIONS OF I_z

We have defined the eigenvalues and eigenfunctions of I_z in (5.6) and we shall discuss a scheme in this section to obtain all such eigenfunctions that correspond to the same total angular momentum $\mathbf{I}^2$. In fact, because I_z^2 and $\mathbf{I}^2$ commute with I_z (see eqn 4.15) it follows that the u_m are also eigenfunctions of these operators:

5.6 $$I_z^2 u_m = m^2 u_m, \tag{1}$$

$$\mathbf{I}^2 u_m = M^2 u_m, \tag{2}$$

where in (2) we have conveniently called M^2 the eigenvalue of the total angular momentum.

The first thing we have to do is to show that m is bounded above and below:

2,1 $$(\mathbf{I}^2 - I_z^2)u_m = (M^2 - m^2)u_m. \tag{3}$$

4.14 $$(I_x^2 + I_y^2)u_m = (M^2 - m^2)u_m. \tag{4}$$

Since $I_x^2 + I_y^2$ is a positive operator, it follows from known results that its eigenvalues must also be positive:

4 $$M^2 - m^2 \geqslant 0 \qquad \Rightarrow \qquad |m| \leqslant M. \tag{5}$$

It follows from this result that, if we label the eigenvalues from j for the top one to k for the lowest, then the functions u_m form an array, ordered from top to bottom as follows,

$$u_j \ldots u_m \ldots u_k. \tag{6}$$

We shall show in this section that this array is discrete, m going down in steps of 1 from j down to $k = -j$, forming a ladder of precisely $2j+1$ steps. Perhaps even more importantly, in proving this result we shall also give rules that determine uniquely all the functions of the ladder.

In order to do this, we shall always assume that the top function of the ladder, u_j, is normalized and that we shall go down the ladder by using the I_- operator, always keeping normalization. Thus, if

$$\langle u_m \mid u_m \rangle = 1, \tag{7}$$

(which we expect to maintain for all m) we shall write

$$I_- u_m = N_m u_{m-1}. \tag{8}$$

In fact, we know from the shift theorem that (R8) must belong to $(m-1)$, i.e. must be u_{m-1} times a constant. This constant, N_m, will be so chosen that u_{m-1} will also be normalized. In the same way, on going up the ladder with I_+ we must have

$$I_+ u_m = P_{m+1} u_{m+1}. \tag{9}$$

(The different indexing rules for P and N have been chosen so as to simplify the result that follows.) The conservation of normalization on shifting requires in (8) that $\langle u_{m-1} \mid u_{m-1} \rangle$ be unity, whence

,8 $$N_m = N_m \langle u_{m-1} \mid u_{m-1} \rangle = \langle u_{m-1} \mid I_- u_m \rangle. \tag{10}$$

We can take I_- to the other side of the bracket as $I_-^\dagger$ (this is the definition of the adjoint operator) which, from (5.10) is I_+. Thus

,9 $$N_m = \langle I_+ u_{m-1} \mid u_m \rangle = P_m^* \langle u_m \mid u_m \rangle = P_m^*. \qquad (11)$$

The basic shifting rules are now, therefore, from (8), (9), and (11),

$$I_- u_m = N_m u_{m-1} \quad \Rightarrow \quad u_{m-1} = N_m^{-1} I_- u_m, \qquad (12)$$

$$I_+ u_m = N_{m+1}^* u_{m+1}. \qquad (13)$$

The condition for the top of the ladder u_j is that, when introduced into (L13), (R13) be zero (so that no further function is generated above u_j). Likewise, u_k on (L12) must lead to the vanishing of (R12). Thus, the top and bottom of the ladder require the conditions

$$N_{j+1} = 0, \qquad N_k = 0. \qquad (14)$$

In order to generate the ladder fully it would be nice to have a recurrence relation on N_m since then, starting from N_{j+1} from (14) and going down, all the N_m would be determined. This, alas, is impossible but, instead, a recurrence relation can be derived for the coefficients

$$C_m =_{\text{def}} N_m N_m^* = |N_m|^2, \qquad (15)$$

which is

$$C_m = C_{m+1} + 2m. \qquad (16)$$

Proof:

13 $$I_+ u_{m-1} = N_m^* u_m. \qquad (17)$$

12|L17 $$I_+ u_{m-1} = N_m^{-1} I_+ I_- u_m \qquad (18)$$

5.3|18 $$= N_m^{-1}(I_- I_+ + 2I_z) u_m \qquad (19)$$

13,5.6|19 $$= N_m^{-1}(I_- N_{m+1}^* u_{m+1} + 2m u_m) \qquad (20)$$

12|20 $$= N_m^{-1}(|N_{m+1}|^2 + 2m) u_m. \qquad (21)$$

17,21 $$|N_m|^2 = |N_{m+1}|^2 + 2m. \qquad (22)$$

The coefficients C_m can readily be obtained from the recurrence relation (16) and the initial value for C_{j+1} from (14). (See Problem 1.) It is found that

$$C_m = j(j+1) - m(m-1). \qquad (23)$$

From (15), the coefficients N_m are related to the C_m as follows:

$$N_m = C_m^{1/2} \exp(\mathrm{i}\alpha), \qquad (24)$$

for arbitrary α. It is a convention, called the *Condon and Shortley convention* (see §8) to take α equal to zero for the whole ladder, i.e. to take all the phase factors in (24) as unity and therefore all the N_m as real.

This convention completely fixes the factors N_m:

23 $$N_m = \{j(j+1) - m(m-1)\}^{1/2}, \tag{25}$$

whence the shifting rules (12) and (13) become

$$I_{\pm} u_m = \{j(j+1) - m(m \pm 1)\}^{1/2} u_{m \pm 1}. \tag{26}$$

Finally, we want to investigate the eigenvalues of $\mathbf{I}^2$. Because I_z commutes with $\mathbf{I}^2$, its eigenfunctions u_m must also be eigenfunctions of $\mathbf{I}^2$ and they should thus carry a second label to indicate the eigenvalue of $\mathbf{I}^2$ to which they belong. Since, as follows below, this eigenvalue depends only on the number j corresponding to the top of the ladder, it is natural to label all the functions in it with a superscript j. The eigenvalue equation for $\mathbf{I}^2$ (see Problem 2) is

$$\mathbf{I}^2 u_m^j = j(j+1) u_m^j. \tag{27}$$

Problems 4-6 (p. 280)

1. Prove (23).
2. Prove (27).

7. THE IRREDUCIBLE BASES FOR SO(3)

Let us consider the array (6.6) of functions of the ladder with a view to counting them. From (6.14), the top function of the ladder, u_j, is such that the coefficient N_{j+1} must vanish. From (6.25) this is indeed the case, as we had to expect. Likewise, from (6.14), the bottom function of the ladder, u_k, is characterized by the condition that N_m vanishes for the given k, which, from (6.25), requires that m be equal to $-j$, thus identifying the bottom of the ladder as u_{-j}. Thus, we can write the ladder as the row vector of $2j+1$ elements

$$\langle u_j^j u_{j-1}^j \dots u_{-j}^j|. \tag{1}$$

Since $2j+1$ must be an integer, the angular momentum quantum number j (see eqn 6.27) must be an integral or half-integral number (the latter meaning, of course, half an odd integer).

We shall now prove that a ladder such as (1) is an irreducible basis for representations of SO(3). In order to do this we must show first that the set (1) forms a basis of this group. This requires that (1) be invariant under all rotations, i.e. that the transform of a function of the set under any rotation be always given in terms of the functions of the same set. Secondly, we must show that no subset of (1) smaller than (1) exists which is invariant in the same manner.

As regards the invariance of the set (1): from (2.15) any rotation can be expressed in terms of I_x, I_y, I_z, and thus, from (5.1), in terms of $I_\pm$ and I_z. I_z merely multiplies each function u^j_m of (1) by m. The $I_\pm$ operators transform such a function into $u^j_{m\pm 1}$ multiplied by some constant. Thus, (1) is invariant under all rotations. That there is no subset of (1) that is invariant under all rotations follows from the fact a rotation $R(\phi\mathbf{n})$ can always be chosen that will transform any given function u^j_m of the ladder into combinations involving $u^j_{m\pm 1}$, so that, by repeated application of this process any u^j_m of the ladder will transform into combinations of all the functions of the ladder.

We shall now consider the irreducible bases for integral j and therefore of odd dimension $1, 3, 5, \ldots$ etc. Since, from Problem 2.2, $\mathbf{I}^2$ is the negative of the Legendrian Λ, that is the Laplacian ∇^2 on the unit sphere, the u^j_m must be eigenfunctions of this operator. The Laplacian, clearly, is invariant under inversion and therefore its eigenfunctions must be eigenfunctions of the inversion, i.e. either *gerade* (even with respect to the inversion) or *ungerade* (odd with respect to the inversion). Since u^0_0 $(j=0)$ must be a constant (basis of one function only) it must be even, whereas the three u^1_m must be odd and so on. That is u^j_m is even or odd for even or odd j respectively. The functions that span these bases are well known, being the eigenfunctions of ∇^2 over the unit sphere. They are the spherical harmonics written, with j replaced by l (l integer), as follows,

$$Y^m_l(\theta\varphi) = \mathrm{i}^{m+|m|} P^{|m|}_l(\cos\theta)\exp(\mathrm{i}m\varphi), \tag{2}$$

$$m \equiv |m| : P^m_l(\cos\theta) = \left\{\frac{(2l+1)(l-m)!}{4\pi(l+m)}\right\}^{1/2} \frac{1}{2^l l!} \sin^m\theta \frac{\mathrm{d}^{l+m}(\cos^2\theta - 1)^l}{\mathrm{d}(\cos\theta)^{l+m}}. \tag{3}$$

(Spherical and solid harmonics are further discussed at the end of this section.) Notice that the position of the labels in Y^m_l is transposed with respect to the u^l_m. More importantly, they satisfy the following relations under conjugation

$$(Y^m_l)^* = (-1)^m Y^{-m}_l, \tag{4}$$

as can readily be verified. This is a result of the Condon and Shortley convention of §6, and will be further discussed in §8.

As expected from the discussion above, the spherical harmonics are gerade or ungerade respectively for l even or odd:

$$\boldsymbol{i}Y^m_l(\theta, \varphi) = Y^m_l(\pi-\theta, \varphi+\pi) = (-1)^l Y^m_l(\theta\varphi). \tag{5}$$

Before we go on to discuss the situation for half-integral j, we shall consider the form of the representative $\hat{R}(\phi\mathbf{n})$ for rotations around the $\mathbf{z}$ axis, which will give us some useful information about the representations themselves. From (2.15),

$$R(\phi\mathbf{z}) = \exp(-\mathrm{i}\phi I_z) = \sum_n (n!)^{-1}(-\mathrm{i}\phi I_z)^n, \tag{6}$$

so that, on using the eigenvalue equation for I_z, (5.6),

$$R(\phi\mathbf{z})u^j_m = \sum_n (n!)^{-1}(-\mathrm{i}\phi m)^n u^j_m = \exp(-\mathrm{i}\phi m)u^j_m. \tag{7}$$

The matrix representation will thus have the form

$$R(\phi\mathbf{z})\langle u^j_j u^j_{j-1} \ldots u^j_{-j}| = \langle u^j_j u^j_{j-1} \ldots u^j_{-j}| \begin{bmatrix} \exp(-\mathrm{i}\phi j) & & & & \\ & \exp\{-\mathrm{i}\phi(j-1)\} & & & \\ & & \cdot & & \\ & & & \cdot & \\ & & & & \exp(\mathrm{i}\phi j) \end{bmatrix}. \tag{8}$$

We shall denote the matrix on (R8) with the symbol $\hat{R}^j(\phi\mathbf{z})$ to indicate the irreducible representation, labelled by j, to which it belongs. We can now consider the case for half-integral j and we shall take the simplest example for $j = \frac{1}{2}$.

$$\hat{R}^{1/2}(\phi\mathbf{z}) = \begin{bmatrix} \exp(-\mathrm{i}\frac{1}{2}\phi) & \\ & \exp(\mathrm{i}\frac{1}{2}\phi) \end{bmatrix}. \tag{9}$$

On adding 2π to ϕ, however, it is clear that the exponentials on (R9) change sign, although $R(\phi + 2\pi, \mathbf{z})$ is identically the same operation as $R(\phi\mathbf{z})$. Thus, we find the disturbing result that the matrices of the alleged irreducible representations of SO(3) are two-valued for half-integral j:

$$\hat{R}^{1/2}(\phi\mathbf{z}) = \pm \begin{bmatrix} \exp(-\mathrm{i}\frac{1}{2}\phi) & \\ & \exp(\mathrm{i}\frac{1}{2}\phi) \end{bmatrix}. \tag{10}$$

It should be clear that we cannot produce out of (10) two representations of SO(3): the negative matrices do not form a group and if we multiply, say, the two positive matrices $+\hat{R}^{1/2}(\pi\mathbf{z})$ and $+\hat{R}^{1/2}(\pi\mathbf{z})$ a negative matrix (for the identity) results. Also, we do not have even one representation, as the latter multiplication example shows. There are two ways out of this. One, due to Bethe (1929) consists in introducing heuristically rotations $R(\phi + 2\pi, \mathbf{z})$ to which the negative matrices are attached. Thus SO(3) is doubled, with as many 'operations' in the new group formed, called the *double group*, as there are matrices. This approach entails serious problems: consider the products $C_{2x}C_{2y}$ and $C_{2y}C_{2x}$ in Table **2**-5.2. Both of them are equal to C_{2z}. How do we know when this result should be interpreted as C_{2z} enhanced or not by a 2π rotation?

There are ways in which, within the double-group approach, the type of uncertainty exemplified above can be resolved (Chapter **13**), but there is a more systematic scheme, which will be developed in this book, in which no additional operations are introduced but, instead, a new type of representation is used called a *projective* or *ray representation* (Chapter **8**).

Spherical and solid harmonics

Solid harmonics are defined as homogeneous polynomials of degree j (j integral here) that satisfy Laplace's equation. Thus, with the additional condition of restriction to the unit sphere ($x^2+y^2+z^2=1$), they coincide with the spherical harmonics. The number of such independent polynomials is in fact $2j+1$ (Problem 5). It is most important to adapt the solid harmonics in such a way that they transform under rotations precisely like the angular momentum basis $\langle u_j^j \ldots u_{\bar{j}}^j|$, for which purpose they must transform precisely as the spherical harmonics defined in (2) and (3). Consider the basis Y_1^1, Y_1^0, Y_1^{-1}:

2,3 $$Y_1^1 = -(\tfrac{3}{8}\pi)^{1/2}\sin\theta\,(\cos\varphi + \mathrm{i}\sin\varphi), \tag{11}$$

2,3 $$Y_1^0 = (\tfrac{3}{4}\pi)^{1/2}\cos\theta, \tag{12}$$

2,3 $$Y_1^{-1} = (\tfrac{3}{8}\pi)^{1/2}\sin\theta\,(\cos\varphi - \mathrm{i}\sin\varphi). \tag{13}$$

The factor $(\frac{3}{4}\pi)^{1/2}$ constant through the basis, can be conveniently disregarded since it will not affect its transformation properties. Thus, we write from (**2**-4.6), (**2**-4.7),

$$\begin{aligned} u_1^1 &= -2^{-1/2}(x+\mathrm{i}y) & u_0^1 &= z & u_{-1}^1 &= 2^{-1/2}(x-\mathrm{i}y) \\ &=_{\text{def}} u_1, & &=_{\text{def}} u_0, & &=_{\text{def}} u_{-1}. \end{aligned} \tag{14}$$

the distinction between x, y and x, y being left implicit as discussed in §**2**-4. In order to build solid harmonics correctly symmetrized we shall change the variables x, y, z into u_1, u_0, u_{-1}. (Notice, however, that when the latter appear in polynomial sets, they must be treated as functions, as explained at the end of §**2**-4.) The Laplacian takes a particularly simple form in these variables (Problem 6) that leads to a simple expansion for the solid harmonics (Problem 7) which shall be denoted as u_m^j since, with the restriction to the unit sphere stated, they coincide with these functions.

Problems 4-7 (p. 281)

1. Obtain the matrix in (8) by acting directly with $R(\phi\mathbf{z})$ on the spherical harmonics (2).
2. Prove that $\chi^j(\phi)$, which is $\operatorname{Tr}\hat{R}^j(\phi\mathbf{n})$ has the values

$$\chi^j(\phi) = \sin(j+\tfrac{1}{2})\phi\,(\sin\tfrac{1}{2}\phi)^{-1}, \qquad \phi \neq 0, \tag{15}$$

$$= 2j+1 \qquad , \qquad \phi = 0. \tag{16}$$

3. Prove that, in the jth representation, the character of an improper rotation of O(3) of the form $\boldsymbol{i}R(\phi\mathbf{n})$ is $(-1)^j\chi^j(\phi)$, with $\chi^j(\phi)$ defined in Problem 2.
4. Prove, on using Table **2**-5.1, that the irreducible representation of

SO(3) corresponding to $j=3$ reduces in $\mathbf{C}_{3v}$ into the representations $2A_1+A_2+2E$. (The numbers in front of the representation symbols indicate how many times the representation appears in the reducible one.)

5. Prove that the number of independent homogeneous polynomials u^j of degree j in x, y, and z, that are solutions of Laplace's equation, $\nabla^2 u^j=0$, is $2j+1$, (j integral).
6. Prove that the Laplacian ∇^2 takes the following form, in the independent variables defined in (14):

$$\nabla^2=-2\,\partial^2/\partial u_1\,\partial u_{-1}+\partial^2/\partial(u_0)^2. \tag{17}$$

7. Write the $2j+1$ independent homogeneous polynomials of degree j in the form

$$u^j_m=\sum_r c_r(u_1)^{m+r}(u_{-1})^r(u_0)^{j-m-2r}, \qquad -j\leqslant m\leqslant j, \tag{18}$$

where r is chosen in the range that makes all exponents in (18) positive or zero, and prove that: (i) (18) leads to the basis (14); (ii) the coefficients c_r satisfy the recurrence relation

$$c_{r+1}=\tfrac{1}{2}(j-m-2r)(j-m-2r-1)\{(r+1)(m+r+1)\}^{-1}c_r. \tag{19}$$

Hence show that

$$c_r=\{(j-m)!\,m!\,c_0\}\{2^r(m+r)!\,r!\,(j-m-2r)!\}^{-1}. \tag{20}$$

Notice that the first curly bracket in (20) denotes a coefficient that depends on j and m, C_{jm} say, that can be taken out of the summation in (18). Its value can be obtained by use of the step-down operator. (See eqn 3.153 in Biedenharn and Louck 1981, which, with a slight change in the basis, agrees with eqn 18.)

8. THE CONDON AND SHORTLEY CONVENTION

It should be clear from the discussion of the last section that the form of the matrix $\hat{R}^j(\phi\mathbf{n})$ on the basis $\langle u^j_j \ldots u^j_{-j}|$ vitally depends on the form of the equations that give the action of the operators $I_\pm$ on the u^j_m. The equations we use are given in (6.26) in the Condon and Shortley (CS) convention. Otherwise, an arbitrary phase factor would appear on (R6.26). What really happens is this: each choice of phase factors will give a representation equivalent to the one chosen with CS factors (all unity). (It must be equivalent because the matrices in eqn 7.8 do not change since I_z does not entail an arbitrary phase.) This is as we must expect, since a single irreducible representation entails really an infinite set of equivalent representations. What the CS convention does is to single out one of the representations of this infinite set as a *canonical representation* to be always used, thus eliminating the inherent indeterminacy of the representation matrices.

The CS convention requires a particular form of behaviour of the basis u^j_m under conjugation, namely that

$$(u^j_m)^* = (-1)^m u^j_{-m}, \tag{1}$$

a result that has already been used (eqn 7.4) in re-defining the spherical harmonics in the CS phase (eqns 7.2, 7.3) and which will now be demonstrated. We shall need in the proof two preliminary results:

$$I_-^* = -I_+, \tag{2}$$

as follows at once from (5.12), and

$$u^j_0 = \text{independent of } \varphi, \tag{3}$$

as follows from applying I_z on u^j_0, that must give zero, while I_z from (2.18) is $-\mathrm{i}\,\partial/\partial\varphi$. The results that follow are independent of j so that we shall write u_m for u^j_m. We shall prove (1), which is obviously true for u_0, by induction, showing first that

$$u_1^* = -u_{-1}. \tag{4}$$

Proof:

6.26 $\quad I_-u_0 = \{j(j+1)\}^{1/2}u_{-1}.$ (5)

5 $\quad u_{-1} = \{j(j+1)\}^{-1/2}I_-u_0$ (6)

5.12, 3|6 $\quad = -\{j(j+1)\}^{-1/2}\exp(-\mathrm{i}\varphi)\,\partial u_0/\partial\theta.$ (7)

On the other hand

6.26 $\quad I_+u_0 = \{j(j+1)\}^{1/2}u_1.$ (8)

8 $\quad u_1 = \{j(j+1)\}^{-1/2}I_+u_0$ (9)

5.12, 3|9 $\quad = \{j(j+1)\}^{-1/2}\exp(\mathrm{i}\varphi)\,\partial u_0/\partial\theta.$ (10)

Eqns (7) and (10) entail (4). We go on:

$$u_2^* = u_{-2}. \tag{11}$$

Proof:

4 $\quad I_+u_1 = -I_+u_{-1}^*$ (12)

2|12 $\quad = I_-^*u_{-1}^*.$ (13)

Also

6.26 $\quad I_+u_1 = \{j(j+1)-2\}^{1/2}u_2,$ (14)

6.26 $\quad I_-u_{-1} = \{j(j+1)-2\}^{1/2}u_{-2}.$ (15)

15 $\quad I_-^*u_{-1}^* = \{j(j+1)-2\}^{1/2}u_{-2}^*.$ (16)

On replacing (14) and (16) on either side of (13), eqn (11) follows and, by induction, (1).

It should be noted that although the CS convention uniquely determines a canonical basis, the representation matrices themselves can be altered by writing the basis as $\langle u^j_{-j}, u^j_{-j+1} \dots u^j_j|$ instead of having the highest m as the first element. Also, by writing the basis as a column rather than a row (which is, of course, a mistake). We agree always to write the canonical basis as shown in (7.8).

Another important feature of the CS convention is that *it involves a definite choice of the* **x**, **y** *axes,* as we shall now demonstrate. It is important to recognize first for this purpose that if the phase factor $\exp(i\alpha)$ is introduced in I_+, then $\exp(-i\alpha)$ must appear in I_-, as it follows from (6.12), (6.13), and (6.24). Let us then consider how such a choice,

$$I'_+ = \exp(i\alpha) I_+, \qquad I'_- = \exp(-i\alpha) I_-, \tag{17}$$

affects the operators I_x and I_y which, from (5.1) are

$$I_x = \tfrac{1}{2}(I_+ + I_-), \qquad I_y = -\tfrac{1}{2}i(I_+ - I_-). \tag{18}$$

When the operators (17) are introduced in (18), one finds

$$I'_x = \tfrac{1}{2}\{\exp(i\alpha) I_+ + \exp(-i\alpha) I_-\} = \cos\alpha\ I_x - \sin\alpha\ I_y, \tag{19}$$

$$I'_y = -\tfrac{1}{2}i\{\exp(i\alpha) I_+ - \exp(-i\alpha) I_-\} = \sin\alpha\ I_x + \cos\alpha\ I_y, \tag{20}$$

that is any choice of phase factor other than the CS choice will result in a rotation of the **x**, **y** axes. It is largely for this reason that the second Euler angle is now always chosen around the **y** and not the **x** axis, since within the CS convention the choice between these two alternatives must be made once for all. (See also Problem **6**-7.1.)

Problem 4.8 (p. 282)

1. Verify that (17) gives the correct behaviour when the $I_\pm$ operators operate on $\langle u_m \mid u_{m'}\rangle$.

9. APPLICATIONS. MATRICES FOR $j = 1$ AND $j = \frac{1}{2}$ (PAULI MATRICES)

We are now in a position to go back to eqn (2.15),

$$R(\phi\mathbf{n}) = \exp(-i\phi\mathbf{n} \cdot \mathbf{I}), \tag{1}$$

with a view to obtaining once more $\hat{R}(\phi\mathbf{n})$ in three-space, now from the matrix for the angular momentum operator **I**. This has to be expressed in terms of I_x, I_y, from (8.18), in which (6.26) is introduced, and of I_z from

(5.6):

$$I_x u_m = \tfrac{1}{2}\{j(j+1)-m(m+1)\}^{1/2}u_{m+1}+\tfrac{1}{2}\{j(j+1)-m(m-1)\}^{1/2}u_{m-1}, \qquad (2)$$

$$I_y u_m = -\tfrac{1}{2}\mathrm{i}\{j(j+1)-m(m+1)\}^{1/2}u_{m+1}+\tfrac{1}{2}\mathrm{i}\{j(j+1)-m(m-1)\}^{1/2}u_{m-1}, \qquad (3)$$

$$I_z u_m = m u_m. \qquad (4)$$

Since we want to form a three-dimensional representation, we take $j=1$ and on using (2) to (4) the following matrices are easily found (Problem 1),

$$\begin{array}{cccc} \text{Basis} & \hat{I}_x & \hat{I}_y & \hat{I}_z \\ \langle u_1^1 u_0^1 u_{-1}^1|, & 2^{-1/2}\begin{bmatrix} & 1 & \\ 1 & & 1 \\ & 1 & \end{bmatrix} & 2^{-1/2}\begin{bmatrix} & -\mathrm{i} & \\ \mathrm{i} & & -\mathrm{i} \\ & \mathrm{i} & \end{bmatrix} & \begin{bmatrix} 1 & & \\ & 0 & \\ & & -1 \end{bmatrix}. \end{array} \qquad (5)$$

The functions u above are the spherical harmonics in the CS convention, so that in order to get matrices related to the **x**, **y**, **z** axes we must take the harmonics into Cartesian form. We do this by using the expressions for u_1^1, u_0^1 and u_{-1}^1 in (7.14) but, because these must now be regarded as functions, the x, y, and z that appear in them are not the independent variables but rather the functions $\langle \mathsf{xyz}|$ of §**2**-4 (eqn **2**-4.11 and discussion below it). From (7.14) the relation between these two bases is as follows:

$$\langle u_1^1 u_0^1 u_{-1}^1| = \langle \mathsf{xyz}|\begin{bmatrix} -2^{-1/2} & 0 & 2^{-1/2} \\ -\mathrm{i}2^{-1/2} & 0 & -\mathrm{i}2^{-1/2} \\ 0 & 1 & 0 \end{bmatrix} =_{\text{def}} \langle \mathsf{xyz}|\, A, \qquad (6)$$

where A is clearly unitary. On the new basis $\langle \mathsf{xyz}|$ the matrices $\hat{I}$ in (5) are transformed in $A\hat{I}A^{\dagger}$, which are (Problem 2),

$$\begin{array}{cccc} \text{Basis} & \hat{I}_x & \hat{I}_y & \hat{I}_z \\ \langle \mathsf{xyz}| & \begin{bmatrix} 0 & 0 & 0 \\ 0 & 0 & -\mathrm{i} \\ 0 & \mathrm{i} & 0 \end{bmatrix} & \begin{bmatrix} 0 & 0 & \mathrm{i} \\ 0 & 0 & 0 \\ -\mathrm{i} & 0 & 0 \end{bmatrix} & \begin{bmatrix} 0 & -\mathrm{i} & 0 \\ \mathrm{i} & 0 & 0 \\ 0 & 0 & 0 \end{bmatrix}. \end{array} \qquad (7)$$

When these matrices are fed into (1) and thus multiplied by $(-\mathrm{i})$ the correct matrices $\hat{\mathbf{J}}$ of (**3**-5.11) appear, as given in (**3**-5.10).

Various important points have to be noticed. First, the work leading to (**3**-5.11) was done in the configuration space $\mathbb{R}^3$, i.e. on the basis $|xyz\rangle$. We have seen in §**2**-4, however, that as long as one changes correctly the column or row nature of the bases, the matrices should not change when going from configuration to function space. Secondly, the $\hat{\mathbf{J}}$ matrices have already been correctly checked by the infinitesimal matrix generators in §3 with a formula, eqn (3.7), that looks exactly like (1) but with the opposite sign. It should be clear now, however, that the meaning of the matrices $\hat{I}$ when working in $\mathbb{R}^3$ as we did in §3, is quite different from the one they have here. Thirdly, it must be appreciated that the agreement obtained in (7) depends entirely on the use of the CS convention. The extraordinary importance of this convention will now, I hope, be fully appreciated.

The Pauli matrices, $j=\frac{1}{2}$

In exactly the same form as for $j=1$, eqns (2), (3), (4) will give the transforms of the basis $\langle u_{1/2}^{1/2} u_{-1/2}^{1/2}|$ and thus the corresponding matrices $\hat{I}_x, \hat{I}_y, \hat{I}_z$, displayed below:

$$\begin{array}{cccc} \text{Basis} & \hat{I}_x & \hat{I}_y & \hat{I}_z \\ \langle u_{1/2}^{1/2} u_{-1/2}^{1/2}| & \frac{1}{2}\begin{bmatrix} & 1 \\ 1 & \end{bmatrix} & \frac{1}{2}\begin{bmatrix} & -\mathrm{i} \\ \mathrm{i} & \end{bmatrix} & \frac{1}{2}\begin{bmatrix} 1 & \\ & -1 \end{bmatrix}. \end{array} \tag{8}$$

It is customary to take the $\frac{1}{2}$ factors away from the matrices and to write the following *Pauli matrices* in the CS convention

$$\begin{array}{ccc} \boldsymbol{\sigma}_x & \boldsymbol{\sigma}_y & \boldsymbol{\sigma}_z \\ \begin{bmatrix} & 1 \\ 1 & \end{bmatrix} & \begin{bmatrix} & -\mathrm{i} \\ \mathrm{i} & \end{bmatrix} & \begin{bmatrix} 1 & \\ & -1 \end{bmatrix}. \end{array} \tag{9}$$

When the $\hat{I}$ matrices of (8) are introduced into (1) via the Pauli matrices, this equation takes the form

$$\hat{R}^{1/2}(\phi\mathbf{n}) = \exp(-\tfrac{1}{2}\mathrm{i}\phi\mathbf{n}\cdot\boldsymbol{\sigma}). \tag{10}$$

Problems 4.9 (p. 282)

1. Verify the matrices in (5).
2. Verify the matrices in (7).
3. Verify the matrices in (9).
4. From the properties of the matrix $\mathbf{n}\cdot\boldsymbol{\sigma}$ prove that (see eqn 10)

$$\hat{R}^{1/2}(\phi\mathbf{n}) = \exp(-\tfrac{1}{2}\mathrm{i}\phi\mathbf{n}\cdot\boldsymbol{\sigma}) = (\cos\tfrac{1}{2}\phi)\mathbf{1}_2 - \mathrm{i}(\sin\tfrac{1}{2}\phi)(\mathbf{n}\cdot\boldsymbol{\sigma}). \tag{11}$$

Hence, that

$$\hat{R}^{1/2}(\phi\mathbf{n}) = \begin{bmatrix} \cos\frac{1}{2}\phi - in_z\sin\frac{1}{2}\phi & -(n_y+in_x)\sin\frac{1}{2}\phi \\ (n_y-in_x)\sin\frac{1}{2}\phi & \cos\frac{1}{2}\phi + in_z\sin\frac{1}{2}\phi \end{bmatrix}. \tag{12}$$

5

TENSOR BASES: INTRODUCTION TO SPINORS

> Tensor Calculus is one of the most satisfactory chapters of mathematics.
>
> C. Lanczos (1970). *Space through the ages*. Academic Press, London, p. 99.

We shall hardly do justice in this book, alas, to Lanczos' delight in tensors. The reader will find nothing here about tensors *à la* Bourbaki, as beautifully introduced in the book by Schutz (1980). Neither, on the more applied side, shall we handle the technicalities of co- and contravariant bases as developed for angular-momentum problems by Fano and Racah (1959). All that we want to do is to develop a modicum of expertise required to handle correctly the bases of the irreducible representations of SO(3). We need for this purpose some ideas that played a crucial part in the early development of tensors. Given a basis – that is, a one-dimensional array of, say, vector components, functions of the configuration-space variables, etc. – its transformation properties under linear transformations of the configuration space are fully described by their corresponding matrices. (These matrices are, for us, the matrices that represent all rotations of SO(3) and the inversion, these operations determining all linear transformations of the configuration space in which we are interested.) We also know that these matrices firmly depend on the basis used. Our tensorial programme will be based on two principles. The first is a building-up principle by means of which we propose to construct a hierarchy of bases, such that each level of the hierarchy will be called a tensor of a given *rank*, the rank indicating the level in the hierarchy. Thus, *scalars* will be tensors of rank zero and *vectors* tensors of rank one, the building-up principle providing the machinery to create the higher levels of the hierarchy. The second and most important principle of this tensor scheme is that a basis will be said to belong to a given level in the hierarchy, i.e. to be a tensor of a given rank, if and only if the matrices under which the basis transforms under all linear transformations of configuration space are precisely those that are accepted as the standard transformation matrices for the rank given. In other words, the *names* given to different kinds of bases, such as scalars, vectors, tensors of rank two and so on, are nothing more than a shorthand or a key that uniquely specifies the precise transformation matrices that correspond to the bases.

The CS or canonical bases $\langle u^j_m|$ of §**4**-8 provide us with a good set of models as regards bases with precise transformation laws. Take, e.g., the one-dimensional basis $\langle u^0_0|$. From (**4**-7.8) and Problem **4**-7.3 we know that the transformation matrix for all rotations and for the inversion is +1. A basis with this property is called a *scalar.* A one-dimensional basis (still to be found) that transforms by the matrix +1 under all rotations and the matrix −1 under inversion is called a *pseudoscalar.*

We shall continue along this argument in the rest of the chapter, defining bases of various types, like vectors, tensors and spinors.

1. VECTORS AND SPHERICAL VECTORS

We call a vector a basis $|xyz\rangle$ of $\mathbb{R}^3$ that transforms by the matrix $\hat{R}^1_{\mathbf{r}}(\phi\mathbf{n})$ of (**3**-3.11) under rotations and by the matrix $-\mathbf{1}_3$ under inversion. Naturally, in each case, the matrix pre-multiplies the basis (because we write it as a column). We also know that to this basis there corresponds a *dual basis,* that transforms under exactly the same matrices as long as the basis is written as a row basis. Such bases can be triples from the function space, such as the functions $\langle \mathsf{xyz}|$, or the unit vectors spanning the configuration space, $\langle \mathbf{ijk}|$ (see §**2**-4 and the note following eqn **3**-1.15). The important point to note is that the same matrix serves to denote two dual bases, one in configuration space (column basis) and the other in function space (row basis). Thus, given a triple in function space with agreed transformation properties we automatically define a triple in configuration space; and vice versa.

All this is very good but we have departed from our plan to stick to the representation matrices of the canonical bases $\langle u^j_m|$. We have seen, however, that the canonical basis $\langle u^1_1 u^1_0 u^1_{-1}|$, that is, the spherical harmonics $\langle Y^1_1 Y^0_1 Y^{-1}_1|$, are related by a simple linear transformation to a basis $\langle \mathsf{xyz}|$ the dual of which is precisely given by the components $|xyz\rangle$ of a vector $\mathbf{r}$ of $\mathbb{R}^3$.

What we propose to do is to bypass the basis $\langle \mathsf{xyz}|$, to take straightaway the basis $\langle Y^1_1 Y^0_1 Y^{-1}_1|$ as our function-space or row-basis, and to define a 'vector' as the dual of it, say $|\mathbb{u}_1, \mathbb{u}_0, \mathbb{u}_{-1}\rangle$, i.e. as a triple that transforms column-wise under rotations and the inversion by the same matrices that transform $\langle Y^1_1 Y^0_1 Y^{-1}_1|$. Two remarks now: first, as we must expect from (**4**-9.6), and will soon be evident, the triple $|\mathbb{u}\rangle$ is made up of complex numbers and thus belongs to $\mathbb{C}^3$ (the set of all triples of complex numbers) rather than $\mathbb{R}^3$. Secondly, so far, we only know the canonical matrix for two operations: the inversion and $R(\phi\mathbf{z})$ so, as we shall see, we shall have to be very careful in constructing $|\mathbb{u}_1\mathbb{u}_0\mathbb{u}_{-1}\rangle$ so that it transforms correctly.

This vector $|\mathbb{u}_1\mathbb{u}_0\mathbb{u}_{-1}\rangle$ dual to the canonical basis $\langle Y_1^1 Y_1^0 Y_1^{-1}|$ is called a *spherical vector*, in a new configuration space $\mathbb{C}^3$ that replaces $\mathbb{R}^3$. This slight complication is more than offset by the fact that the necessary matrices of transformation become much simpler. We shall now determine these spherical vectors.

From (**4**-7.8), the spherical harmonics for $l=1$ transform as follows under a rotation around the **z** axis:

$$R(\phi\mathbf{z})\langle Y_1^1 Y_1^0 Y_1^{-1}| = \langle Y_1^1 Y_1^0 Y_1^{-1}| \begin{bmatrix} e^{-i\phi} & & \\ & 1 & \\ & & e^{i\phi} \end{bmatrix}. \tag{1}$$

Under inversion, the corresponding matrix is $-\mathbf{1}_3$, (see eqn **4**-7.5). From (**4**-7.11) to (**4**-7.13) the spherical harmonics can be written in terms of the Cartesian functions† x, y, z, in the same manner as in (**4**-7.14):

$$Y_1^1 = -2^{-1/2}(\mathsf{x}+i\mathsf{y}), \qquad Y_1^0 = \mathsf{z}, \qquad Y_1^{-1} = 2^{-1/2}(\mathsf{x}-i\mathsf{y}). \tag{2}$$

Our problem is to find combinations of the *variables* x, y, z, that, column-wise, behave like the spherical-harmonics basis. We cannot expect, however, $x \pm iy$ to behave like $\mathsf{x} \pm i\mathsf{y}$. It is easy to verify, in fact, that, when acting on those independent variables with the *configuration space* operator $R(\phi\mathbf{z})$, the result is

$$R(\phi\mathbf{z})(\mathsf{x} \pm i\mathsf{y}) = \exp(\pm i\phi)(x \pm iy). \tag{3}$$

(See Problem 1.) We can thus guess that $\mathbb{u}_1$ and $\mathbb{u}_{-1}$ must respectively be $2^{-1/2}(x-iy)$ and $-2^{-1/2}(x+iy)$, the factors being taken from (2). It is tempting to guess that, likewise, $\mathbb{u}_0$ must be z but this is not so. The problem is that, in principle, we should satisfy the transformation matrix not just for $R(\phi\mathbf{z})$ as in (1) but for all $R(\phi\mathbf{n})$. Let us now consider $R(\beta\mathbf{y})$ acting on Y_1^0 from (**4**-7.12), disregarding the irrelevant normalization factor:

$$R(\beta\mathbf{y})Y_1^0 = R(\beta\mathbf{y})\cos\theta = \cos(\theta-\beta) = \cos\theta\,\cos\beta + \sin\theta\,\sin\beta \tag{4}$$

$$= \mathsf{z}\cos\beta + \mathsf{x}\sin\beta. \tag{5}$$

We can easily obtain z and x from the harmonics in (**4**-7.11) to (**4**-7.13) (disregarding numerical factors) and introduce them in (5):

$$R(\beta\mathbf{y})Y_1^0 = -\sin\beta Y_1^1 + \cos\beta Y_1^0 + \sin\beta Y_1^{-1}. \tag{6}$$

† *Warning.* The distinction made here between the independent variable $x+iy$ and the function $\mathsf{x}+i\mathsf{y}$ will not always be made. Elements of row-vector bases must always be understood as functions and when $x+iy$, say, appears in such a basis, it must be understood as a polynomial of degree one and not as the independent variable.

If we call A the matrix of transformation of the basis $\langle Y_1^1 Y_1^0 Y_1^{-1}|$, the coefficients in (6) give the *second column* of A. In order to consider the transform $R(\beta\mathbf{y})z$, however, we need the *second row* of A since, first, z being an independent variable it must transform as a member of a column basis and, second, we want this basis to transform, column-wise, by the identical matrix A. The second row of A is easy to obtain, fortunately, since A being orthogonal, A^{T} equals A^{-1}. Thus, the second row of A must be the second column of A^{-1}, that is the column in (6) but with β replaced by $-\beta$:

$$\langle \sin\beta, \cos\beta, -\sin\beta|. \tag{7}$$

We can now transform the variable z by means of the matrix for the rotation $R(\beta\mathbf{y})$ in (**3**-1.14):

$$R(\beta\mathbf{y})z = z\cos\beta - x\sin\beta. \tag{8}$$

We want to express this as the transform $R(\beta\mathbf{y})\,|\mathbb{u}_1\mathbb{u}_0\mathbb{u}_{-1}\rangle$, with $\mathbb{u}_{\pm1} = \pm(x\mp \mathrm{i}y)$ from our previous guess (disregarding again numerical factors). Expressing x in terms of these variables and substituting the result into (8), we get

$$R(\beta\mathbf{y})z = -\sin\beta\,\mathbb{u}_1 + \cos\beta\, z + \sin\beta\,\mathbb{u}_{-1}. \tag{9}$$

The coefficients here, of course, should be the elements of the second row of A given in (7), but they are not them. If, instead, we take the second element of the basis, $\mathbb{u}_0$, to be $-z$, we have, on changing the sign on both sides of (9),

$$R(\beta\mathbf{y})(-z) = \sin\beta\,\mathbb{u}_1 + \cos\beta(-z) - \sin\beta\,\mathbb{u}_{-1}, \tag{10}$$

which agrees with (7). We therefore define the spherical vector as the basis $|\mathbb{u}_1\mathbb{u}_0\mathbb{u}_{-1}\rangle$ in configuration space with elements

$$\mathbb{u}_1 = 2^{-1/2}(x - \mathrm{i}y), \qquad \mathbb{u}_0 = -z, \qquad \mathbb{u}_{-1} = -2^{-1/2}(x + \mathrm{i}y). \tag{11}$$

In using the spherical vector, the configuration space, previously $\mathbb{R}^3$, is taken to be $\mathbb{C}^3$. The coefficients in (11) have been chosen to make the transformation from $|xyz\rangle$ to $|\mathbb{u}_1\mathbb{u}_0\mathbb{u}_{-1}\rangle$ a unitary one, as was the case in (**4**-9.6) between $\langle \mathsf{xyz}|$ and $\langle u_1^1 u_0^1 u_{-1}^1|$.

The reader should verify that the basis (11) transforms under the matrix in (1). (Problem 2.) That it also transforms correctly under all $R(\phi\mathbf{n})$ will be seen in §**7**-3.

Warning. Although (5) is absolutely correct, eqn (4) is not a general expression for the transform of Y_1^0. We are assuming in this equation that the harmonic is defined only on the **xz** plane, since the transformation law for θ therein used is valid only in that plane, the correct rule of transformation actually depending on φ. The correct transformation

(Problem 3) agrees with (5) and it will also be obtained in a simpler way in Problem **7**-3.6.

Problems 5-1 **(p. 283)**

1. Verify eqn (3).
2. Verify that the basis $|\mathbb{u}_1 \mathbb{u}_0 \mathbb{u}_{-1}\rangle$ of (11) transforms under the matrix given in (1).
3. Obtain correctly the transform of Y_1^0 (given as $\cos\theta$) by acting on it with the function–space operator for $R(\beta\mathbf{y})$. Hence derive (5).

2. TENSOR BASES AND TENSOR PRODUCTS

When we realize the canonical bases $\langle u^j|$ through the spherical harmonics $\langle Y_l^m|$, bases for all irreducible representations of SO(3) for integral j are obtained (i.e. all the odd-dimensional representations appear). The basis $\langle Y_l^m|$, however, is gerade for even l and ungerade for odd l (§**4**-7). As a result, when forming the irreducible representations of odd dimension of O(3), half of them appear to be missing, namely the ungerade bases for even l and the gerade bases for odd l. (Remember that since $O(3) = SO(3) \otimes \mathbf{C}_i$, with $\mathbf{C}_i = E \oplus \boldsymbol{i}$, then each irreducible representation of SO(3) should split in two irreducible representations of O(3), one gerade and the other ungerade, adapted to the two corresponding irreducible representations of $\mathbf{C}_i$.)

It is clear that we need some machinery for constructing new bases, and we shall use for this purpose a scheme which creates a whole hierarchy of bases in such a way that each basis belongs to a clearly stated representation (at first, not necessarily irreducible). We shall first work entirely with the representations for integral j. A canonical basis $\langle u^j|$ for j integral will be called a *tensor of rank one* and dimension $2j+1$. It must be clearly understood that such an apellation is entirely linked to the transformation matrices of these objects under all the operations of O(3):

$$g\,\langle u^j| = \langle u^j|\,\hat{G}^j(g). \tag{1}$$

Here, if g is a proper rotation, $\hat{G}^j(g)$ must be the precise rotation matrix corresponding to the canonical bases (we do not have them yet, except for $j=1$, but they will be fully specified later on). If g is the inversion $\hat{G}^j(g)$ must be $(-1)^j \mathbf{1}_{2j+1}$. Here $(-1)^j$ gives the gerade–ungerade property of the basis and it is called its *parity*. Any basis which transforms in this way, irrespectively of where it comes from (i.e. it need not be the stated spherical harmonics) will be a tensor of rank one and dimension $2j+1$. A spherical vector as defined in §1 will be, in fact, a tensor of rank one and dimension three.

We now define a *tensor of rank two* and dimension $(2j+1)(2j'+1)$ as any array that transforms by the same matrices as $\langle u^j| \otimes \langle u^{j'}|$ (see eqn **2**-5.22). In fact, the matrices corresponding to such bases are well defined, being the direct products $\hat{G}^j \otimes \hat{G}^{j'}$ of the matrices in (1). Such bases are in general reducible. Within SO(3), it can be proved (see Problem 1) that the reduction goes by the famous *Clebsch–Gordan series:*

$$\langle u^j| \otimes \langle u^{j'}| = \langle u^{j+j'}| \oplus \langle u^{j+j'-1}| \oplus \ldots \oplus \langle u^{j-j'}|, \qquad j \geq j'. \tag{2}$$

If we consider this reduction as performed in O(3) the parity of the basis under inversion must be considered. Since the basis on (L2) is made up of products of the form $u^j_m u^{j'}_m$ it is clear that its parity, and thus that of each of the bases on (R2) must be $(-1)^{j+j'}$. Consider for example

$$\langle u^1| \otimes \langle u^1| = \langle u^2| \oplus \langle u^1| \oplus \langle u^0|. \tag{3}$$

The basis $\langle u^1|$ on (R3) behaves under SO(3) exactly as the canonical basis $\langle u^1|$ in (1.1) (see Problem 2), which must be the case; otherwise the notation in (R2) and (R3) would be nonsense. In O(3), however, it does not behave in the same way, since, being a tensor of rank *two* it is gerade. We have thus obtained the missing basis of O(3) corresponding to j equal to unity but gerade. This basis, which behaves like a vector under rotations but is gerade, is often called an *axial vector* although this name falls outside the present hierarchy. It is a gerade tensor of rank two and dimension three (or, as we shall see, an antisymmetric tensor of rank two and dimension three). It should be clear that in the same way all the gerade bases for odd j and the ungerade bases for even j are formed in O(3). (A more systematic treatment of the irreducible bases of O(3) is given in §**11**-9.) The nature of this process, based on reductions such as the one effected on (R3), is made clearer by a process called the symmetrization of the tensor bases.

Symmetrization of tensors

Call $\langle w|$ the tensor $\langle u^j| \otimes \langle u^{j'}|$ of components $w_{mm'}$ equal to $u^j_m u^{j'}_{m'}$. We shall say that w is *symmetrical* or *antisymmetrical* respectively when

$$w_{mm'} = \pm w_{m'm}. \tag{4}$$

This nomenclature is introduced not just to cover special cases but, rather, because it is valid for all tensors, in the sense that any arbitrary tensor w can always be written as a sum of a symmetrical and an antisymmetrical tensor. This is so because

$$w_{mm'} = \tfrac{1}{2}(w_{mm'} + w_{m'm}) + \tfrac{1}{2}(w_{mm'} - w_{m'm}). \tag{5}$$

Clearly, the tensors with the components given by the first and second brackets on (R5) are symmetric and antisymmetric respectively.

There is, moreover, a most important conservation law of the symmetry or antisymmetry of a tensor with respect to all rotations and the inversion, which we shall now prove: if $w_{mm'}$ equals $w_{m'm}$ we assert that the same is valid for the transformed tensor gw under any operation of O(3): $(gw)_{mm'}$ equals $(gw)_{m'm}$.

Proof: From the ordinary matrix rule, but for double indices, $(gw)_{mm'}$ can be written in either of the two identical ways given in the following equation:

$$(gw)_{mm'} = \sum w_{rs}\hat{G}(g)_{rs,mm'} \equiv \sum w_{sr}\hat{G}(g)_{sr,mm'} \tag{6}$$

6
$$= \sum w_{rs}\hat{G}(g)_{sr,mm'} \tag{7}$$

2-5.28|7
$$= \sum w_{rs}\hat{G}(g)_{sm}\hat{G}(g)_{rm'} \equiv \sum w_{rs}\hat{G}(g)_{rm'}\hat{G}(g)_{sm} \tag{8}$$

2-5.28|8
$$= \sum w_{rs}\hat{G}(g)_{rs,m'm} = (gw)_{m'm}, \tag{9}$$

where, in (7), we use the symmetry of w_{sr} in (6).

This result allows us immediately to reduce a tensor basis $\langle w|$ equal to $\langle u^{j}|\otimes\langle u^{j'}|$, because if we write this basis as in (5), then the symmetrical and antisymmetrical parts of the tensor must belong to different irreducible bases, since the theorem just proved shows that there is no operation of O(3) that can transform a symmetric into an antisymmetric component or vice versa.

Let us take as an example a tensor product of two bases belonging to $j=1$, like in (3), but we shall want to distinguish between them with symbols $\langle u^1|$ and $\langle v^1|$ respectively.

$$\langle u^1|\otimes\langle v^1| = \langle u^1_1 v^1_1,\, u^1_1 v^1_0,\, u^1_1 v^1_{-1},\, \ldots,\, u^1_{-1} v^1_{-1}|, \tag{10}$$

(compare with eqn 1.1 for the suffices, which correspond to the superscripts in that equation). We shall see that there can be only three linearly independent antisymmetric components on (R10). For $u^1_i v^1_i$ will give a null antisymmetric term, and $u^1_i v^1_j$ and $u^1_j v^1_i$ will both give the same antisymmetric term except for the sign. These three antisymmetric components must form an irreducible tensor basis on their own, corresponding to $\langle u^1|$ on (R3), whereas the six symmetric components form a basis reducible into $\langle u^2|$ (five components) and $\langle u^0|$ (one component), as shown in (R3). The antisymmetric components are

$$\langle u^1_1 v^1_0 - u^1_0 v^1_1,\, u^1_1 v^1_{-1} - u^1_{-1} v^1_1,\, u^1_{-1} v^1_0 - u^1_0 v^1_{-1}|. \tag{11}$$

They are clearly gerade, and it is now easy to prove that, under rotations, they behave precisely as a tensor of rank one (spherical vector). (Problem 2.) The 'vector' (11) is a so-called axial vector, in our hierarchy an antisymmetric tensor of rank two. (This vector can in fact be identified with the cross product of the vectors u and v.)

Problems 5-2 (p. 283)

1. Obtain the Clebsch–Gordan series (2).
2. Prove that the basis (11) transforms under rotations as a tensor of rank one and dimension three (spherical vector).

3. HALF-INTEGRAL BASES: SPINORS

We have seen in §2 how to form a hierarchy of tensors in function space by forming tensor products of the bases $\langle u^j|$. We can, of course, do the same in configuration space, forming a hierarchy of which the spherical vector basis $|\mathbb{u}_1\mathbb{u}_0\mathbb{u}_{-1}\rangle$ is at the lowest level. We know from (1.1) that under rotations around the **z** axis, this basis transforms as follows:

$$R(\phi\mathbf{z})\,|\mathbb{u}_1\mathbb{u}_0\mathbb{u}_{-1}\rangle=\begin{bmatrix} e^{-i\phi} & & \\ & 1 & \\ & & e^{i\phi}\end{bmatrix}|\mathbb{u}_1\mathbb{u}_0\mathbb{u}_{-1}\rangle. \tag{1}$$

Under inversion, instead,

$$\boldsymbol{i}\,|\mathbb{u}_1\mathbb{u}_0\mathbb{u}_{-1}\rangle=-\mathbf{1}_3\,|\mathbb{u}_1\mathbb{u}_0\mathbb{u}_{-1}\rangle. \tag{2}$$

The following question was posed by Elie Cartan towards the turn of the century: are vectors the simplest building bricks of the tensor hierarchy or are they themselves tensor products of some simpler entities? The latter, of course, must be arrays of dimension two, since one-dimensional arrays (scalars) do not increase in dimension when forming tensor products.

It is extremely easy to realize that the answer to the above question must be affirmative, since all we need is to guess a matrix that, on the tensor product with itself, leads to the matrix in (1). Let us call μ_1 and μ_2 the elements of the corresponding basis and assume the following transformation rule:

$$R(\phi\mathbf{z})\,|\mu_1\mu_2\rangle=\begin{bmatrix} e^{-i\frac{1}{2}\phi} & \\ & e^{i\frac{1}{2}\phi}\end{bmatrix}|\mu_1\mu_2\rangle, \tag{3}$$

and similar rules for another such basis like $|\nu_1\nu_2\rangle$. Under the rule for the tensor product of matrices the basis

$$|\mu_1\mu_2\rangle\otimes|\nu_1\nu_2\rangle=|\mu_1\nu_1,\ \mu_1\nu_2,\ \mu_2\nu_1,\ \mu_2\nu_2\rangle \tag{4}$$

transforms under the matrix

2-6.54
$$\begin{bmatrix} e^{-i\frac{1}{2}\phi} & \\ & e^{i\frac{1}{2}\phi}\end{bmatrix}\otimes\begin{bmatrix} e^{-i\frac{1}{2}\phi} & \\ & e^{i\frac{1}{2}\phi}\end{bmatrix}=\begin{bmatrix} e^{-i\phi} & & & \\ & 1 & & \\ & & 1 & \\ & & & e^{i\phi}\end{bmatrix}, \tag{5}$$

and it is pretty immediate that on symmetrization the basis (4) separates into a symmetrical part

$$|\mu_1\nu_1,\ \mu_1\nu_2+\mu_2\nu_1,\ \mu_2\nu_2\rangle, \tag{6}$$

which transforms as the vector in (1) and an antisymmetrical part

$$|\mu_1\nu_2-\mu_2\nu_1\rangle, \tag{7}$$

which transforms under the remaining diagonal element in (R5), i.e. unity. (See Problem 1.)

In order to show that (6) is a spherical vector we must still consider its behaviour under the inversion, but before we do this let us go back to the canonical bases of SO(3) in order to consider the so-far unused half-integral bases. First, from (**4**-7.9), the canonical basis $\langle u^{1/2}_{1/2}u^{1/2}_{-1/2}|$ (or more precisely its dual) transforms exactly as $|\mu_1\mu_2\rangle$ in (3). Secondly, from the Clebsch–Gordan series, (2.2),

$$\langle u^{1/2}|\otimes\langle u^{1/2}|=\langle u^{1}|\oplus\langle u^{0}|, \tag{8}$$

precisely as in going from (4) to (6) and (7). The objects that transform under the transformation rules that we are defining (because they are not complete yet) are called *spinors*. In (3) we define the spinors as 2-tuples from $\mathbb{C}^2$ whereas their duals, which behave like the functions $\langle u^{1/2}_{1/2}u^{1/2}_{-1/2}|$, are 2-tuples of complex functions on $\mathbb{C}^2$. Rather than using this cumbersome notation, they will be also denoted by Greek components such as $\langle\mu_1\mu_2|$ or $\langle\nu_1\nu_2|$. Strictly speaking, we should here establish the same difference in notation between configuration-space vectors $|xy\rangle$ and function-space bases $\langle \mathrm{xy}|$ that we made before, but there is no serious danger of confusion with spinors and we shall depend on the brackets $\langle\ |$, $|\ \rangle$ to denote the space.

We must now address the question of the behaviour of spinors under inversion. Just as in (5), in order to get the right matrix $-\mathbf{1}_3$ in (2), we guess the following matrix:

$$\begin{bmatrix} \mathrm{i} & \\ & \mathrm{i}\end{bmatrix}\otimes\begin{bmatrix} \mathrm{i} & \\ & \mathrm{i}\end{bmatrix}=\begin{bmatrix}-1 &&&\\ &-1&&\\ &&-1&\\ &&&-1\end{bmatrix}. \tag{9}$$

There is a point here that must be considered. In eqn (9), as in (5), the matrix on the left can be multiplied by ±1 without affecting the right-hand side. For reasons of consistency with phase factors to be introduced later (see **6**-6.8 to **6**-6.10), we shall take the negative of the matrix in (9), that is

$$i\,|\mu_1\mu_2\rangle=\begin{bmatrix}-\mathrm{i} & \\ & -\mathrm{i}\end{bmatrix}|\mu_1\mu_2\rangle. \tag{10}$$

(In reading eqn 10, remember that the whole of the left-hand side must be understood as the single symbol of a new basis, namely the transform of $|\mu_1\mu_2\rangle$. Thus, it makes no sense to cancel the bases on both sides of the equation, a misreading which would lead to the wrong identification of the operator with its matrix.) With this rule it can immediately be verified that (6) is ungerade as behoves a vector. The antisymmetric component (7) is then a *pseudoscalar.*

The status of the inversion, alas, is not as simple as all this and the reasons for the situation as it obtains in practice are partly historical. The spinors, as we know, are nothing else than the canonical functions $\langle u^{1/2}_{1/2} u^{1/2}_{-1/2}|$, which are angular-momentum eigenfunctions. Their behaviour under inversion is often discussed by an argument which apparently goes back to Pauli, although the first use of eqn (11) below that I found is that of van der Waerden (1932), p. 82. The argument runs as follows. Angular momentum is an axial vector (gerade) and therefore its eigenfunctions, the spinors, must be taken to be also gerade:

$$\boldsymbol{i}\,|\mu_1\mu_2\rangle = \mathbf{1}_2\,|\mu_1\mu_2\rangle. \tag{11}$$

This argument, however, cannot be taken too seriously, because we have seen in Problem **2**-7.4 that the classical classification in gerade and ungerade under the inversion does not necessarily hold in quantum mechanics. We shall see later, however, that the law of transformation (10) can be written, under what is called a *gauge transformation,* in the form (11). Eventually (see §**11**-9) we shall conclude that spinors can be chosen to transform under the inversion by eqn (10), which we shall call the *Cartan gauge,* or (11), which we shall call the *Pauli gauge.*

Even then, the spinor story is not concluded, because the argument presented here is grossly over-simplified. On analysing the behaviour of vectors under rotations we contented ourselves with considering rotations around the **z** axis only (eqn 1). We would have to prove, really, that the symmetrized tensor product of spinors (6) transforms like a vector under *all* rotations. This will be done in §**7**-3.

Problem 5-3 (p. 284)

1. Verify that the basis (7) transforms under eqn (1) like $\mathbb{u}_0$. Show that it is a pseudoscalar.

6

THE BILINEAR TRANSFORMATION: INTRODUCTION TO SU(2), SU′(2), AND ROTATIONS. MORE ABOUT SPINORS

> L'un des buts principaux de cet Ouvrage est de développer systématiquement la théorie des spineurs en donnant de ces êtres mathématiques une définition purement géométrique: grâce à cette origine géométrique, les matrices qu'utilisent les physiciens en Mécanique quantique se présentent d'elles-mêmes . . .
>
> Élie Cartan (1938), p. 4.

> One of the principal purposes of this work is to develop systematically the theory of spinors, by giving an entirely geometrical definition of these mathematical entities: thanks to this geometrical origin, the matrices used by the physicists in quantum mechanics turn up by themselves . . .

We have now finished the revision of the material that will be needed in order to develop the main argument of the book, which is the study of the spinor representations – such as they might be, since we have seen (§**4**-7) that the matrices corresponding to $j=\frac{1}{2}$ do not form a representation of SO(3) at all.

Having by now reinforced the reader's knowledge on a variety of useful subjects, we shall in a way start from scratch. The problem we have in mind is the following one. In (**4**-7.9) we obtained the matrix $\hat{R}^{1/2}(\phi\mathbf{z})$ and we noticed in the following equation that this matrix is two-valued. It is this property of these matrices that prevents their forming a representation of SO(3). We want to understand very clearly how this double-valuedness of the matrices arises. And we want to understand how 2×2 matrices can possibly 'represent' (in quotes because the word is not quite yet usable in a proper group-theoretical sense) three-dimensional rotations.

We shall make a start in this chapter into the study of these problems and the next one will deliver the final results. The method that we shall use here will be that of the *bilinear transformation*, also called a *projective* or a *Möbius transformation*, after its inventor, although it was also very much developed by Cayley. For those with no previous experience of it, this subject will look like a very strange game for quite a while until we begin to make contact with reality, reality here meaning rotations in SO(3).

1. THE BILINEAR TRANSFORMATION

Given any two-dimensional matrix

$$B = \begin{bmatrix} B_{11} & B_{12} \\ B_{21} & B_{22} \end{bmatrix}, \tag{1}$$

the following transformation of x (a complex number in general) into $\bar{x}$ is called a *bilinear transformation*

$$\bar{x} = \frac{B_{11}x + B_{12}}{B_{21}x + B_{22}}, \tag{2}$$

and will be denoted with the symbol

$$\bar{x} = B \circ x. \tag{3}$$

Notice that the operation symbol $\circ$ in (3) is necessary since, otherwise, x would have to be a two-dimensional column vector, in which case (L3) would be something entirely different from (2). The validity of the matrix notation here depends on the fact that the composition rule of successive transformations (2) will entail a composition rule of the corresponding matrices that is precisely the ordinary matrix multiplication rule. We shall now prove this. Consider a second transformation

$$\bar{\bar{x}} = A \circ \bar{x} = \frac{A_{11}\bar{x} + A_{12}}{A_{21}\bar{x} + A_{22}}. \tag{4}$$

On combining (4) and (2) we are transforming x into $\bar{\bar{x}}$ and what we want to prove is that

$$\bar{\bar{x}} = C \circ x = \frac{C_{11}x + C_{12}}{C_{21}x + C_{22}}, \tag{5}$$

with $C = AB$, which is the same as saying that

$$\bar{\bar{x}} = A \circ (B \circ x) = AB \circ x. \tag{6}$$

The proof is elementary:

2|4

$$\bar{\bar{x}} = \frac{A_{11}(B_{11}x + B_{12})(B_{21}x + B_{22})^{-1} + A_{12}}{A_{21}(B_{11}x + B_{12})(B_{21}x + B_{22})^{-1} + A_{22}} \tag{7}$$

$$= \frac{(A_{11}B_{11} + A_{12}B_{21})x + A_{11}B_{12} + A_{12}B_{22}}{(A_{21}B_{11} + A_{22}B_{21})x + A_{21}B_{12} + A_{22}B_{22}}, \tag{8}$$

which agrees with (5). Notice that we have assumed in the proof that B_{21}

and B_{22} are not both zero, in which case the determinant of the matrix vanishes and the matrix is singular. Thus, *we must require of the matrices used in the bilinear transformation that they be non singular,*

$$\det A \neq 0. \tag{9}$$

The bilinear transformation has a very important property: the transformation of x into $\bar{x} = A \circ x$ is unaltered if A is multiplied by a constant λ, thus forming

$$\lambda A = \begin{bmatrix} \lambda A_{11} & \lambda A_{12} \\ \lambda A_{21} & \lambda A_{22} \end{bmatrix}, \tag{10}$$

as follows immediately from, e.g., (4).

On the other hand, the mapping $x \mapsto \bar{x}$ becomes pathological when the matrix is of the form

$$A = \begin{bmatrix} A_{11} & A_{12} \\ kA_{11} & kA_{12} \end{bmatrix}, \tag{11}$$

in which case it is clear (compare with eqn 4) that $\bar{x} = k^{-1}$ for all x. A matrix such as (11), however, is singular, so that it is ruled out by condition (9).

The inverse

Given

$$\bar{x} = (A_{11}x + A_{12})(A_{21}x + A_{22})^{-1} = A \circ x, \tag{12}$$

the inverse transformation can be obtained by elimination, and it is also bilinear:

12 $$x = \frac{A_{22}\bar{x} - A_{12}}{-A_{21}\bar{x} + A_{11}} =_{\text{def}} A' \circ \bar{x}; \qquad A' = \begin{bmatrix} A_{22} & -A_{12} \\ -A_{21} & A_{11} \end{bmatrix}. \tag{13}$$

From (12) and (13), on using (6),

$$x = A'A \circ x \qquad \Rightarrow \qquad A'A = \mathbf{1}_2. \tag{14}$$

A', of course, should be the inverse A^{-1} of A, but we have not yet assumed this, since we have to satisfy ourselves that eqn (14) is possible, owing to the fact that A' and A in that relation are both completely defined, this not being the case when the relation $A^{-1}A = \mathbf{1}$ is written, in which case the matrix A^{-1} is still to be found. Moreover, remember that in (12) and (13) the matrices can be multiplied by an arbitrary constant. It will be sufficient for generality to do so for A', which will be multiplied by

λ. Thus, we have in (14)

$$A'A = \begin{bmatrix} \lambda A_{22} & -\lambda A_{12} \\ -\lambda A_{21} & \lambda A_{11} \end{bmatrix} \begin{bmatrix} A_{11} & A_{12} \\ A_{21} & A_{22} \end{bmatrix}$$

$$= \begin{bmatrix} \lambda(A_{11}A_{22} - A_{12}A_{21}) & 0 \\ 0 & \lambda(A_{11}A_{22} - A_{12}A_{21}) \end{bmatrix} = \begin{bmatrix} 1 & \\ & 1 \end{bmatrix}. \quad (15)$$

Thus, the necessary and sufficient condition for the inversion of the bilinear transformation to be possible is, from (15), on recognizing that the factors of λ in it are det A,

$$\lambda \det A = 1, \quad (16)$$

a condition which can always be satisfied as long as det $A \neq 0$, i.e. as long as the matrix is non-singular, as we have already required.

2. SPECIAL UNITARY MATRICES. THE SU(2) GROUP

When dealing with proper-rotation matrices we found that two conditions had to be imposed on them: they had to be orthogonal or unitary (according to whether the space on which they act is real or complex) and they had to have unit determinant. Matrices which satisfy this last condition shall be called *unimodular.* With a view to making later contact between the bilinear transformation and rotations, we shall consider now bilinear transformations the matrices of which are *unimodular and unitary*. Unimodular matrices are often called 'special' whence the matrices in question are *special unitary matrices* of dimension two, in short SU(2). It is clear that the set of all such matrices forms a group, which is called the SU(2) group.

Because for any bilinear transformation matrix A its inverse A^{-1} is given, disregarding for the time being any constant factor, by the matrix A' in (1.13),

$$A^{-1} = \begin{bmatrix} A_{22} & -A_{12} \\ -A_{21} & A_{11} \end{bmatrix}, \quad (1)$$

a very strong condition on the matrix elements of a SU(2) matrix is imposed by the unitary condition, which requires $A^{\dagger} = A^{-1}$:

$$A^{\dagger} = \begin{bmatrix} A_{11}^* & A_{21}^* \\ A_{12}^* & A_{22}^* \end{bmatrix} = \begin{bmatrix} A_{22} & -A_{12} \\ -A_{21} & A_{11} \end{bmatrix} = A^{-1}. \quad (2)$$

Necessary and sufficient conditions for the SU(2) property are therefore,

$$A_{22} = A_{11}^*, \qquad A_{22}^* = A_{11}, \quad (3)$$

$$A_{21}^* = -A_{12}, \qquad -A_{21} = A_{12}^*. \quad (4)$$

The equations on the right of each line here are equivalent to those on the left. To verify the SU(2) condition it is thus sufficient to write $A_{11}=a$, $A_{12}=b$, whereupon the second row of the matrix is determined by (3) and (4):

$$A=\begin{bmatrix} a & b \\ -b^* & a^* \end{bmatrix}. \tag{5}$$

Naturally, a and b have to be chosen so as to satisfy the unimodular condition

$$\det A = aa^* + bb^* = 1. \tag{6}$$

Two points have to be noticed. First, every unitary matrix A ($|\det A| = 1$) can be *normalized* so that it becomes unimodular ($\det A = 1$) by multiplying it with a constant, which, as we know, does not affect the bilinear transformation:

$$A=\begin{bmatrix} a & b \\ c & d \end{bmatrix} \mapsto \begin{bmatrix} \lambda a & \lambda b \\ \lambda c & \lambda d \end{bmatrix}, \qquad \lambda = \pm(\det A)^{-1/2}. \tag{7}$$

Secondly, it must be realized that every SU(2) matrix can still be multiplied by ± 1. This does not affect the bilinear transformation entailed by it or the unimodular condition. That is:

$$A \text{ and } -A \text{ are interchangeable,} \tag{8}$$

a statement which we display because it has remarkable consequences in the rotation group. It is tempting to regard this double-valuedness of the matrices as a nuisance that one would like to sweep under the carpet, but not only can this not be done: on the contrary, with a little bit of tender care (8) will provide a lot of important results.

One small matter of convention. Because of the rule (8), we would always have to write a SU(2) matrix preceded by the $\pm$ sign. We shall agree to leave this as implicit, the rule (8) always carefully kept in mind. For the same reason, when normalizing a matrix by (7) we shall take only the positive sign of λ. This is, of course, only a temporary measure since the time will come when the two signs will have to be made explicit.

3. ROTATIONS AND SU(2): A FIRST CONTACT

In order to show how the bilinear transformation with SU(2) matrices gives three-dimensional rotations we must first discuss a strange but useful concept, that of an *isotropic vector* first introduced by Cartan in 1913 (see Cartan 1938, Temple 1960).

Consider two unit orthogonal vectors, $\mathbf{p}$ and $\mathbf{q}$,

$$\mathbf{p}^2 = \mathbf{q}^2 = 1, \qquad \mathbf{p}\cdot\mathbf{q} = 0. \tag{1}$$

An orthogonal transformation (proper or improper rotation) is one that keeps relations (1) invariant. This can be formulated in a more compact way by inventing a vector

$$\mathbf{v}=\mathbf{p}+i\mathbf{q}, \tag{2}$$

for which $\mathbf{v}^2$, (in the ordinary rather than the Hermitian sense) is

$$\mathbf{v}^2=\mathbf{p}^2-\mathbf{q}^2+2i\mathbf{p}\,.\,\mathbf{q}=0. \tag{3}$$

Such vectors of zero length are called *isotropic vectors* and their main use is that they embody the two conditions in (1) in a single equation, so that an orthogonal transformation must always transform an isotropic vector into another isotropic vector.

Let us now write an isotropic vector of components x, y, z (these must be, of course, complex numbers in which both real and imaginary parts of x must be understood to belong to the x axis and similarly for the others):

$$x^2+y^2+z^2=0. \tag{4}$$

It is very convenient, as we shall see, to define an *isotropic parameter*

$$w=(x-iy)z^{-1}, \tag{5}$$

for which

5,4 $$ww^*=(x^2+y^2)z^{-2}=(-z^2)z^{-2}=-1, \tag{6}$$

whence

$$w^*=-w^{-1}. \tag{7}$$

The condition for a SU(2) matrix to be a rotation is that (6) (and therefore eqn 4) be invariant under the rotation. Thus, given

$$\bar{w}=\begin{bmatrix} a & b \\ -b^* & a^* \end{bmatrix}\circ w, \qquad \bar{w}^*=\begin{bmatrix} a^* & b^* \\ -b & a \end{bmatrix}\circ w^*, \tag{8}$$

$$\bar{w}\bar{w}^*=\frac{aw+b}{-b^*w+a^*}\,\frac{a^*w^*+b^*}{-bw^*+a}=\frac{aa^*ww^*+ab^*w+a^*bw^*+bb^*}{bb^*ww^*-ab^*w-a^*bw^*+aa^*}, \tag{9}$$

we have the condition

6|9 $$\frac{-aa^*+ab^*w+a^*bw^*+bb^*}{-bb^*-ab^*w-a^*bw^*+aa^*}=-1, \tag{10}$$

which can be satisfied in either of the two following cases only:

Case 1 $$aa^*=1, \qquad b=0. \tag{11}$$

Case 2 $$a=0, \qquad bb^*=1. \tag{12}$$

These conditions are particularly simple to apply to three types of rotations: (i) all rotations $R(\phi\mathbf{z})$ around the $\mathbf{z}$ axis; (ii) a binary rotation $R(\pi\mathbf{x})$ around the $\mathbf{x}$ axis; (iii) a binary rotation $R(\pi\mathbf{y})$ around the $\mathbf{y}$ axis. We shall now show how these rotations actually emerge from the SU(2) transformations when cases 1 and 2 above are considered.

Case 1. On introducing (11) into (8), we have

,11 $$\bar{w} = aw(a^*)^{-1} = a^2 w. \tag{13}$$

This is precisely the eigenvalue equation for the variable $\mathbb{u}_1$ (which equals $x - \mathrm{i}y$, compare with eqn 5) under a rotation $R(\phi\mathbf{z})$. In fact, from (**5**-1.1) and Problem **5**-1.2,

$$\bar{\mathbb{u}}_1 = R(\phi\mathbf{z})\mathbb{u}_1 = \exp(-\mathrm{i}\phi)\mathbb{u}_1. \tag{14}$$

On comparing (13) and (14) we obtain a^2 and from it, with (11), the corresponding SU(2) matrix, which we shall denote with $\hat{R}(\phi\mathbf{z})$:

$$\hat{R}(\phi\mathbf{z}) = \begin{bmatrix} \mathrm{e}^{-\mathrm{i}\frac{1}{2}\phi} & \\ & \mathrm{e}^{\mathrm{i}\frac{1}{2}\phi} \end{bmatrix}, \tag{15}$$

and which, lo and behold, is identically the matrix $\hat{R}^{1/2}(\phi\mathbf{z})$ in (**4**-7.9).

Case 2. On introducing (12) into (8) we have

,12,7 $$\bar{w} = b(-b^* w)^{-1} = -b^2 w^{-1} = b^2 w^*. \tag{16}$$

Compare this result with the following two transformations

,5 $$\bar{w} = R(\pi\mathbf{x})\{(x - \mathrm{i}y)z^{-1}\} = (x + \mathrm{i}y)(-z)^{-1} = -w^*, \tag{17}$$

,5 $$\bar{w} = R(\pi\mathbf{y})\{(x - \mathrm{i}y)z^{-1}\} = (-x - \mathrm{i}y)(-z)^{-1} = w^*. \tag{18}$$

Clearly, they are both of the form (16) with $b^2 = -1, +1$ respectively; that is, $b = \pm\mathrm{i}$ for $R(\pi\mathbf{x})$ and $b = \pm 1$ for $R(\pi\mathbf{y})$. We thus obtain the matrices

$$\hat{R}(\pi\mathbf{x}) = \begin{bmatrix} & -\mathrm{i} \\ -\mathrm{i} & \end{bmatrix}, \qquad \hat{R}(\pi\mathbf{y}) = \begin{bmatrix} & -1 \\ 1 & \end{bmatrix}, \tag{19}$$

where we have chosen the sign of the matrices so as to agree with our choice of phases, as will be seen in §4.

The reader may think that all that we have done in this section is to deal with rotations in the $\mathbf{x}$, $\mathbf{y}$ plane (in a fancy complex variable) so that it is no wonder that we have got two-dimensional matrices as was the case, say, in (**2**-4.11). The situation now, however, is different. In order to obtain (**2**-4.11) we had to transform *two* variables, whereas SU(2) allows us to get a two-dimensional matrix out of the transformation of a *single* variable.

Even then, the matrices that we have obtained appear to pertain to suspiciously highly special cases of rotations and thus to be of no general significance. It is indeed a quite extraordinary fact that the matrices which we have obtained do generate all rotations in three-dimensional space, as we shall now prove.

4. BINARY ROTATIONS AS THE GROUP GENERATORS

From (3.15), with ϕ equal to π, and (3.19) we have three SU(2) matrices corresponding to binary rotations around the orthogonal axes **x**, **y**, **z**, which, from §**2**-1, form a set of three bilateral binary rotations. For simplicity, we shall call these matrices $\boldsymbol{I}_x$, $\boldsymbol{I}_y$, $\boldsymbol{I}_z$:

$$\begin{matrix} \boldsymbol{I}_x & \boldsymbol{I}_y & \boldsymbol{I}_z \\ \begin{bmatrix} & -\mathrm{i} \\ -\mathrm{i} & \end{bmatrix} & \begin{bmatrix} & -1 \\ 1 & \end{bmatrix} & \begin{bmatrix} -\mathrm{i} & \\ & \mathrm{i} \end{bmatrix} \end{matrix} \tag{1}$$

(remember that their signs, which are arbitrary, have been chosen with foresight).

We shall now prove that for the spinor representations these matrices are the infinitesimal generators of the rotation group, **I** in (**4**-2.15) as determined in matrix form in (**4**-9.8) and (**4**-9.9). As is common in this field, the infinitesimal generators will be written as $\frac{1}{2}\boldsymbol{\sigma}$, as in eqn (**4**-9.10),

$$\hat{R}^{1/2}(\phi\mathbf{n}) = \exp(-\mathrm{i}\tfrac{1}{2}\phi\mathbf{n}\cdot\boldsymbol{\sigma}), \tag{2}$$

where $\boldsymbol{\sigma}$ stands for the vector of components $\boldsymbol{\sigma}_x$, $\boldsymbol{\sigma}_y$, $\boldsymbol{\sigma}_z$, the Pauli matrices.

For ϕ equal to π and **m** either **x**, **y**, or **z**, the matrix in (2) must become the corresponding $\boldsymbol{I}$ matrix in (1):

$$\hat{R}^{1/2}(\pi\mathbf{m}) = \exp(-\mathrm{i}\tfrac{1}{2}\pi\boldsymbol{\sigma}_m) = \boldsymbol{I}_m, \tag{3}$$

whence

$$\boldsymbol{\sigma}_m = 2\mathrm{i}\pi^{-1}\ln\boldsymbol{I}_m. \tag{4}$$

It is easy to verify, on the other hand, that for all matrices $\boldsymbol{I}_m$ in (1),

$$\boldsymbol{I}_m^2 = -\mathbf{1}. \tag{5}$$

This is a very important relation that could not be expected to hold for ordinary representations of SO(3) (i.e. representations of odd dimension). This is so because the square of the matrix of a binary rotation (as the $\boldsymbol{I}$ are) must be the matrix of the identity. In the present case, however, owing to the sign duality of SU(2) the matrix of the identity may appear with negative sign, as in (R5). It will be shown in §**11**-5 that, in fact, such a sign *must* appear for the square of binary rotation matrices in spinor representations, that is in those constructed from SU(2). It is easy to prove (see Problem 1) that the multiplication rule (5) entails the following relation:

$$\ln \boldsymbol{I} = \tfrac{1}{2}\pi \boldsymbol{I}, \tag{6}$$

which, introduced in (4) with the correct identification of the corresponding axis, gives

$$\boldsymbol{\sigma}_m = \mathrm{i}\boldsymbol{I}_m, \qquad m = x, y, z. \tag{7}$$

Equation (7) confirms our assertion that the $\boldsymbol{I}_m$ are, except for a constant factor i, the infinitesimal group generators for the spinor representations of SO(3). Moreover, comparison of $\boldsymbol{\sigma}_m$ as derived from (1) and (7), with the Pauli matrices in (**4**-9.9) shows complete agreement with our previously determined generators. It will now be seen that the phases in (1) were chosen so as to give the Pauli matrices in the correct CS phase.

It is now clear that the three matrices in (1) obtained from the very simple properties of SU(2) must allow us through (2) to obtain the matrix of this group corresponding to any arbitrary rotation $R(\phi\mathbf{n})$, precisely as it was done in (**4**-9.12), which is, of course, a SU(2) matrix. Thus, the most general SU(2) matrix A in (2.5) can be associated with a rotation $R(\phi\mathbf{n})$. (See Problem 3.)

It is remarkable that, apparently, we have been able to telescope the whole of three-dimensional space into the plane of the w variable that we used. The reason behind this feat will be clear in Chapter **7**, when it will be demonstrated that this is precisely what the stereographic projection does.

Problems 6-4 (p. 285)

1. Prove that for a matrix $\boldsymbol{I}$ such that $\boldsymbol{I}^2 = -1$, then

$$\exp(\tfrac{1}{2}\pi \boldsymbol{I}) = \boldsymbol{I}. \tag{8}$$

2. Prove that the infinitesimal generators of SU(2) must be *traceless matrices*, i.e. matrices of zero trace.
3. Identify, in terms of the angle of rotation ϕ and the axis of rotation $\mathbf{n}$ the matrix elements of the general SU(2) matrix (2.5):

$$a = \cos\tfrac{1}{2}\phi - \mathrm{i}n_z \sin\tfrac{1}{2}\phi, \qquad b = -(n_y + \mathrm{i}n_x)\sin\tfrac{1}{2}\phi. \tag{9}$$

(These are the Cayley–Klein parameters of a rotation. See §**7**-3.)

5. DO WE HAVE A REPRESENTATION OF SO(3)?

Equation (**4**-9.12) provides two SU(2) matrices (with opposite signs) for every rotation of SO(3), so that it should be fairly clear that we do not have a true representation of the group, but we had better hammer this point home. Consider the subgroup of rotations around the **z** axis, $R(\phi\mathbf{z})$. Let us abbreviate them as $R(\phi)$ and the corresponding matrices (3.15) as $\hat{R}(\phi)$. Clearly, the multiplication rule of the rotations

$$R(\phi)R(\phi')=R(\phi+\phi'), \tag{1}$$

is paralleled by that of the matrices (3.15):

$$\hat{R}(\phi)\hat{R}(\phi')=\hat{R}(\phi+\phi'). \tag{2}$$

Yet, we do not have a representation, for the following reason. A matrix of SU(2) can always be replaced by its negative, whence (R2) can perfectly well be rewritten as follows,

$$\hat{R}(\phi)\hat{R}(\phi')=-\hat{R}(\phi+\phi'), \tag{3}$$

whereupon it is seen that the multiplication rule (1) is not conserved. Neither is it possible to produce a representation by stamping out of existence the negative matrices. Take, e.g., always the positive matrices $\hat{R}(\phi)$ from (3.15). We shall then have

$$\hat{R}(\pi)=\begin{bmatrix}-\mathrm{i} & \\ & \mathrm{i}\end{bmatrix}, \tag{4}$$

leading to

$$\hat{R}(\pi)\hat{R}(\pi)=\begin{bmatrix}-1 & \\ & -1\end{bmatrix}=-\hat{R}(2\pi)\equiv-\hat{R}(0), \tag{5}$$

with the negative sign raising its head again.

Of course, we already knew about the duality of the rotation matrices for half-integral j in §**4**-7. But this duality appears now in a much more fundamental way as an inherent property of the bilinear transformation which can be demonstrated without having to take ϕ outside the range from $-\pi$ to π, as we had to do in that section.

6. SU(2) PLUS THE INVERSION: SU′(2)

A question of terminology first: if we add the inversion to SO(3) the 'special' property of the orthogonal matrices (i.e. determinant +1) is dropped, and so is the letter S, thus designating with O(3) the group of all orthogonal three-dimensional matrices with determinant ±1. This, in principle, is what we want to do in this section for SU(2) but, unfortunately, the name U(2) is pre-empted: it denotes the group of all two-dimensional unitary matrices whatever the value of their determinants (which, because of the unitary condition must be a phase factor $e^{i\omega}$). The group of all two-dimensional unitary matrices with determinant ±1 appears, in fact, to have been so neglected in the past that it did not have an agreed name. Altmann and Herzig (1982) called it SU′(2).

Most of the major properties of SU(2) obtain for SU′(2) but, for this reason, one must be very careful with such changes as are required. The vital existence of the inverse is ensured in eqn (1.16), when $\det A = -1$, by taking λ as -1. Thus the inverse of A, A^{-1} is given by

1.15 $$A^{-1} = \begin{bmatrix} -A_{22} & A_{12} \\ A_{21} & -A_{11} \end{bmatrix} = A^{\dagger} = \begin{bmatrix} A_{11}^{*} & A_{21}^{*} \\ A_{12}^{*} & A_{22}^{*} \end{bmatrix}. \tag{1}$$

As in (2.2), we equate A^{-1} here with the inverse, $A^{\dagger}$, as follows from the unitary condition, whence, as in (2.3) and (2.4) we obtain the following condition on the matrix elements:

$$A_{22} = -A_{11}^{*}, \qquad A_{21} = A_{12}^{*}. \tag{2}$$

Thus, a SU′(2) matrix has the form

$$A = \begin{bmatrix} a & b \\ b^{*} & -a^{*} \end{bmatrix}, \qquad \det A = -1 \quad \Rightarrow \quad aa^{*} + bb^{*} = +1. \tag{3}$$

Because the matrix here differs from that on (L3.8) inasmuch as its second row is multiplied by -1, it is clear that, in finding the condition for A to be an improper rotation, eqns (3.9) and (3.10) are entirely unaltered, leading to precisely the same cases 1 and 2 as before:

$$\text{Case 1: } aa^{*} = 1,\ b = 0; \qquad \text{Case 2: } aa^{*} = 0,\ bb^{*} = 1. \tag{4}$$

In the same manner as in (3.13) and (3.16), but now making use of the matrix A in (3) we obtain the following relations for these cases:

$$\textit{Case 1.} \quad \bar{w} = aw(-a^{*})^{-1} = -a^{2}w. \tag{5}$$

,3.7 $$\textit{Case 2.} \quad \bar{w} = b(b^{*}w)^{-1} = -b^{2}w^{*}. \tag{6}$$

Case 1 will clearly pertain to the inversion, since for this operation

3.5 $$\boldsymbol{i}w = \bar{w} = (-x + \mathrm{i}y)(-z)^{-1} = w, \tag{7}$$

whence, on comparing with (5),

$$a^{2} = -1, \qquad a = \pm\mathrm{i}. \tag{8}$$

We write the matrix, which we shall call $\boldsymbol{I}_{i}$, in the form

$$\boldsymbol{I}_{i} = \begin{bmatrix} -\mathrm{i} & \\ & -\mathrm{i} \end{bmatrix}, \tag{9}$$

which is precisely the matrix for the inversion that we have guessed in (**5**-3.10), the sign having been chosen for reasons discussed below.

It is significant to remark that the inversion acts like the identity on w, from (7), but, of course, it could not be represented by $\mathbf{1}_2$ since this is a unimodular matrix.

It can easily be seen that case 2 corresponds to improper rotations, such as reflections. (Problem 1.)

Inversion and parity

We must recall that the inversion, being a symmetry operation, is a unitary and not a Hermitian operator and that is why its eigenvalues (see eqn **5**-3.10) are not necessarily real. On the other hand, we have been able to correlate the rotations $\boldsymbol{I}_x$, $\boldsymbol{I}_y$, $\boldsymbol{I}_z$ with the corresponding observables $\boldsymbol{\sigma}$, by using eqn (4.2) that relates a rotation $R(\phi\mathbf{n})$ with its corresponding observable. This relation, of course, is only strictly valid for a proper rotation, but we can extend it also to the inversion by the following argument. In two dimensions, the inversion is identical to a binary rotation around an axis orthogonal to the plane. In three dimensions $\boldsymbol{i}$ can be considered as a binary rotation around an axis along a fourth orthogonal dimension. Moreover, the matrix $\boldsymbol{I}_i$ in (9) is such that its square is $-\mathbf{1}_2$ and it therefore satisfies the same multiplication rules given in (4.5) for the rotation matrices. Thus, the relation that we have given in (4.7) between the binary rotations and their observables is also valid for the inversion:

$$\boldsymbol{\sigma}_i = \mathrm{i}\boldsymbol{I}_i. \tag{10}$$

This observable that corresponds to the inversion we call *parity*. From (10) and (9) its matrix is

$$\boldsymbol{\sigma}_i = \begin{bmatrix} 1 & \\ & 1 \end{bmatrix}. \tag{11}$$

It can now be seen that we have taken the phase in (9) in such a way that (11) is the unit matrix, thus asserting that the angular-momentum eigenfunctions are even with respect to parity, as we must expect in analogy with axial vectors. It should be stressed that the matrix in (11) need not be the identity: if we had taken the opposite phase in (9), $-\mathbf{1}_2$ would appear in (11) thus defining odd functions. It should be clear that inversion and parity are not the same thing. The angular-momentum eigenfunctions are invariant under parity but, as in (**5**-3.10), they are multiplied by $-\mathrm{i}$ under the inversion.

Problems 6-6 (p. 285)

1. Prove that the reflection planes σ_x and σ_y normal to the axes **x** and **y** respectively, give *Case* 2 transformations and obtain their corresponding matrices.

2. Obtain the matrices of Problem 1 by writing σ_x and σ_y as $\boldsymbol{i}R(\pi\mathbf{x})$ and $\boldsymbol{i}R(\pi\mathbf{y})$ respectively.

7. SPINORS AND THEIR INVARIANTS

We have seen that a variable w transforms as follows under SU(2):

2.5
$$\bar{w} = (aw+b)(-b^*w+a^*)^{-1}. \tag{1}$$

Let us write w as the quotient of two new variables μ_1 and μ_2 (these are called the *homogeneous coordinates* of the bilinear transformation in eqn 1, the meaning of which will be clearer in §**7**-2). Thus

1
$$\frac{\bar{\mu}_1}{\bar{\mu}_2} = \frac{a\mu_1\mu_2^{-1}+b}{-b^*\mu_1\mu_2^{-1}+a^*} = \frac{a\mu_1+b\mu_2}{-b^*\mu_1+a^*\mu_2}, \tag{2}$$

(an expression which will be indeterminate for $\mu_2 \to 0$). Thus, while w transforms under a bilinear transformation, its homogeneous coordinates μ_1 and μ_2 transform linearly:

$$|\bar{\mu}_1\bar{\mu}_2\rangle = \begin{bmatrix} a & b \\ -b^* & a^* \end{bmatrix} |\mu_1\mu_2\rangle. \tag{3}$$

We have seen in Chapter **5** that it is the transformation matrix that gives a basis its name. A basis that transforms under (3) is called a *spinor.* Our original first-approach definition, through eqn (**5**-3.3) is, of course, a particular case of (3). We thus know two things about spinors. One is that they are useful entities in our tensor hierarchy, since tensors of higher rank (like vectors and the like) are obtained by tensor products of spinors. Also that they are duals of the canonical basis $\langle u^{1/2}_{1/2} u^{1/2}_{-1/2}|$ from (**4**-9.8).

It should be remembered that the SU(2) matrix in (3) is the rotation matrix $\hat{R}^{1/2}(\phi\mathbf{n})$ in (**4**-9.12). A property of this matrix that is very easy to prove is that, on incrementing ϕ by 2π, it changes sign, which means that the spinor basis itself changes sign under a rotation by 2π.

The *complex conjugate spinor* $|\mu_1^*\mu_2^*\rangle$ has a very remarkable transformation property, which is given by taking complex conjugates on both sides of (3):

$$|\bar{\mu}_1^*\bar{\mu}_2^*\rangle = \begin{bmatrix} a^* & b^* \\ -b & a \end{bmatrix} |\mu_1^*\mu_2^*\rangle. \tag{4}$$

The matrices in (3) and (4) have the same trace, so that the conjugate spinor transforms under the same representation (within an equivalence) as the normal spinor. It is easy to guess the transformation of the basis $|\mu_1^*\mu_2^*\rangle$ in (4) that will bring the matrix in (4) into precisely that in (3). It can easily be verified, in fact, that

4 $$|\bar{\mu}_2^*, -\bar{\mu}_1^*\rangle = \begin{bmatrix} a & b \\ -b^* & a^* \end{bmatrix} |\mu_2^*, -\mu_1^*\rangle. \tag{5}$$

That is, if we use the symbol $\leftrightarrow$ to denote exchangeability as regards transformation properties, then:

$$|\mu_2^*, -\mu_1^*\rangle \leftrightarrow |\mu_1\mu_2\rangle \qquad \text{or} \qquad -\mu_1^* \leftrightarrow \mu_2,\ \mu_2^* \leftrightarrow \mu_1. \tag{6}$$

We can now study the *invariants* under spinor transformations. If we call m and l the spinors $|\mu_1\mu_2\rangle$ and $|\lambda_1\lambda_2\rangle$ respectively, the matrix transformation (3), because it is unitary, must leave invariant the inner product $m^\dagger l$; that is

$$\bar{\mu}_1^*\bar{\lambda}_1 + \bar{\mu}_2^*\bar{\lambda}_2 = \mu_1^*\lambda_1 + \mu_2^*\lambda_2. \tag{7}$$

The invariant

$$\mathscr{I} = \mu_1^*\lambda_1 + \mu_2^*\lambda_2, \tag{8}$$

can be rewritten as follows, on introducing (6),

$$\mathscr{I} = \mu_1\lambda_2 - \mu_2\lambda_1. \tag{9}$$

That this is an invariant agrees with Problem **5**-3.1 where it was shown that this quantity is a pseudoscalar. (In this section we are not concerned with the inversion; hence both scalars and pseudoscalars are considered invariants, being invariant under rotations.)

When, in the above work, l and m are identical the invariant $\mathscr{I} = \mu_1^*\mu_1 + \mu_2^*\mu_2$ can be regarded as the square of the modulus of the spinor m. From (9), however, $\mathscr{I}$ vanishes in this case, that is spinors can be considered to have zero modulus, as long as the relations in (6) are taken as identities.

There is one final point which we must make about conjugate spinors. We have seen that $|\mu_1^*\mu_2^*\rangle$ and $|\mu_1\mu_2\rangle$ must belong to the same representation (within a similarity) when we are dealing with proper rotations; that is, with SO(3). Under the inversion, however, we find, on using (6.9),

$$\boldsymbol{i}\,|\mu_1\mu_2\rangle = \begin{bmatrix} -\mathrm{i} & \\ & -\mathrm{i} \end{bmatrix} |\mu_1\mu_2\rangle \tag{10}$$

and correspondingly,

$$\boldsymbol{i}\,|\mu_1^*\mu_2^*\rangle = \begin{bmatrix} \mathrm{i} & \\ & \mathrm{i} \end{bmatrix} |\mu_1^*\mu_2^*\rangle, \tag{11}$$

and the matrices do not have the same trace. It will be seen later that although the representations to which spinors and their conjugates belong are not equivalent in O(3), they are *gauge equivalent* (see §**8**-2 and eqns **11**-9.7, **11**-9.8) which amounts to almost the same thing.

Problem 6-7 (p. 286)

1. Prove that complex conjugation of a spinor is the same as a binary rotation around the **y** axis. (This is the reason why the Euler angle β is now chosen around **y** and not **x**. Complex conjugation is deeply connected with the important operation of time reversal.)

7

ROTATIONS AND SU(2). THE STEREOGRAPHIC PROJECTION

> Erwähnen muss ich dabei jedoch eines Gedankens der mir aus Riemann's Vorlesungen durch mündliche Ueberlieferung zu Ohren kam, und der auf meine Darstellung von nicht geringen Einfluss wurde. Dieser Gedanke besteht in der Projection der auf der Horizontalebene ausgebreiteten Functionswerthe nach einer Kugelfläche hin.
>
> C. Neumann (1865), p. vi.

> I must mention also, however, a concept from Riemann's lectures about which I heard by oral tradition and which had a not negligible influence on my presentation. This concept consists in the projection on a spherical surface of values of functions displayed over the horizontal plane.

There are several points which we shall want to tidy up in this chapter, in particular the precise nature of the relation between SU(2) and SO(3). It is clear from the work of Chapter **6** that all rotations of SO(3) are given by SU(2) matrices, but we must define in more detail the mapping between the two groups. Another question that we shall handle at the same time is the completion of the proof, sketched in Chapter **5**, that (spherical) vectors are symmetrized direct products of spinors.

It is important to realize that although the general proofs that we are seeking are easy to obtain, there is a great gap between existential and constructive proofs, and that we aim at the second type: we want to establish some unequivocal rules of transformation which require a great deal of care and attention in the work that follows. The rewards, however, will not be negligible, for we shall be able not only to construct some perfectly sound matrices but also to obtain explicit expressions for the so far mysterious spinor components μ_1 and μ_2. The method which we shall use, that of the stereographic projection, appears to have been first described by Carl Neumann (1865, Fourth Lecture), although it must have been used by Riemann before, as the epigraph shows.

1. THE STEREOGRAPHIC PROJECTION

The secret will now be revealed as to how, as was apparent in Chapter **6**, three-dimensional rotations can actually be performed in a plane. This is done by means of the *stereographic projection*, in which each point p of a sphere (*sphere of projection*) is projected onto a plane. This is done by taking one fixed point of the sphere, called the *projection pole* Ω. The *projection plane* Π is a plane normal to the diameter of the sphere through Ω and we shall take it to be the equatorial plane of the sphere, although other choices are possible and indeed used. The projection point P of a point p of the sphere is defined by the intersection of Ωp with Π. (See Fig. 1*a*.)

The fundamental property of the stereographic projection is that the projection of a circle (given, of course, on the surface of the sphere) is always a circle, as illustrated in Fig. 1*b*. We shall merely sketch a proof of this important result; a full proof, as well as a complete treatment of the projection can be found in Hilbert and Cohn-Vossen (1952). The circle γ to be projected is shown in Fig. 2, which has been drawn so that the great circle of the figure is a symmetry plane of γ, only one-half of which is visible. Since the lines that determine the projection are generators of the cone through γ, the projection Γ of the circle, being a section of the cone, can only be either a circle or an ellipse. In either case, the plane of the figure (i.e. the plane that contains the great circle) must be a symmetry plane of Γ and therefore, if Γ were an ellipse, PQ must be an axis of it. It can then be seen, because of fairly obvious equalities of sides and angles, that if we perform a binary rotation around Ωc, then Γ will be taken to a plane parallel to γ, and the latter, instead, becomes parallel to the projection plane. It is sufficient for this purpose to verify that the points p and q of γ transform into p', q', respectively, in such a way that $q'p'$ is now

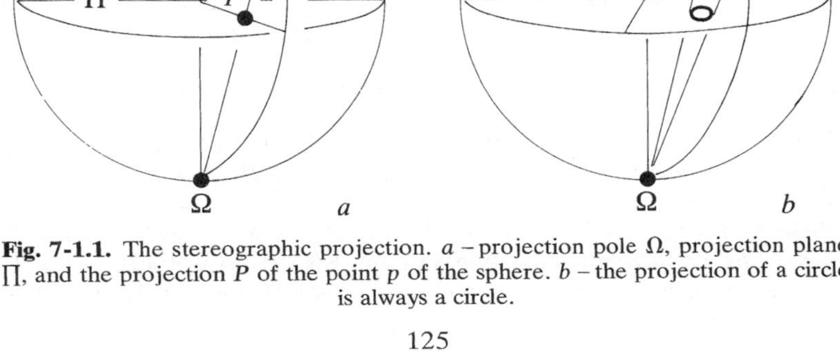

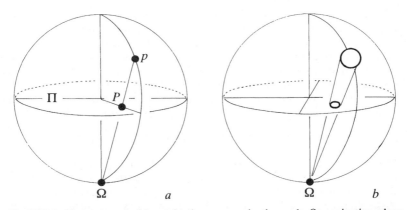

Fig. 7-1.1. The stereographic projection. *a* – projection pole Ω, projection plane Π, and the projection P of the point p of the sphere. *b* – the projection of a circle is always a circle.

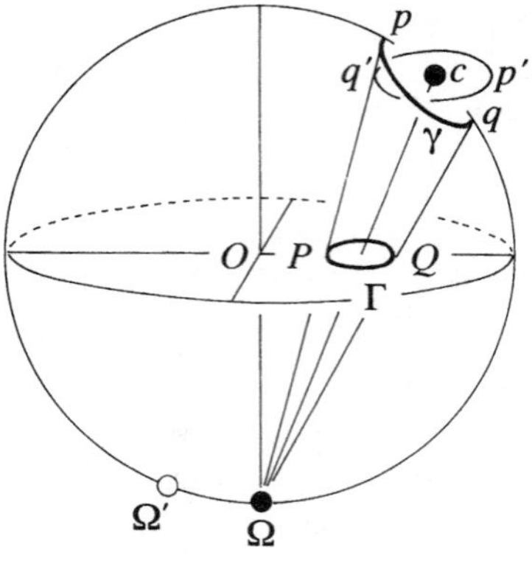

Fig. 7-1.2. Stereographic projection of a circle.

parallel to PQ. (It is easy to believe that under the binary rotation around Ωc, p, and q should be exchanged. Think about the equatorial plane of the figure: rotation of it by π around $\Omega' O$ will not exchange east and west; rather, it will bring this circle into an orientation parallel to γ.) Since two parallel sections of a cone must be either both circles or both ellipses, γ being a circle entails Γ also being one.

The importance of the result that we have just proved is self-evident. When a SO(3) operation is performed on the sphere, a point p on it will describe a circle round the rotation axis, which will be projected as a circle on the projection plane: it will be shown at the end of Theorem 1 in §3 that the projected point P actually rotates around some fixed point on the projection plane (although this result appears rather obvious, its proof is not trivial). This is equivalent to saying that every rotation of SO(3) is mapped to a rotation of SU(2). The converse result will be proved first; namely, that when the point on the projection circle rotates, then its pre-image on the sphere also rotates – that is, that every rotation of the plane (SU(2) rotation) maps to a rotation of SO(3). Before we do this the geometry of the projection, and the coordinates, require a little attention.

2. GEOMETRY AND COORDINATES OF THE PROJECTION

Before we start, we must repeat that we are not merely interested in proving the theorems just stated, important as they are. We consider them as vehicles for obtaining matrices for SO(3) and SU(2) and we want these matrices to be unimpeachable, i.e. such that they transform identified objects in an unambiguous fashion. Clearly, in dealing with SO(3) we are, in principle, interested in transforming vectors $|xyz\rangle$, these being the coordinates of a point on the unit sphere (the sphere of projection):

$$x^2+y^2+z^2=1. \tag{1}$$

We must carefully bear in mind, however, that we prefer to standardize our basis in configuration space in terms of the spherical vector $|ш_1 ш_0 ш_{-1}\rangle$ of components

5-1.11 $$ш_{\pm 1}=\pm 2^{-1/2}(x \mp iy), \qquad ш_0=-z. \tag{2}$$

Bearing this in mind, we take the projection pole Ω as shown in Fig. 1, a choice which has been made so that the points for negative z, i.e. in the range $-1 \leqslant z \leqslant 0$, project onto the projection circle, whereas the points for positive z (in which, from eqn 2, we are not interested) project on points of the projection plane outside this circle. We now relate x and y with X and Y and then express z in terms of the latter variables. From the figure (remember that z is negative),

$$x=(1-z)X, \qquad y=(1-z)Y. \tag{3}$$

3|1 $$(1-z)^2(X^2+Y^2)+z^2=1. \tag{4}$$

4 $$z^2(1+X^2+Y^2)-2z(X^2+Y^2)+X^2+Y^2-1=0. \tag{5}$$

5 $$z=(X^2+Y^2 \pm 1)(1+X^2+Y^2)^{-1}. \tag{6}$$

Since z ranges from 0 to -1, we must take the negative sign in (6):

$$z=(X^2+Y^2-1)(1+X^2+Y^2)^{-1}. \tag{7}$$

With a view to introducing the spherical vector components defined in

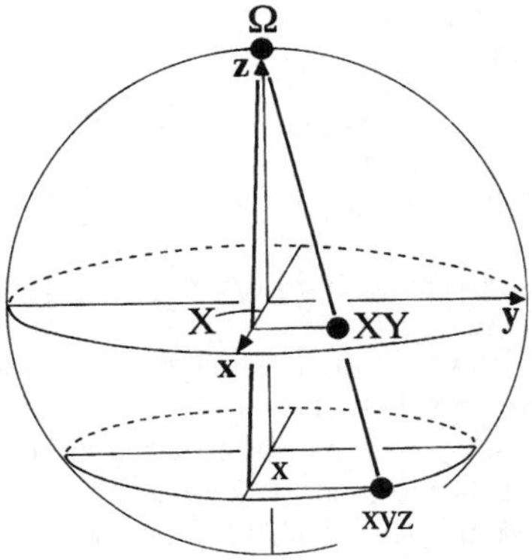

Fig. 7-2.1. Coordinates in the stereographic projection. The projection sphere is of unit radius.

(2), we replace x, y and X, Y by the complex variables

$$u = x - \mathrm{i}y, \qquad U = X - \mathrm{i}Y. \tag{8}$$

3|8 $$u = (1 - z)U. \tag{9}$$

8|7 $$z = (UU^* - 1)(1 + UU^*)^{-1}. \tag{10}$$

10|9 $$u = 2U(1 + UU^*)^{-1}. \tag{11}$$

We saw in §6-7 that it was advantageous to replace the complex variable w by the quotient $\mu_1\mu_2^{-1}$ of two *homogeneous coordinates* μ_1, μ_2 (which turned out to be spinor components). We shall do this for the variable U on the projection plane, but the relation between the homogeneous coordinates μ_1, μ_2 and the point U on the projection plane is now more geometrical, since the ray $\mu_1 : \mu_2$ denotes the projection ray of which U is the projection point. Since μ_1 and μ_2 can be multiplied by any constant, we can choose them so that $|\mu_1\mu_2\rangle$ be normalized in the Hermitian sense

$$U = \mu_1\mu_2^{-1}, \qquad \mu_1\mu_1^* + \mu_2\mu_2^* = 1, \tag{12}$$

and with this convention we easily obtain the sphere variables z, u in terms of the homogeneous coordinates on the projection plane:

12|10 $$z = \mu_1\mu_1^* - \mu_2\mu_2^*, \tag{13}$$

12|11 $$u = 2\mu_1\mu_2^*. \tag{14}$$

We are now ready to establish the homomorphism between SU(2) and SO(3). Before we do this, however, it is important to remark that eqns (12) to (14) lead to explicit expressions for the spinor components μ_1 and μ_2, as discussed at the end of §3.

Problem 7-2 (p. 286)

1. Verify that U is a projection point of the sphere for any values of μ_1, μ_2.

3. THE HOMOMORPHISM BETWEEN SU(2) AND SO(3)

Because a circle on the projection plane is the image of a circle on the projection sphere, we expect that a rotation of U, a point of the projection plane, will map to a rotation of the sphere. Since U has been expressed in terms of homogeneous coordinates μ_1, μ_2 (which are, of course, nothing else than spinor components), we express a rotation of U as a rotation of μ_1, μ_2 under a SU(2) transformation

$$|\bar{\mu}_1\bar{\mu}_2\rangle = \mathsf{A}\,|\mu_1\mu_2\rangle, \tag{1}$$

with

6-2.5 $$A=\begin{bmatrix} a & b \\ -b^* & a^* \end{bmatrix}, \qquad \det A = aa^*+bb^*=1. \qquad (2)$$

(In fact, since A is unitary eqn 1 is undoubtedly a rotation.) Since the expressions obtained in §2 entailed the components of the complex spinor, we shall also need its transformation rule

6-7.5 $$|\bar{\mu}_2^*, -\bar{\mu}_1^*\rangle = A\,|\mu_2^*, -\mu_1^*\rangle. \qquad (3)$$

We are now ready to enunciate and prove our first result.

Theorem 1. *Mapping of* SU(2) *into* SO(3)

Each rotation A of the homogeneous coordinates maps to a proper rotation $R(A)$ of the sphere. That is we have a mapping $\mathrm{SU}(2)\to\mathrm{SO}(3)$ such that

$$A \mapsto R(A), \qquad \forall A\in \mathrm{SU}(2), \qquad \text{some } R(A)\in \mathrm{SO}(3). \qquad (4)$$

Proof. In order to map SU(2) to SO(3) we must go from the two-dimensional bases in (1) and (3) of SU(2) to a three-dimensional basis. We do this by the method discussed in §**5**-2, in which we form the direct product basis

$$|\mu_1\mu_2\rangle \otimes |\mu_2^*, -\mu_1^*\rangle = |\mu_1\mu_2^*, -\mu_1\mu_1^*, \mu_2\mu_2^*, -\mu_2\mu_1^*\rangle, \qquad (5)$$

and then symmetrize it. This work is straightforward and runs as follows. The matrix corresponding to the basis (5) is from (1) and (3),

2-6.54 $$A\otimes A=\begin{bmatrix} a^2 & ab & ba & b^2 \\ -ab^* & aa^* & -bb^* & ba^* \\ -b^*a & -b^*b & a^*a & a^*b \\ b^{*2} & -b^*a^* & -a^*b^* & a^{*2} \end{bmatrix} = B. \qquad (6)$$

This matrix can be reduced (symmetrized) by forming $L^{-1}BL$ (which we shall call C), with the unitary matrix, (which symmetrizes R5),

$$L=\begin{bmatrix} 1 & & & \\ & 2^{-1/2} & 2^{-1/2} & \\ & 2^{-1/2} & -2^{-1/2} & \\ & & & 1 \end{bmatrix}, \qquad (7)$$

when

$$C=\begin{bmatrix} a^2 & 2^{1/2}ab & 0 & b^2 \\ -2^{1/2}ab^* & aa^*-bb^* & 0 & 2^{1/2}a^*b \\ 0 & 0 & aa^*+bb^* & 0 \\ b^{*2} & -2^{1/2}a^*b^* & 0 & a^{*2} \end{bmatrix}, \qquad (8)$$

and the corresponding basis (obtained by pre-multiplying R5 with L^{-1}) is

$$|\mu_1\mu_2^*, 2^{-1/2}(\mu_2\mu_2^* - \mu_1\mu_1^*), -2^{-1/2}(\mu_2\mu_2^* + \mu_1\mu_1^*), -\mu_1^*\mu_2\rangle. \tag{9}$$

Thus, the symmetrized matrix, which we shall call $\hat{R}^1(A)$ (the superscript denoting the dimension $2\times 1+1$ of the matrix), is

8 $$\hat{R}^1(A) = \begin{bmatrix} a^2 & 2^{1/2}ab & b^2 \\ -2^{1/2}ab^* & aa^* - bb^* & 2^{1/2}a^*b \\ b^{*2} & -2^{1/2}a^*b^* & a^{*2} \end{bmatrix}, \tag{10}$$

with the corresponding symmetrized basis from (9),

$$|\mu_1\mu_2^*, 2^{-1/2}(\mu_2\mu_2^* - \mu_1\mu_1^*), -\mu_1^*\mu_2\rangle. \tag{11}$$

(Notice that the second element of this basis is, correctly, the sum of the second and third elements of the basis in R5.)

Because C is the product of unitary matrices it is unitary and so is $\hat{R}^1(A)$. Also, it is easy to verify (Problem 1) that

$$\det \hat{R}^1(A) = 1. \tag{12}$$

In order to prove that $\hat{R}^1(A)$ entails a rotation of SO(3) we must prove that when the basis (11) is transformed into $\mathbf{r} = |xyz\rangle$ then the corresponding matrix transform of $\hat{R}^1(A)$, which we shall call $\hat{R}_{\mathbf{r}}(A)$, is orthogonal and of unit determinant. We must first identify the basis (11), which we can do most simply by forming, from (2.13) and (2.14) our canonical basis in $\mathbb{C}^3$, i.e. the spherical vector basis $|\mathbb{u}_1, \mathbb{u}_0, \mathbb{u}_{-1}\rangle$, of (2.2):

2.2,2.8,2.14 $$\mathbb{u}_1 = 2^{-1/2}u \quad = 2^{1/2}\mu_1\mu_2^*, \tag{13}$$

2.2,2.13 $$\mathbb{u}_0 = -z \quad = \mu_2\mu_2^* - \mu_1\mu_1^*, \tag{14}$$

2.2,2.8,2.14 $$\mathbb{u}_{-1} = -2^{-1/2}u^* \quad = -2^{1/2}\mu_1^*\mu_2. \tag{15}$$

This basis, in fact, is identical with (11) except for an irrelevant constant factor. Thus, we need go no further: because of the way in which we constructed the basis $|\mathbb{u}_1\mathbb{u}_0\mathbb{u}_{-1}\rangle$ in §5-1 we know that its matrix $\hat{R}^1(A)$ must transform correctly into some orthogonal matrix $\hat{R}_{\mathbf{r}}^1(A)$ when the basis is changed into $|xyz\rangle$. For future reference, though, we shall sketch this work. The matrix $\hat{R}^1(A)$ transforms the spherical vectors as follows

$$|\bar{\mathbb{u}}_1\bar{\mathbb{u}}_0\bar{\mathbb{u}}_{-1}\rangle = \hat{R}^1(A)\,|\mathbb{u}_1\mathbb{u}_0\mathbb{u}_{-1}\rangle. \tag{16}$$

Then, the matrix $\hat{R}_{\mathbf{r}}^1(A)$ that transforms $|xyz\rangle$ in the same way is

$$\hat{R}_{\mathbf{r}}^1(A) = M^{-1}\,\hat{R}^1(A)M, \tag{17}$$

where M is the matrix that pre-multiplied with $|xyz\rangle$ transforms this basis into the spherical vectors of (2.2). From this expression it can be proved that $\hat{R}_{\mathbf{r}}^1(A)$ must belong to SO(3), it can be obtained and its orthogonality directly verified. (Problems 2 to 4.)

Comments

(1) Because the matrix (10) transforms the basis $|\text{ш}_1\text{ш}_0\text{ш}_{-1}\rangle$ under a rotation, it must also transform its dual, the spherical harmonics $\langle Y_1^1 Y_1^0 Y_1^{-1}|$ as will be directly verified in Problem **14**-2.2. (See also §**5**-1.) Thus (10) is a very useful matrix. Moreover, $\hat{R}^1(A)$ leads immediately into $\hat{R}_{\mathbf{r}}^1(A)$ (Problem 3), which transforms the basis $|xyz\rangle$ and this matrix can readily be seen to agree with $\hat{R}_{\mathbf{r}}^1(\phi\mathbf{n})$, as directly obtained in (**3**-3.11). (Problem 4.) This is a strong check on all the work so far. (2) The reader may feel that our matrices are left in the air, not apparently being related to the rotation axes and angles or any other useful parameters like the Euler angles. This is not so, since the a and b parameters that characterize a SU(2) matrix A are related to $R(\phi\mathbf{n})$ by (**6**-4.9). These parameters, a, b that fully identify a rotation matrix A, are called the *Cayley–Klein parameters.* $\hat{R}^1(A)$ and $\hat{R}_{\mathbf{r}}^1(A)$ can more properly now be written as $\hat{R}^1(ab)$ and $\hat{R}_{\mathbf{r}}^1(ab)$ respectively. (3) The work we have done here shows very clearly that the basis obtained by forming the symmetrized tensor product of two spinors transforms by the same matrix as a spherical vector under some rotation of the sphere. That this is true for all rotations of SO(3) will, in fact, be proved in Theorem 2 below, thus completing the work sketched in §**5**-3.

We have proved so far that there is a mapping of SU(2) into SO(3), that is that each matrix of SU(2) has an image in SO(3). The converse result states that every rotation $R(\phi\mathbf{n})$ of SO(3) maps to a rotation of SU(2) (a rotation of SU(2) being realized by two matrices $\pm A$). The proof of this statement is immediate since, from (**6**-4.9) $R(\phi\mathbf{n})$ uniquely determines the Cayley–Klein a, b parameters and the latter uniquely determine the pair $\pm A$ corresponding to a SU(2) rotation, that is a rotation of the projection point P in the projection plane Π. (See eqn 1 and Problem 2.1.) It follows that SU(2) maps to the whole of SO(3) or, as it is said, onto SO(3). We shall prove in Theorem 2 that this mapping is homomorphic.

Theorem 2. *The homomorphism*

The mapping of SU(2) onto SO(3) is a two-to-one homomorphism, the two matrices $+A$ and $-A$ of SU(2) being uniquely mapped to the single matrix $\hat{R}_{\mathbf{r}}^1(A)$ of SO(3). Vice versa, the pre-images of $\hat{R}_{\mathbf{r}}^1(A)$ in the mapping are, and only are, $+A$ and $-A$.

Proof. The matrix $\hat{R}^1(A)$ in (10) is, from (6), the symmetrized product of A with A. We shall denote this symmetrized product with the symbol $\bar{\otimes}$ and we shall assume that the multiplication rules which we obtained for the ordinary direct product of matrices are still valid for it, which is easy to verify. With this notation in (17) we write

$$\hat{R}^1_{\mathbf{r}}(A) = M^{-1}(A \bar{\otimes} A)M, \qquad A \in \mathrm{SU}(2), \qquad \hat{R}^1_{\mathbf{r}}(A) \in \mathrm{SO}(3). \quad (18)$$

In order to prove that the mapping $A \mapsto \hat{R}^1_{\mathbf{r}}(A)$ is homomorphic we must show that the multiplication rules are preserved. From (18)

$$\hat{R}^1_{\mathbf{r}}(A_i)\hat{R}^1_{\mathbf{r}}(A_j) = M^{-1}(A_i \bar{\otimes} A_i)MM^{-1}(A_j \bar{\otimes} A_j)M \quad (19)$$

$$= M^{-1}(A_i \bar{\otimes} A_i)(A_j \bar{\otimes} A_j)M \quad (20)$$

2-6.55|20 $$= M^{-1}(A_iA_j \bar{\otimes} A_iA_j)M \quad (21)$$

18|21 $$= \hat{R}^1_{\mathbf{r}}(A_iA_j). \quad (22)$$

It follows from (18) that $+A$ and $-A$ both map to the same $\hat{R}^1_{\mathbf{r}}(A)$ in SO(3). In order to prove that the mapping is two-to-one we prove that A and $-A$ are the only matrices of SU(2) that have $\hat{R}^1_{\mathbf{r}}(A)$ as their image. Suppose that, on the contrary,

$$\hat{R}^1_{\mathbf{r}}(A) = \hat{R}^1_{\mathbf{r}}(A'), \qquad A' \neq \pm A. \quad (23)$$

23 $$\hat{R}^1_{\mathbf{r}}(A)^{-1}\hat{R}^1_{\mathbf{r}}(A') = \mathbf{1}_3. \quad (24)$$

It is not difficult to verify from (18) that $\hat{R}^1_{\mathbf{r}}(A)^{-1}$ equals $\hat{R}^1_{\mathbf{r}}(A^{-1})$ whence, on using (22),

$$\hat{R}^1_{\mathbf{r}}(A^{-1})\hat{R}^1_{\mathbf{r}}(A') = \hat{R}^1_{\mathbf{r}}(A^{-1}A') = \mathbf{1}_3. \quad (25)$$

Thus, on substituting $A^{-1}A'$ for A into (R18), (L18) must become $\mathbf{1}_3$, which requires that $A^{-1}A'$ be equal to $\pm\mathbf{1}_2$, whence $A' = \pm A$, thus verifying the theorem.

The spinor components

We defined in (**6**-7.3) a spinor as a basis $|\mu_1\mu_2\rangle$ with μ_1 and μ_2 in $\mathbb{C}^2$, which is used as a convenient substitute for $\mathbb{R}^3$, the configuration space. Its dual will be written as $\langle\mu_1\mu_2|$ making no typographical distinction to indicate that μ_1 and μ_2 refer to functions over $\mathbb{C}^2$. We are doing this because, for future reference (§**14**-3), it will be useful to identify the spinor components μ_1 and μ_2 in function space. The dual of the basis in (13) to (15) is $\langle u^1_1 u^1_0 u^1_{-1}|$ and the spinor components μ_1 and μ_2 go over to the dual changing nothing else than their meaning. Thus:

13,14,15 $$\langle u^1_1 u^1_0 u^1_{-1}| = \langle 2^{1/2}\mu_1\mu_2^*, \mu_2\mu_2^* - \mu_1\mu_1^*, -2^{1/2}\mu_1^*\mu_2| \quad (26)$$

4-7.14 $$= \langle -2^{-1/2}(x+\mathrm{i}y), z, 2^{-1/2}(x-\mathrm{i}y)|. \quad (27)$$

Bearing in mind that the spinor components must also satisfy condition (R2.12), we have with (26) and (27), and the notation

$$v = x + \mathrm{i}y, \quad (28)$$

the conditions

26,27 $$\mu_1\mu_2^* = -\tfrac{1}{2}v, \tag{29}$$

26,27 $$\mu_1\mu_1^* - \mu_2\mu_2^* = -z, \tag{30}$$

2.12 $$\mu_1\mu_1^* + \mu_2\mu_2^* = 1. \tag{31}$$

(Notice that these are mere manifestations of the eqns 2.12 to 2.14 that were obtained from the stereographic projection. The work that follows could have been done directly from those equations but we left it until now, first to refine the numerical constants, and secondly to work in function space, the latter being the cause of the apparent discrepancies with the just-mentioned equations.) It is straightforward to obtain μ_1 and μ_2 from (29) to (31) and this is done in Problems 7 to 9. In Problem 10 we verify that the basis $\langle \mu_1\mu_2|$, which must coincide with $\langle u^{1/2}_{1/2} u^{1/2}_{-1/2}|$, actually behaves like this basis under rotations $R(\phi\mathbf{z})$. Problem 11 produces an expansion of the solid harmonics u^j_m in terms of the spinor components.

Problems 7-3 (p. 286)

1. Prove that $\det \hat{R}^1(A)$ given by (10) is unity.
2. Verify that the matrix $\hat{R}^1_{\mathbf{r}}(A)$ of (17) is orthogonal and of unit determinant.
3. Show that

$$\hat{R}^1_{\mathbf{r}}(A) = \begin{bmatrix} \tfrac{1}{2}(a^2-b^2+a^{*2}-b^{*2}) & -\tfrac{1}{2}\mathrm{i}(a^2+b^2-a^{*2}-b^{*2}) & -(ab+a^*b^*) \\ \tfrac{1}{2}\mathrm{i}(a^2-b^2-a^{*2}+b^{*2}) & \tfrac{1}{2}(a^2+b^2+a^{*2}+b^{*2}) & -\mathrm{i}(ab-a^*b^*) \\ ab^*+a^*b & -\mathrm{i}(ab^*-a^*b) & aa^*-bb^* \end{bmatrix}. \tag{32}$$

(This matrix transforms $\mathbf{r} = |xyz\rangle$.)

4. Verify that (32) agrees with (**3**-3.11).
5. Show that the matrix that transforms $\langle Y_1^1 Y_1^0 Y_1^{-1}|$ under $R(\beta\mathbf{y})$ is

$$\hat{R}^1(\beta\mathbf{y}) = \begin{bmatrix} \cos^2\tfrac{1}{2}\beta & -2^{-1/2}\sin\beta & \sin^2\tfrac{1}{2}\beta \\ 2^{-1/2}\sin\beta & \cos\beta & -2^{-1/2}\sin\beta \\ \sin^2\tfrac{1}{2}\beta & 2^{-1/2}\sin\beta & \cos^2\tfrac{1}{2}\beta \end{bmatrix}. \tag{33}$$

6. Transform Y_1^0 under the matrix (33) and show that

$$R(\beta\mathbf{y})Y_1^0 = \cos\theta\ \cos\beta + \sin\beta\ \sin\theta\ \cos\varphi. \tag{34}$$

Why is this result different from (**5**-1.4)?

7. Verify that v from (28) can be written as

$$v = \{(1+z)(1-z)\}^{1/2}\exp(\mathrm{i}\varphi), \tag{35}$$

where φ is the usual polar angle.

8. Verify that the functions μ_1 and μ_2 that satisfy eqns (29) to (31) are

$$\mu_1 = -2^{-1/2}(1-z)^{1/2} \exp(\mathrm{i}\tfrac{1}{2}\varphi), \qquad \mu_2 = 2^{-1/2}(1+z)^{1/2} \exp(-\mathrm{i}\tfrac{1}{2}\varphi). \quad (36)$$

Verify that μ_1 and μ_2 in (36) can be multiplied by -1, in agreement with (**6**-7.2), that defines μ_1 and μ_2 to within a constant factor.

9. Obtain the following Cartesian form for the spinor components

$$\begin{aligned} \mu_1 &= -2^{-1/2}(1+z)^{-1/4}(1-z)^{1/4}(x+\mathrm{i}y)^{1/2}, \\ \mu_2 &= 2^{-1/2}(1+z)^{1/4}(1-z)^{-1/4}(x-\mathrm{i}y)^{1/2}. \end{aligned} \quad (37)$$

10. Obtain the matrix $\hat{R}^{1/2}(\phi\mathbf{z})$ on the basis $\langle\mu_1\mu_2|$ and show that it agrees with (**4**-7.9).

11. Introduce u_1, u_0, u_{-1} in terms of the spinor components of (26) into the expansion (**4**-7.18) of the spherical harmonics u^j_m and obtain

$$u^j_m = \mathbb{C}_{jm} \sum_{k=r}^{j-m-r} \sum_r (-1)^k \frac{(\mu_1)^{m+k}(\mu_2^*)^{j-k}(\mu_2)^{j-m-k}(\mu_1^*)^k}{(m+r)!\, r!\, (k-r)!\, (j-m-k-r)!}. \quad (38)$$

Here, $\mathbb{C}_{jm} = 2^{\frac{1}{2}m} C_{jm}$, the latter being defined under (**4**-7.20), and r must be chosen so that r, $m+r$, $j-m-2r$ are all positive or zero.

8

PROJECTIVE REPRESENTATIONS

Umgekehrt entspricht auch jedem System von Matrizen, deren Determinanten sämtlich von Null verschieden sind, und welche die Eigenschaft besitzen, daß für je zwei Elemente A, B der Gruppe eine Gleichung der Form $(A)(B) = r_{A,B}(AB)$ besteht, wo die Größen $r_{A,B}$ Konstanten bedeuten, eine Darstellung der Gruppe durch gebrochene lineare Substitutionen.

Isaiah Schur (1904), p. 21.

Conversely, there also corresponds a representation of the group by a linear fractional substitution to every system of matrices, the determinants of which are all different from zero, and such that they satisfy the property that for every two elements A, B of the group, an equality of the form $(A)(B) = r_{A,B}(AB)$ obtains.

We have seen that, though SU(2) is a group of matrices homomorphic to SO(3), the mapping of SU(2) onto SO(3) is two-to-one and thus cannot be a representation of SO(3). (See also §**6**-5.) We shall define in this chapter a new type of representation, called a *projective or ray representation*, which will allow us to treat the mapping SU(2) → SO(3) as a representation of SO(3). As we shall see, projective representations are essential in the treatment of quantal symmetries, but the subject is somewhat neglected in the elementary literature, despite an early and excellent introduction in Hamermesh (1962). A fairly detailed account can be found in Altmann (1977). A very complete treatment of the mathematical theory is given by Asano and Shoda (1935) and a useful review is provided by Morris (1971). Recent results for finite groups are given by Haggarty and Humphreys (1978) and in the book by Karpilovsky (1985). Before we define the projective representations we shall consider a simple example.

1. THE GROUP $\mathbf{D}_2$ AND ITS SU(2) MATRICES. DEFINITION OF PROJECTIVE REPRESENTATIONS

The group $\mathbf{D}_2$ consists of the identity E and three bilateral binary rotations, C_{2x}, C_{2y}, C_{2z} (see Fig. **2**-5.1). It must be remembered, (§**2**-1), that a *bilateral binary rotation* is a rotation $R(\pi\mathbf{n})$ such that there exists in

Table **8**-1.1. SU(2) matrices for $\mathbf{D}_2$.

Operation	E	C_{2x}	C_{2y}	C_{2z}
Matrix symbol	$\boldsymbol{I}$	$\boldsymbol{I}_x$	$\boldsymbol{I}_y$	$\boldsymbol{I}_z$
Matrix	$\begin{bmatrix} 1 & \\ & 1 \end{bmatrix}$	$\begin{bmatrix} & -i \\ -i & \end{bmatrix}$	$\begin{bmatrix} & -1 \\ 1 & \end{bmatrix}$	$\begin{bmatrix} -i & \\ & i \end{bmatrix}$

the group to which it belongs another operation $R(\pi\mathbf{n}')$ with $\mathbf{n}' \perp \mathbf{n}$. Bilateral binary rotations will play a fundamental rôle in our work: we have already seen (§**6**-4) that they are, except for a factor, the infinitesimal generators of SU(2) for the spinor representations. We have written in (**6**-4.1) SU(2) matrices for the operations C_{2x}, C_{2y}, C_{2z}, which we now copy in Table 1 together with the matrix for E. Of course, we should actually write eight SU(2) matrices, each matrix in Table 1 appearing also as its negative. We want to keep only the matrices in Table 1, however, so as to explore how far can we go in trying to establish a one-to-one mapping between them and $\mathbf{D}_2$. It must also be appreciated that the signs chosen in Table 1 have been carefully taken in order to agree with phase conventions (such as the CS convention).

The multiplication table for $\mathbf{D}_2$ has already been given in Table **2**-5.2 and it is repeated in Table 2, together with the multiplication table for the $\boldsymbol{I}$ matrices, which can easily be obtained from their value in Table 1.

It is clear from the table that the $\boldsymbol{I}$ matrices do not even form a group, since the set does not close, but, if we were to ignore the signs in the table, then the $\boldsymbol{I}$ matrices would form a representation. What happens is that, e.g., C_{2z} is the image both of $\boldsymbol{I}_z$ and of $-\boldsymbol{I}_z$ and when it appears as the product $C_{2x}C_{2y}$, $\boldsymbol{I}_z$ is the pre-image but, when it appears as $C_{2y}C_{2x}$, $-\boldsymbol{I}_z$ turns up instead. This sounds pretty horrible: $C_{2x}C_{2y}$ commute, but on multiplying the SU(2) matrices the commutation is lost, owing to their sign uncertainty. A nuisance undoubtedly, but, like many nuisances, if one learns to live with them, they become eventually not just acceptable

Table **8**-1.2. $\mathbf{D}_2$ multiplication tables.
A product ab appears in the intersection of the a row with the b column.

	Operations					$\boldsymbol{I}$ matrices			
	E	C_{2x}	C_{2y}	C_{2z}		$\boldsymbol{I}$	$\boldsymbol{I}_x$	$\boldsymbol{I}_y$	$\boldsymbol{I}_z$
E	E	C_{2x}	C_{2y}	C_{2z}	$\boldsymbol{I}$	$\boldsymbol{I}$	$\boldsymbol{I}_x$	$\boldsymbol{I}_y$	$\boldsymbol{I}_z$
C_{2x}	C_{2x}	E	C_{2z}	C_{2y}	$\boldsymbol{I}_x$	$\boldsymbol{I}_x$	$-\boldsymbol{I}$	$\boldsymbol{I}_z$	$-\boldsymbol{I}_y$
C_{2y}	C_{2y}	C_{2z}	E	C_{2x}	$\boldsymbol{I}_y$	$\boldsymbol{I}_y$	$-\boldsymbol{I}_z$	$-\boldsymbol{I}$	$\boldsymbol{I}_x$
C_{2z}	C_{2z}	C_{2y}	C_{2x}	E	$\boldsymbol{I}_z$	$\boldsymbol{I}_z$	$\boldsymbol{I}_y$	$-\boldsymbol{I}_x$	$-\boldsymbol{I}$

but positively delightful. This will be the case here, since we shall be able to obtain many illuminating results from these properties of the matrices.

In order to learn how to cope with this situation, we shall start by doing a cosmetic exercise on the multiplication table of the $\boldsymbol{I}$ matrices in Table 2. Since, if we forget their signs, they multiply in the same manner as the operations of $\mathbf{D}_2$, we shall give their multiplication rules exactly from the multiplication table of $\mathbf{D}_2$, read in conjunction with a table that gives the ± 1 factors that appear on multiplying the matrices. We display this in Table 3, where the factor ± 1 corresponding to the product of g_i and g_j is denoted with the symbol $[g_i, g_j]$. It must be clear that, although we label everything in Table 3 in terms of the group operations, what this table does is to give us the products of the $\boldsymbol{I}$ matrices. For example: the matrix corresponding to the product $C_{2x}C_{2y}$ is the matrix for $C_{2z}(\boldsymbol{I}_z)$ with + sign. The matrix corresponding to the product $C_{2y}C_{2x}$ is the matrix corresponding to $C_{2z}(\boldsymbol{I}_z)$ with − sign, and so on.

We are thus led to the idea of a new type of matrix representation of a group. Let us recall first how a matrix representation $\hat{G}(g)$ of G is defined:

$$g_i g_j = g_k \qquad \Rightarrow \qquad \hat{G}(g_i)\hat{G}(g_j) = \hat{G}(g_k). \tag{1}$$

If we call $\check{G}(g)$ the matrices of a new type of representation that behaves like the set of the four $\boldsymbol{I}$ matrices explicit in Table 2 and implicit in Table 3, then we must have:

$$g_i g_j = g_k \qquad \Rightarrow \qquad \check{G}(g_i)\check{G}(g_j) = [g_i, g_j]\check{G}(g_k). \tag{2}$$

A mapping of a group G of order n to a set of n matrices $\check{G}$ that satisfy the multiplication rules for some system of stated n^2 factors $[g_i, g_j]$ is called a *projective or ray representation* of the group G. (Naturally, like for any representation, the mapping must satisfy the conditions that the image of an element g_i of G be one and only one element $\check{G}(g_i)$ of $\check{G}$, and that every element $\check{G}(g_i)$ of $\check{G}$ be the image of some element g_i of G.) The factors $[g_i, g_j]$ are called the projective factors and the set of n^2

Table **8**-1.3. $\mathbf{D}_2$ multiplication and factor tables.
A product $g_i g_j$ or a factor $[g_i, g_j]$ appears in the intersection of the g_i row with the g_j column.

$g_i g_j$	E	C_{2x}	C_{2y}	C_{2z}
E	E	C_{2x}	C_{2y}	C_{2z}
C_{2x}	C_{2x}	E	C_{2z}	C_{2y}
C_{2y}	C_{2y}	C_{2z}	E	C_{2x}
C_{2z}	C_{2z}	C_{2y}	C_{2x}	E

$[g_i, g_j]$	E	C_{2x}	C_{2y}	C_{2z}
E	1	1	1	1
C_{2x}	1	−1	1	−1
C_{2y}	1	−1	−1	1
C_{2z}	1	1	−1	−1

such factors is called a *factor system*. The factors $[g_i, g_j]$ are, in general, complex numbers (although in our example we have taken them as ± 1) and they must be uniquely defined for each pair of elements of G, g_i and g_j.

With this definition, the four $\boldsymbol{I}$ matrices given for $\mathbf{D}_2$, although they do not form a group, form a projective representation and we shall see, eventually, that despite some differences, the work with projective representations is surprisingly similar to the work with ordinary representations. In order to avoid confusion, ordinary representations (which must satisfy eqn 1) are now called *vector representations.*

We have been led here to the definition of projective representations purely on heuristic grounds. In the next section we shall see that they must play a fundamental rôle in quantum mechanics.

2. BASES OF THE PROJECTIVE REPRESENTATIONS

Projective representations (under the name of linear fractional transformations; see the epigraph of this chapter) were invented in 1904 by Isaiah Schur well before quantum mechanics existed. Indeed, in a series of four papers (Schur 1904, 1907, 1910, 1927), he gave the theory of these representations in astonishing completeness. It was Hermann Weyl (1925, 1926) who first applied projective representations to the rotation group and who in 1928 in his seminal book on group theory (Weyl 1931) pointed out the importance of these representations in quantum mechanics. Surprisingly, it was not until much later that a definition of the bases of projective representations was attempted (Rudra 1965) and even then, this definition was later found to require modification. We shall follow here the definition proposed by Altmann (1979).

In physical applications, as we have seen in §**2**-5, if we want to construct a representation of a group G we start with a *basis* of functions $\langle \varphi_1(\mathbf{r}) \ldots \varphi_n(\mathbf{r})|$ which are invariant under the group ($\mathbf{r}$ being the coordinate in configuration space). This is true whether we deal with vector or projective representations. In §**2**-5, however, we assumed that we transformed the functions of the basis under the function space operators $\dot{g}$ (henceforth written as g and identified from the context) by means of eqn (**2**-3.1). This, however, is not right in quantum mechanics, the correct rule of transformation being

2-7.1 $$g\varphi(\mathbf{r}) = \omega\varphi(g^{-1}\mathbf{r}), \tag{1}$$

where ω is a phase factor, for which $\omega\omega^*$ equals unity. We saw in §**2**-7 that these phase factors are independent of the operand of g. There is a temptation, therefore, to take them to be dependent on g alone, ω_g say. This, however, will not do. We have seen in §1, e.g., that C_{2z} realized as the product $C_{2x}C_{2y}$ entails a factor of $+1$, whereas the same operation realized as the product $C_{2y}C_{2x}$ entails a factor of -1. This is rather awkward, because for us, so far, the symbol g_ig_j and the result of the product, say g_k, are identically the same thing, so that we do not even have the means to distinguish the symbols $C_{2x}C_{2y}$ or $C_{2y}C_{2x}$ from C_{2z}. The result of all this is that the phase factor ω in (1) must be defined in relation not to a single operation g_i but to a pair of operations g_i and g_j, and that this pair of operations cannot be written as g_ig_j. So, we adopt the following convention:

g_ig_j: product of two operations, identical with the operation g_k that is its result. (2)

g_i, g_j: the operation g_i performed after g_j but this transformation is not identified with the result g_k of the composite operation. (3)

Warning. The alert reader will notice here what appears to be a departure from the fundamental view of a linear transformation or symmetry operation given in Chapter **1**, as being entirely described by the initial and final configurations and entirely unconcerned with the history of how the transformation came about. The symbols C_{2x}, C_{2y}; C_{2y}, C_{2x} and C_{2z} are now different whereas $C_{2x}C_{2y}$, $C_{2y}C_{2x}$ and C_{2z} are all the self-same linear transformation: thus we are introducing an element of 'history' in our notation. The precise meaning of this will become clearer in §§**10**-5 and **10**-6, where it will be seen that, after all, we are not reneging on the fundamental view of linear transformations just described.

Be that as it may, notation (3) does precisely the job we need: whereas ω cannot be made dependent on g_k or on g_ig_j, it depends on g_i, g_j and for simplicity we shall denote it with the symbol $[g_i, g_j]$ already used in §1. Thus, the phase factors in (1) must be defined as follows:

$$g_i, g_j\varphi(\mathbf{r}) = [g_i, g_j]\varphi\{(g_ig_j)^{-1}\mathbf{r}\}. \tag{4}$$

Correspondingly, when we act on the basis (remembering that it is invariant under G), we have

$$g_i, g_j \langle\varphi| = [g_i, g_j]\langle\varphi|\check{G}(g_ig_j). \tag{5}$$

The matrix $\check{G}$ here is uniquely defined for each group element since, from (4), it depends only on g_ig_j (i.e. a given element g_k, say) and not on g_i, g_j.

It should be realized that, as shown by the definition (4), it is not possible to express fully the action of a *single* operation on the basis. If we want to act on $\langle\varphi|$ with g_j alone, the nearest thing we can do is to act on it with the pair e, g_j (here we shall write the group identity always as

e):

5 $$e,\, g_j \langle\varphi| = [e, g_j]\langle\varphi|\, \check{G}(g_j). \tag{6}$$

We shall see in §3, however, that the projective factor system can always be chosen so that $[e, g]$ equals unity for all $g \in G$. (Such projective factor systems are said to be *standardized* and we shall always assume that only standardized factor systems are used. The one in Table 1.3 is an example.) Because of this property, it is possible to rewrite (6) as follows,

$$g_j \langle\varphi| = \langle\varphi|\, \check{G}(g_j). \tag{7}$$

If we now act on (7) with g_i we have, on using (7) again,

$$g_i,\, g_j \langle\varphi| = \langle\varphi|\, \check{G}(g_i)\check{G}(g_j), \tag{8}$$

whence, from (5) and (8),

$$\check{G}(g_i)\check{G}(g_j) = [g_i, g_j]\check{G}(g_i g_j), \tag{9}$$

thus proving, on comparison with (1.2), that the basis $\langle\varphi|$ under the rules given spans a projective representation.

It will be seen that the work is surprisingly similar to that required for vector representations. Indeed, (7) is formally identical with the definition of the matrix representative for such representations, but attention is required in (5) to the use of the correct phase factors when multiplying operations.

Bases and energy levels

In dealing with vector representations, it is generally accepted that an irreducible basis $\langle\varphi_1 \ldots \varphi_n|$ is such that the functions in it form a complete set of degenerate eigenfunctions of the Hamiltonian H, corresponding to an energy eigenvalue E:

$$H \langle\varphi| = E \langle\varphi|. \tag{10}$$

This statement is sometimes called the *irreducibility postulate*, although it is more in the nature of a convention, the convention being that whenever a basis of degenerate functions as in (10) is not irreducible then the degeneracy is called *accidental*. Let us henceforth qualify our work with the condition that accidental degeneracy is excluded and let us see what happens to the above statement when phase factors and projective representations are admitted. We must accept that the concept of an irreducible basis and representation, which is very much the same thing as for vector representations, is still valid for projective ones.

Given the assumption of completeness which we have stated, any other basis $\langle\psi|$ that satisfies (10) must be a linear combination of the functions in $\langle\varphi|$. Apply now $g_j \in G$ (the group of the Schrödinger equation, all the elements of which commute with H) on both sides of (10) and then commute $g_j H$:

$$Hg_j \langle\varphi| = Eg_j \langle\varphi|. \tag{11}$$

It follows that the basis $g_j \langle\varphi|$ is degenerate with $\langle\varphi|$ so that it must be a linear combination of it. Call $\check{G}(g_j)$ the corresponding matrix: we then have,

$$g_j \langle\varphi| = \omega_{g_j} \langle\varphi| \check{G}(g_j), \tag{12}$$

where, for generality, we have included the arbitrary phase factor by which the basis can always be multiplied. (Notice that this is not a projective factor: it is correlated to a single operation and not to a product.) Again for generality, we shall continue using such tedious factor every time we operate on a basis. On operating with g_i on (12), as in (8), we have

$$g_i, g_j \langle\varphi| = \omega_{g_i}\omega_{g_j} \langle\varphi| \check{G}(g_i)\check{G}(g_j). \tag{13}$$

On the other hand, operating directly with g_i, g_j on $\langle\varphi|$, as in (5), we have to include the projective factor, as well as an arbitrary factor $\omega_{g_i g_j}$ pertaining to the operation $g_i g_j$. It will be convenient to call the projective factor $[g_i, g_j]'$, for reasons which will soon be clear:

$$g_i, g_j \langle\varphi| = \omega_{g_i g_j}[g_i, g_j]' \langle\varphi| \check{G}(g_i g_j). \tag{14}$$

On comparing (13) and (14),

$$\check{G}(g_i)\check{G}(g_j) = \omega_{g_i g_j}(\omega_{g_i}\omega_{g_j})^{-1}[g_i, g_j]' \check{G}(g_i g_j). \tag{15}$$

Thus, the matrices $\check{G}(g)$ form a projective representation of G with factor system

$$[g_i, g_j] = \omega_{g_i g_j}(\omega_{g_i}\omega_{g_j})^{-1}[g_i, g_j]'. \tag{16}$$

At the same time, the degenerate basis $\langle\varphi|$ in (10) corresponding to the energy level E spans a projective representation of G and, just as for the vector representations, we accept that, barring accidental degeneracies, this representation must be irreducible. So, the situation is very much the same as for the vector representations. One has to be a bit more careful, however, because of the fact, illustrated by the proof above, that there may be a change of factor system, as we shall now discuss. If we consider eqns (12), (13), and (14) we could have absorbed the phase factor ω_g into the corresponding matrix by defining

$$\check{G}'(g) = \omega_g \check{G}(g), \qquad \forall g \in G. \tag{17}$$

Clearly, these matrices, from (15), form a projective representation with projective factor $[g_i, g_j]'$, given from (16) by

$$[g_i, g_j]' = \omega_{g_i}\omega_{g_j}\omega^{-1}_{g_ig_j}[g_i, g_j]. \tag{18}$$

It is clear, thus, that given the basis $\langle\varphi|$ belonging to the eigenvalue E, the corresponding matrix representation is defined within a free-phase factor as shown in (17). Such a phase factor, defined for each matrix, is called a *gauge*. A change of gauge will effect a change in the factor system as shown in (18), which is called a *gauge equivalence*. As it can be seen from the above proof, a gauge equivalence will not affect the relation between an irreducible basis and its eigenvalue, so that a gauge transformation (as in eqn 17) is often regarded as trivial and to be ignored. You may believe this at your peril: when it is a question of uniquely defining the symmetry properties of functions no phase factor can be disregarded. After all, this is the whole basis, e.g., of the CS convention and whereas it is true that any gauge could be used, unless a gauge is firmly chosen and maintained errors will ensue. The extraordinary advantage, in fact, of the projective-representation method is that it permits the user to keep all phase factors firmly under control.

Naturally, the definition of a projective representation depends entirely on that of a given factor system. In our example in Table 1.3 the factor system was provided to us by the matrices we chose from SU(2). A great deal of the work that will come in the next two or three chapters will in fact be concerned with the consistent and unique choice of a factor system in order to deal with the spinor representations. What we must do next, however, is to study the properties of projective factor systems.

3. THE FACTOR SYSTEM

The factors $[g_i, g_j]$ form a set of $|G|^2$ constants, if $|G|$ is the order of the group. Not every such set, though, is eligible as a factor system since the factor system must satisfy a necessary condition, called the *associativity condition*:

$$[g_i, g_j][g_ig_j, g_k] = [g_i, g_jg_k][g_j, g_k]. \tag{1}$$

Condition (1) must be satisfied for all i, j, k, so that $|G|^3$ equations must be formed. (This is a great advantage in practice, since it allows for a thorough check of any factor system one proposes to use.) In (1), every product of the form g_ig_j must be understood in the sense defined in (2.2). The proof of (1) follows from the requirement of associativity for the matrix products:

$$\check{G}(g_i)\{\check{G}(g_j)\check{G}(g_k)\} = \{\check{G}(g_i)\check{G}(g_j)\}\check{G}(g_k). \tag{2}$$

We express (L2) on using (1.2) twice:

$$\check{G}(g_i)\{\check{G}(g_j)\check{G}(g_k)\} = \check{G}(g_i)[g_j, g_k]\check{G}(g_jg_k)$$
$$= [g_j, g_k][g_i, g_jg_k]\check{G}(g_ig_jg_k). \quad (3)$$

The same is done on (R2) and (1) follows. It can be proved (see, e.g., Altmann 1977) that the associativity condition is sufficient as well as necessary.

Given a factor system, a new one can be obtained by a choice of gauge, that is a set of $|G|$ phase factors ω_g for each $g \in G$, as shown in (2.18). It can be proved (see Altmann 1977) that the associativity conditions are still satisfied after a gauge transformation. It is also easy to prove (Problem 2) that the gauge-equivalence relation is transitive whence, given one factor system, an infinite class of gauge-equivalent factor systems exists such that any two members of this class are gauge equivalent.

Some important identities for the projective factors must be noted:

$$[e, e] = [e, g] = [g, e]. \quad (4)$$

$$[g, g^{-1}] = [g^{-1}, g]. \quad (5)$$

We shall prove the first of (4) and leave the rest to Problem 3. From (1),

$$[e, e][ee, g] = [e, eg][e, g]. \quad (6)$$

6 $$[e, e][e, g] = [e, g][e, g] \quad \Rightarrow \quad [e, e] = [e, g]. \quad (7)$$

If $[e, e] \neq 1$, on choosing in (2.18) the gauge $\omega_g = [e, e]^{-1}$, $\forall g \in G$, we have

2.18 $$[e, e]' = [e, e]^{-1}[e, e]^{-1}[e, e][e, e] = 1, \quad (8)$$

and, as already stated, we assume that this choice of gauge has always been made. Such a factor system is said to be *standardized.* For such a system

4 $$[e, e] = [e, g] = [g, e] = 1, \quad (9)$$

as we already assumed in going from (2.6) to (2.7). A choice of gauge in which (5) is also taken as unity is traditional in some fields of work (Backhouse 1973) but it has the dreadful consequence in the rotation group that the value of -1 for $[C_2, C_2]$ that appears in all three cases of Table 1.3 is not permitted, whereas we shall see that this is an essential result which simplifies the work enormously. Notice, on the other hand, that Table 1.3 is correctly standardized.

Problems 8-3 (p. 287)

1. Verify from Table 1.3 the following associativity relation for $\mathbf{D}_2$:

$$[C_{2x}, C_{2y}][C_{2x}C_{2y}, C_{2z}] \doteq [C_{2x}, C_{2y}C_{2z}][C_{2y}, C_{2z}]. \tag{10}$$

2. Prove that the gauge-equivalence relation (2.18) is transitive.
3. Prove the identities (4) and (5).

4. THE REPRESENTATIONS

Projective representations can be reduced in exactly the same manner as vector representations, or they can be irreducible, in which case they have orthogonality relations which, for unitary projective representations (as all our representations will be) are identical to those that obtain for vector representations.

There is one major difference, however, as regards the number of irreducible projective representations that appear for a fixed factor system. Whereas for vector representations their number equals the total number of classes in the group, this is not so for projectives, although the number of projective representations can be inferred from the fact that the sum of the squares of the dimensions of the representations equals the order of the group, as for vector representations. (See Altmann 1977 for these properties.)

Characters

As for vector representations, the character of an operation g in a representation $\check{G}$, $\chi(g \mid \check{G})$, is the trace of the corresponding matrix,

$$\chi(g \mid \check{G}) = \mathrm{Tr}\, \check{G}(g). \tag{1}$$

There is a major difference for projectives, however, since their characters are not class functions, as is the case for vector representations. This entails serious problems when dealing with projective representations, although we shall see that, with sufficient care, the situation reverts to that obtaining for vector representations. An expression for the character obtained by Altmann (1979), for a standardized factor system is very useful for this purpose:

$$\chi(g_i^g \mid \check{G}) = [g_i^g, g][g, g_i]^{-1}\chi(g_i \mid \check{G}), \qquad \forall g \in G, \tag{2}$$

where, as in (**2**-5.2) g_i^g is the conjugate of g_i by g, gg_ig^{-1}. Clearly, it is only when the product of the two projective factors on (R2) is unity, that the character of g_i equals that of its conjugate g_i^g. (A lot of work will later on be done to ensure that this is always the case for all point groups.) The proof of (2) follows.

1 $$\chi(g_i^g \mid \check{G}) = \operatorname{Tr} \check{G}(g g_i g^{-1}) \tag{3}$$

2.9|3 $$= \operatorname{Tr}\{[g, g_i g^{-1}]^{-1} \check{G}(g) \check{G}(g_i g^{-1})\} \tag{4}$$

$$= [g, g_i g^{-1}]^{-1} [g_i, g^{-1}]^{-1} \operatorname{Tr}\{\check{G}(g) \check{G}(g_i) \check{G}(g^{-1})\}. \tag{5}$$

Since the trace of the product of *two* factors is independent of their order,

5 $$\chi(g_i^g \mid \check{G}) = [g, g_i g^{-1}]^{-1} [g_i, g^{-1}]^{-1} \operatorname{Tr}\{\check{G}(g^{-1}) \check{G}(g) \check{G}(g_i)\} \tag{6}$$

2.9|6 $$= [g, g_i g^{-1}]^{-1} [g_i, g^{-1}]^{-1} [g^{-1}, g] \operatorname{Tr}\{\check{G}(e) \check{G}(g_i)\} \tag{7}$$

2.9|7 $$= [g, g_i g^{-1}]^{-1} [g_i, g^{-1}]^{-1} [g^{-1}, g][e, g_i] \operatorname{Tr} \check{G}(g_i) \tag{8}$$

3.9|8 $$= [g, g_i g^{-1}]^{-1} [g_i, g^{-1}]^{-1} [g^{-1}, g] \operatorname{Tr} \check{G}(g_i). \tag{9}$$

We now use the associativity condition (3.1) in the following form:

$$[g g_i, g^{-1}][g g_i g^{-1}, g] = [g g_i, g^{-1} g][g^{-1}, g]. \tag{10}$$

The first factor on (R10) is of the form $[g', e]$ and equals unity from the standardization condition, whence

10 $$[g g_i, g^{-1}][g_i^g, g] = [g^{-1}, g]. \tag{11}$$

We introduce (R11) into (9):

$$\chi(g_i^g \mid \check{G}) = [g, g_i g^{-1}]^{-1} [g_i, g^{-1}]^{-1} [g g_i, g^{-1}][g_i^g, g] \chi(g_i \mid \check{G}), \tag{12}$$

and use the associativity condition (3.1) once more, now in the form

$$[g, g_i][g g_i, g^{-1}] = [g, g_i g^{-1}][g_i, g^{-1}], \tag{13}$$

whence

$$[g g_i, g^{-1}] = [g, g_i g^{-1}][g_i, g^{-1}][g, g_i]^{-1}. \tag{14}$$

On introducing (14) into (12), eqn (2) follows.

5. DIRECT PRODUCTS OF REPRESENTATIONS

Given a group M with two subgroups G and H such that g and h commute for all g and h, then an element of M can uniquely be written as $m = g_i h_k$ and M is said to be the direct product of G and H, $G \otimes H$. Assume that projective factors for G and H are given, as well as for every pair g_i, h_k, and that two projective representations of G and H are available,

$$\check{G}(g_i)\check{G}(g_j) = [g_i, g_j]\check{G}(g_i g_j); \qquad \check{H}(h_k)\check{H}(h_l) = [h_k, h_l]\check{H}(h_k h_l). \tag{1}$$

We shall then prove that

$$\check{M}(g_i h_k) = [g_i, h_k]^{-1} \check{G}(g_i) \otimes \check{H}(h_k), \tag{2}$$

is a projective representation of M with factor system

$$[g_ih_k, g_jh_l] = [g_i, g_j][h_k, h_l][g_ig_j, h_kh_l][g_i, h_k]^{-1}[g_j, h_l]^{-1}. \quad (3)$$

In fact,

2 $$\check{M}(g_ih_k)\check{M}(g_jh_l) =$$

$$= [g_i, h_k]^{-1}[g_j, h_l]^{-1}\{\check{G}(g_i) \otimes \check{H}(h_k)\}\{\check{G}(g_j) \otimes \check{H}(h_l)\} \quad (4)$$

2-6.55|4 $$= [g_i, h_k]^{-1}[g_j, h_l]^{-1}\{\check{G}(g_i)\check{G}(g_j) \otimes \check{H}(h_k)\check{H}(h_l)\} \quad (5)$$

1|5 $$= [g_i, h_k]^{-1}[g_j, h_l]^{-1}[g_i, g_j][h_k, h_l]\{\check{G}(g_ig_j) \otimes \check{H}(h_kh_l)\} \quad (6)$$

2|6 $$= [g_i, h_k]^{-1}[g_j, h_l]^{-1}[g_i, g_j][h_k, h_l][g_ig_j, h_kh_l]\check{M}(g_ig_jh_kh_l). \quad (7)$$

On the other hand, the projective factor for the product in (L4) is defined from

$$\check{M}(g_ih_k)\check{M}(g_jh_l) = [g_ih_k, g_jh_l]\, \check{M}(g_ig_jh_kh_l), \quad (8)$$

and comparison of (8) and (7) gives (3).

It should be noticed that the above proof would still be valid if the projective factor in (R2) were eliminated, as well as, correspondingly, the last two projective factors on (R3). The associativity conditions for (3), however, must be satisfied and these factors are essential for this purpose (Problem 1).

Another important direct product occurs between two projective representations of the same group but corresponding to different factor systems:

$${}^r\check{G}(g_i)\, {}^r\check{G}(g_j) = {}^r[g_i, g_j]\, {}^r\check{G}(g_ig_j); \qquad {}^s\check{G}(g_i)\, {}^s\check{G}(g_j) = {}^s[g_i, g_j]\, {}^s\check{G}(g_ig_j), \quad (9)$$

in which case their direct product

$$\check{G}(g) = {}^r\check{G}(g) \otimes {}^s\check{G}(g) \quad (10)$$

is a projective representation with factor system

$$[g_i, g_j] = {}^r[g_i, g_j]\, {}^s[g_i, g_j]. \quad (11)$$

The proof is along the same lines as for the previous result (Problem 2).

Problems 8-5 (p. 288)

1. Prove that the projective factors (3) satisfy the associativity condition.
2. Prove (10) and (11).

6. THE COVERING GROUP

We shall here be concerned with the following problem: given a group G with a factor system, to find the irreducible projective representations of

Table 8-6.1. Projective representation for $\mathbf{D}_2$.

$\mathbf{D}_2$ $\check{D}_2$	E $\boldsymbol{I}$	C_{2x} $\boldsymbol{I}_x$	C_{2y} $\boldsymbol{I}_y$	C_{2z} $\boldsymbol{I}_z$
χ	2	0	0	0

G. There are very precise methods to do this job (see Altmann 1977) but we wish to use a fairly heuristic (i.e. brute force) approach.

Let us consider the group $\mathbf{D}_2$, with multiplication table and factor system in Table 1.3. We know, in fact, what to expect, since the SU(2) matrices $\boldsymbol{I}$, $\boldsymbol{I}_x$, $\boldsymbol{I}_y$, $\boldsymbol{I}_z$ in Table 1.1 form a projective representation of $\mathbf{D}_2$ with the given factor system. (The latter, in fact, was obtained from the multiplication rules of these matrices.) Thus, one at least of the representations that we should obtain is the one listed in Table 1. (The characters follow from Table 1.1.)

The procedure that we shall adopt consists in obtaining a *covering group*. Given a group G, a covering group G' is another group that contains the operations of G and such that the irreducible projective representations of G can be obtained by extracting from the irreducible vector representation of G', $\hat{G}'(g')$, those matrices for which g' belongs to G. (Technically, it is said that G' is then *subduced* to G.) In order to fulfil this programme we need a group that contains the operations of $\mathbf{D}_2$ but in which the multiplication rules of these operations are altered in order to incorporate the projective factors. That is, whereas in $\mathbf{D}_2$

$$C_{2x}C_{2z} = C_{2y}, \tag{1}$$

we need in the new group something like

$$C_{2x}C_{2z} = -C_{2y} \tag{2}$$

(see Table 1.3). We therefore need a group that contains all the operations of $\mathbf{D}_2$ plus new 'operations' such as $-C_{2y}$. To make things easier, it is convenient to denote such a new group element with the symbol $\tilde{C}_{2y}$. The covering group, which we shall call $\tilde{\mathbf{D}}_2$, will thus contain eight elements:

$$\tilde{\mathbf{D}}_2: \quad E \quad C_{2x} \quad C_{2y} \quad C_{2z} \quad \tilde{E} \quad \tilde{C}_{2x} \quad \tilde{C}_{2y} \quad \tilde{C}_{2z} \tag{3}$$

The necessary multiplication rules are very easy to guess. When we multiply two operations without the tilde we must add the tilde to the result in Table 1.3 whenever a factor -1 appears in the factor system. We can thus obtain quickly that part of the multiplication table of $\tilde{\mathbf{D}}_2$ that pertains to the operations without tilde. The rest of the table follows

Table **8**-6.2. Multiplication table of $\tilde{\mathbf{D}}_2$.
(The quaternion group)
The element $g_i g_j$ is at the intersection of the row g_i with the column g_j.

	E	C_{2x}	C_{2y}	C_{2z}	$\tilde{E}$	$\tilde{C}_{2x}$	$\tilde{C}_{2y}$	$\tilde{C}_{2z}$
E	E	C_{2x}	C_{2y}	C_{2z}	$\tilde{E}$	$\tilde{C}_{2x}$	$\tilde{C}_{2y}$	$\tilde{C}_{2z}$
C_{2x}	C_{2x}	$\tilde{E}$	C_{2z}	$\tilde{C}_{2y}$	$\tilde{C}_{2x}$	E	$\tilde{C}_{2z}$	C_{2y}
C_{2y}	C_{2y}	$\tilde{C}_{2z}$	$\tilde{E}$	C_{2x}	$\tilde{C}_{2y}$	C_{2z}	E	$\tilde{C}_{2x}$
C_{2z}	C_{2z}	C_{2y}	$\tilde{C}_{2x}$	$\tilde{E}$	$\tilde{C}_{2z}$	$\tilde{C}_{2y}$	C_{2x}	E
$\tilde{E}$	$\tilde{E}$	$\tilde{C}_{2x}$	$\tilde{C}_{2y}$	$\tilde{C}_{2z}$	E	C_{2x}	C_{2y}	C_{2z}
$\tilde{C}_{2x}$	$\tilde{C}_{2x}$	E	$\tilde{C}_{2z}$	C_{2y}	C_{2x}	$\tilde{E}$	C_{2z}	$\tilde{C}_{2y}$
$\tilde{C}_{2y}$	$\tilde{C}_{2y}$	C_{2z}	E	$\tilde{C}_{2x}$	C_{2y}	$\tilde{C}_{2z}$	$\tilde{E}$	C_{2x}
$\tilde{C}_{2z}$	$\tilde{C}_{2z}$	$\tilde{C}_{2y}$	C_{2x}	E	C_{2z}	C_{2y}	$\tilde{C}_{2x}$	$\tilde{E}$

easily from this basic sector, by adding a tilde to the result when only one of the factors contains a tilde, or by using the straight result of the basic sector when both factors contain a tilde. That is: after the basic sector is formed, tildes are manipulated very much as products of -1. The multiplication table of $\tilde{\mathbf{D}}_2$ thus obtained is given in Table 2.

It can readily be seen how the table is constructed. The basic sector is the top left quadrant. Then the top right quadrant is obtained by adding a tilde to the basic sector where there is no tilde, or removing it where there is one. The top left and bottom right quadrants must, by the mode of construction, be identical, and the same obtains for the top right and bottom left quadrants.

It is clear that $\tilde{\mathbf{D}}_2$ closes and hence that it is a group. In order to reduce it we must obtain its class structure. Since $\mathbf{D}_2$ is Abelian, each operation is its own class. In $\tilde{\mathbf{D}}_2$, on the other hand, this changes. It is easy to see that C_{2x} and $\tilde{C}_{2x}$ belong in the same class: from the multiplication table, the conjugate of C_{2x} under C_{2y} is

$$C_{2y}C_{2x}C_{2y}^{-1} = C_{2y}C_{2x}\tilde{C}_{2y} = \tilde{C}_{2z}\tilde{C}_{2y} = \tilde{C}_{2x}. \tag{4}$$

We thus must have three classes of the form C_{2r}, $\tilde{C}_{2r}$ ($r = x$, y, or z). On the other hand, E and $\tilde{E}$ commute with all elements and thus each form a class of their own. Thus, $\tilde{\mathbf{D}}_2$ is a group of order eight, with five classes. It follows that we must have five irreducible vector representations, four of which are one-dimensional and one two-dimensional. From the orthogonality conditions of the characters they are easy to guess and they are displayed in Table 3.

When the characters for the operations E, C_{2x}, C_{2y}, C_{2z} are extracted from Table 3, the first four representations yield the four vector represen-

Table **8**-6.3. The irreducible representations of $\tilde{\mathbf{D}}_2$.
(The quaternion group)

Rep.	E	$\tilde{E}$	$C_{2x}, \tilde{C}_{2x}$	$C_{2y}, \tilde{C}_{2y}$	$C_{2z}, \tilde{C}_{2z}$
1	1	1	1	1	1
2	1	1	−1	−1	1
3	1	1	−1	1	−1
4	1	1	1	−1	−1
5	2	−2	0	0	0

Table **8**-6.4. The two-dimensional representation of $\tilde{\mathbf{D}}_2$.
(The quaternion group)

E	$\tilde{E}$	C_{2x}	$\tilde{C}_{2x}$	C_{2y}	$\tilde{C}_{2y}$	C_{2z}	$\tilde{C}_{2z}$
$\begin{bmatrix} 1 & \\ & 1 \end{bmatrix}$	$\begin{bmatrix} -1 & \\ & -1 \end{bmatrix}$	$\begin{bmatrix} & -i \\ -i & \end{bmatrix}$	$\begin{bmatrix} & i \\ i & \end{bmatrix}$	$\begin{bmatrix} & -1 \\ 1 & \end{bmatrix}$	$\begin{bmatrix} & 1 \\ -1 & \end{bmatrix}$	$\begin{bmatrix} -i & \\ & i \end{bmatrix}$	$\begin{bmatrix} i & \\ & -i \end{bmatrix}$

tations of $\mathbf{D}_2$, as can immediately be seen by comparison with Table **2**-5.3. The last representation does not yield in the same manner any irreducible representation of $\mathbf{D}_2$ (neither can it reduce into a sum of such representations). A full matrix representation can be constructed from the matrices in Table 1.1 and it is shown in Table 4.

It can be seen that these matrices verify the characters in Table 3. Moreover, on multiplication, they satisfy the multiplication rules of the group $\tilde{\mathbf{D}}_2$, and they form therefore a vector representation of it. When the matrices of $\mathbf{D}_2$ are extracted from this representation their characters agree with those of the projective representation expected from Table 1.

Remarks

(1) It should be noticed that the relation between the group G and its covering group $\tilde{G}$ is a subtle one. $\tilde{G}$ contains G but it is not a supergroup of G. This is so because, although the operations of G are in $\tilde{G}$, *their multiplication rules are altered*, as is obvious from the top left quadrant of Table 2. Moreover, under the multiplication rules of $\tilde{G}$, the operations of G do not close and thus do not form a subgroup. (See the quadrant mentioned.)

(2) For the same reasons, inverses change when going from G to $\tilde{G}$. Thus, while the inverse of C_{2y} in $\mathbf{D}_2$ is C_{2y}, it follows from Table 2 that it is $\tilde{C}_{2y}$ in $\tilde{\mathbf{D}}_2$. (It is frightfully easy for this reason to make awful mistakes in work such as that in eqn 4.)

(3) It can be shown (see Altmann 1977) that our heuristic procedure yields all the irreducible projective representations of the given group with the given factor system.

(4) The covering group as defined above is often called the *double group*. It should be noticed, though, that for other factor systems covering groups must be constructed with more than double the number of operations of the original group.

(5) As shown in the headings, the double group of $\mathbf{D}_2$, $\tilde{\mathbf{D}}_2$ is called the *quaternion group*. The reason for this is that the matrices of $\mathbf{D}_2$ in Table 1 are, as we shall see later, quaternion units. They do not close, of course (see Table 1.2), and they do not form a group, so that their double group is the nearest one can get to constructing a group out of the quaternion units. $\tilde{\mathbf{D}}_2$ can in fact be realized by taking the four quaternion units in Table 1 plus their four negatives.

Finally: We shall soon see that the projective representations obtained are at the very core of nature since they are the spin representations. They could never have been obtained if we had not had the factor system in Table 1.3. This factor system came from the SU(2) matrices. If we want to obtain these representations for all point groups, say, we shall need a different and more general method for generating the factor systems. This will be the work of the next few chapters.

Problems 8-6 (p. 288)

1. On multiplying out a few pairs of matrices of the representations 1 to 4 in Table 3, corresponding to elements of $\mathbf{D}_2$, verify that they form vector representations of $\mathbf{D}_2$.
2. Verify that the matrices of $\mathbf{D}_2$ in Table 4 form a projective representation with the projective factor system of Table 1.3.

9

THE GEOMETRY OF ROTATIONS

And for the sake of all things in general let us recall to mind that nothing can be known concerning the things of this world without the power of geometry, . . .

Roger Bacon (ca. 1266). *The Opus Majus of Roger Bacon* (trans. R. B. Burke) 2 vols. University of Pennsylvania Press, Philadelphia, 1928, vol. I, p. 234.

We have seen in Chapter **8** that in order to define a projective factor we must make a distinction between a sequence of two operations g_i, g_j and their result g_k. In order to do this we must first learn how two rotations g_j and g_i carried out in succession determine the rotation g_k: this will be the main task of this chapter, which will thus prepare the ground for Chapter **10**, where the distinction between g_i, g_j and their result g_k will be finally made, thus allowing us to determine projective factors uniquely.

The whole of this chapter consists thus of a purely classical study of the geometry of rotations. Although the methods used are therefore pedestrian, the consequences reach much further than one might expect. In particular, a new parametrization of the rotations will appear, through the Euler–Rodrigues parameters, which will prove to be the most powerful tool available for determining the composition of rotations and the projective factors. This parametrization, moreover, will lead in a most natural way into the study of quaternions.

A very important concept in this work will be that of a *pole* of a rotation. The earliest use which I have found of this concept in English is in Sylvester (1850). Klein was probably the first to exploit rotation poles in the study of point groups in his book on the icosahedron (Klein 1884, trans. p. 9). More importantly, in p. 35 of this reference, although poles are not mentioned by name, the crucial distinction between the two poles corresponding to a rotation axis is made by the sign of the rotation associated with them when looking at the rotation sphere from outside. The first careful application of poles to the question of conjugation and to the study of point groups that I know of is the book by Zassenhaus published in German in 1937 (see Zassenhaus 1949), although he does not make the distinction between the two poles of an axis, which will prove below to be vital. (See Altmann 1977.)

1. THE UNIT SPHERE AND THE ROTATION POLES

We saw in §**2**-1 that a rotation of a unit sphere around its centre is fully specified by the coordinates of two of its points. (From now on, as is current practice, a sphere will always be understood as a surface, whereas the solid sphere will be called a *ball.*) Thus the major tool for specifying rotations will be the unit sphere (sphere of unit radius). In practice, a rotation will be specified by the axis $\mathbf{n}$ and angle of rotation ϕ and we have seen in Chapter **1** that the angle of rotation can never be larger than 2π (remember that rotating and spinning the sphere are not the same thing!). We have already seen in eqn (2) at the beginning of Chapter **3** that it is best to take the angle of rotation ϕ in the range

$$-\pi<\phi\leqslant\pi, \tag{1}$$

a choice which will prove essential in Chapter **10** (see §**10**-1). The range is open below because the rotation by $-\pi$ is identical to the rotation by π. The range (1), compared with the range from 0 to 2π which is often favoured, appears to have the disadvantage that negative angles have to be considered. We shall see, however, that our system of conventions – which must be carefully adhered to if consistent results are to be obtained – ensures that only positive rotations need be considered.

The problem is that the distinction between positive and negative rotations is purely conventional, because $R(\phi\mathbf{n})$ and $R(-\phi, -\mathbf{n})$ are one and the same rotation (see eqn **3**-2.1). We must remove this ambiguity, which we do by introducing two concepts, namely that of the *pole* $\Pi(\mathrm{g})$ of a rotation g and that of the positive hemisphere $\mathfrak{H}$. (For compactness of notation we shall often refer to a rotation with the symbol g, an operation of $G\equiv SO(3)$.)

The *pole* of a rotation is the point on the sphere that is invariant under the given rotation and *such that the rotation is seen as counterclockwise from outside the sphere.* Notice that the italicized proviso is essential to ensure that $R(\phi\mathbf{n})$ has one and only one pole.

At this point, $R(\phi\mathbf{n})$ and $R(-\phi, \mathbf{n})$ will have different poles (one being the *antipole* of the other) but which is which is purely a matter of convention. We notice that the above-mentioned pair of rotations could equally well be written as $R(\phi\mathbf{n})$, $R(\phi, -\mathbf{n})$ respectively, which is more consistent since, in any case, the angle ϕ is always seen as positive, in one case from the end of $\mathbf{n}$ and in the other from the end of $-\mathbf{n}$. What we do in order to distinguish positive from negative rotations unambiguously is this: we divide the sphere in two disjoint areas (not necessarily connected) of exactly the same extension, and by conventional definition we call one area the *positive hemisphere* $\mathfrak{H}$ and the other the *negative hemisphere* $\bar{\mathfrak{H}}$. The vector $\mathbf{n}$ of the axis of rotation will always be taken so that it ends on

$\mathfrak{H}$ for positive rotations and on $\bar{\mathfrak{H}}$ for negative ones. Thus, the poles of positive rotations will always be on $\mathfrak{H}$ and the poles of negative rotations on $\bar{\mathfrak{H}}$. Naturally, we must be careful in defining $\mathfrak{H}$ and $\bar{\mathfrak{H}}$ so that they are disjoint, as they must be, so that no point on the sphere can belong to both areas at the same time. An example is given below.

It will be seen that, with the above definitions, whereas we distinguish between **n** and −**n**, the rotation angle ϕ is always positive in the range

$$0 \leqslant \phi \leqslant \pi. \tag{2}$$

Two remarks about this. First, if we had taken rotations in the range from 0 to 2π, as it is often done, then the inverse C_3^- of C_3^+ has to be understood as $(C_3^+)^2$. If we had stuck to this convention, then both rotations would have been given the same pole, whereas now we have a unique pole for each rotation. Secondly, this most important result is valid even for binary totations although for them the pole of $R(\pi\mathbf{n})$ denotes identically the same rotation as its antipole, the pole of $R(-\pi\mathbf{n})$. The latter operation, however, has already been stamped out of existence since its angle is not in the accepted range (1). Thus binary rotations are all given unique poles, always on $\mathfrak{H}$.

We now give an example of a choice of $\mathfrak{H}$ which will be used for convenience in this chapter and the next one, and which is as follows (see Fig. 1). The point $(xyz) \in \mathfrak{H}$ if any one of the following three alternatives is satisfied:

$$\text{(i) } z>0; \qquad \text{(ii) } z=0, x>0; \qquad \text{(iii) } z=0, x=0, y>0. \tag{3}$$

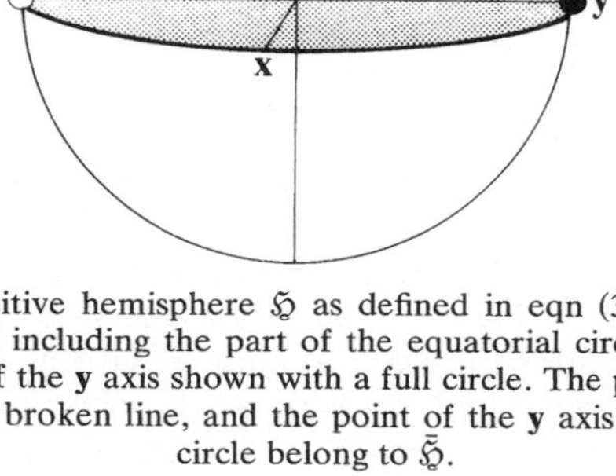

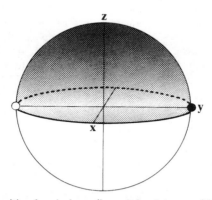

Fig. 9-1.1. The positive hemisphere $\mathfrak{H}$ as defined in eqn (3). $\mathfrak{H}$ is given by the shaded hemisphere, including the part of the equatorial circle shown with a full line and the point of the **y** axis shown with a full circle. The part of the equatorial circle shown with a broken line, and the point of the **y** axis shown with an open circle belong to $\bar{\mathfrak{H}}$.

It must be stressed however that although, when dealing with SO(3), this is the only definition of $\mathfrak{H}$ that is required, in dealing with finite subgroups of it, the point groups, the definition of $\mathfrak{H}$ requires careful attention in each individual case. (See §§**11**-6 and **15**-1.)

Conjugate poles

One extremely useful feature of the theory of poles is that they permit the identification of conjugate operations at a glance, without the need of tedious group multiplications, as we shall now show. The pole $\Pi(g_i)$ of the rotation g_i is, by definition, invariant under g_i:

$$g_i\Pi(g_i)=\Pi(g_i). \tag{4}$$

When another rotation of the group, g, acts on $\Pi(g_i)$ it takes it into another point on the sphere, $g\Pi(g_i)$ which shall be called the *conjugate pole* of $\Pi(g_i)$ under g. This terminology anticipates the result that $g\Pi(g_i)$ is actually the pole of the conjugate operation of g_i, g_i^g:

$$g\Pi(g_i)=\Pi(g_i^g). \tag{5}$$

In order to prove this assertion, we must prove that (L5) is invariant under g_i^g:

2-5.2,4 $$g_i^g g\Pi(g_i)=gg_ig^{-1}g\Pi(g_i)=gg_i\Pi(g_i)=g\Pi(g_i). \tag{6}$$

(The notation of eqn 5 used here will be further refined in §**11**-6.)

This important result explains why we have introduced the concept of *bilateral rotation* (§**2**-1). An axis of rotation is bilateral when there is a binary rotation perpendicular to it which, clearly, exchanges the poles of the rotation $R(\phi\mathbf{n})$, say C^+, and the rotation $R(\phi,-\mathbf{n})$, C^-. Thus, for bilateral rotations, C^+ and C^- are conjugate and belong therefore to the same class. (Such a class, that contains an operation as well as its inverse, is called *ambivalent*.) It should be noticed that this result is void for bilateral binary rotations, since, for them, C^+ and C^- are the self-same operation.

Improper rotations

All improper rotations are of the form $\boldsymbol{i}$g, where g is a proper rotation. An important and simple result is that the pole of $\boldsymbol{i}$g is the same as the pole of g, as is illustrated in Fig. 2. We show in it a cylinder around the rotation axis and, for greater clarity, the centre of inversion is represented outside it. On the left of the figure the pole of the rotation is at Π because from outside the solid this is the end from which the rotation is seen as counter-clockwise. Inversion takes A into B but this cannot be the pole since the rotation now appears clockwise from the corresponding end. The pole has to be placed as shown on the right of the figure and thus it is

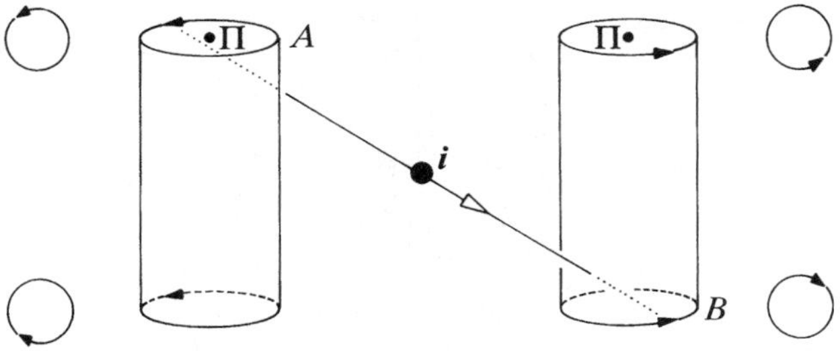

Fig. 9-1.2. The inversion leaves poles invariant. The small circles show how the rotation appears when viewed from *outside* the solid. The centre of inversion is the black circle.

not changed. (Imagine that $\boldsymbol{i}$ is now at the centre of the cylinder and you will see that the result is the same.) With this result, poles can easily be assigned to all operations of O(3).

Problems 9-1 (p. 288)

1. Find all the classes of $\mathbf{C}_{3v}$.
2. Prove that all the C_3 operations of the octahedral group $\mathbf{O}$ (group of the cube) are conjugate, but that this is not so for $\mathbf{T}$ (group of the tetrahedron). Discuss the corresponding classes in these groups.
3. Prove that $\mathbf{D}_2$ is an invariant subgroup of $\mathbf{O}$.

2. THE EULER CONSTRUCTION

We propose here to solve the problem of finding the product of two rotations $R(\alpha\mathbf{n}_A)R(\beta\mathbf{n}_B)$ with poles A and B respectively. We are given the poles of both rotations on the unit sphere and the problem is to find the angle of rotation γ and the pole of the resultant rotation C. The solution of this problem is due to Rodrigues, and it is shown in Fig. 1. Although the pole A has been set at the North Pole, this is only in order to facilitate the reading of the picture: the construction is entirely general for poles A and B in any arbitrary positions, and with any arbitrary rotation angles.

The prescription is as follows. Join A and B with the great circle in between these two points (in our case this great circle is a meridian, which facilitates its identification). The second step in the prescription is to sweep this great circle right and left around each pole by half the rotation angle of the corresponding pole. The four arcs thus swept cross at the points C and C'. We assert that C is the pole of the composite rotation $R(\alpha\mathbf{n}_A)R(\beta\mathbf{n}_B)$. For this to be the case, it must be left invariant by this rotation and we now prove that this is so. When the first rotation $R(\beta\mathbf{n}_B)$

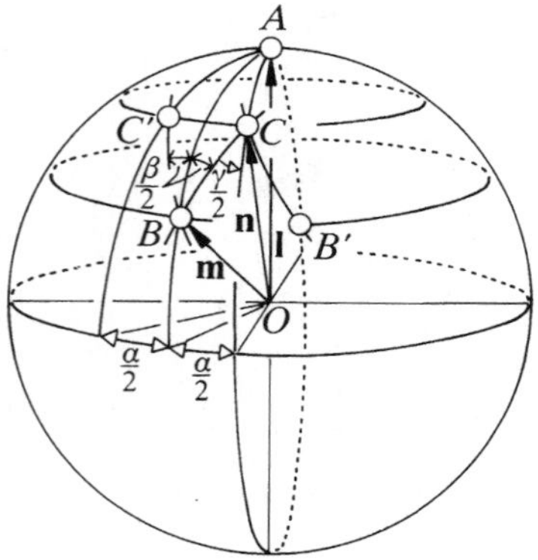

Fig. 9-2.1. The Euler construction. The figure illustrates the product $R(\alpha\mathbf{n}_A)R(\beta\mathbf{n}_B) = R(\gamma\mathbf{n}_C)$, where $\mathbf{n}_A$, $\mathbf{n}_B$, and $\mathbf{n}_C$ are OA, OB, and OC respectively.

operates on C it rotates it counterclockwise (as seen from B) by β until it reaches C'. When $R(\alpha\mathbf{n}_A)$ acts on C' it rotates it by α counter-clockwise (as seen from A) until it reaches C. As regards the angle of the rotation with pole C, we assert that it is γ, where $\frac{1}{2}\gamma$ is displayed in the figure. In order to prove this consider the effect of $R(\alpha\mathbf{n}_A)R(\beta\mathbf{n}_B)$ on B. $R(\beta\mathbf{n}_B)$ leaves it invariant and $R(\alpha\mathbf{n}_A)$ rotates it into B'. Since the rotation around C thus takes B into B', γ is the angle of rotation. Notice that, as it should be, the sense of rotation is counterclockwise as seen from C.

Notice that all the rotation angles appear halved in this construction, a result with notable consequences in the rotation group and which can best be understood by considering a second interpretation of the construction. We have seen (Fig. **2**-1.2) that a rotation by ϕ is the product of two reflections, σ_1 followed by σ_2, on planes through the rotation axis at a dihedral angle $\frac{1}{2}\phi$. The sense of the rotation is from σ_1 to σ_2. We use this result in Fig. 1. The first rotation, around B, is given by the reflection σ_1 on the plane OCB followed by the reflection σ_2 on the plane OAB. The second rotation, around A, is a reflection on the plane OAB (that cancels σ_2), followed by a reflection σ_3 on the plane OAC. The final result of these four reflections is the reflection on OCB followed by the reflection on OAC, that is a rotation around the intersection of these two planes, OC, by twice the angle in between them, that is twice $\frac{1}{2}\gamma$. The sense of the composite rotation is from OCB to OAC, as shown in the figure and previously proved. A deeper reason for the relation between rotations and mirrors will be given in §**12**-10. The present construction is at the basis of Biedenharn's turns, also discussed in §**12**-10.

It will also be seen that we have completed the proof of the *Euler Theorem* (§**2**-1) which asserts that the most general transformation of a sphere with fixed centre is a rotation $R(\phi\mathbf{n})$ around some axis $\mathbf{n}$ by some angle ϕ. This is so because in §**2**-1 we showed that the transformation described can always be given as the product of two rotations which, the Euler construction shows, is always expressible as a single rotation.

We must now solve the problem treated by the Euler construction in such a way as to obtain formulae that determine the angle and vector axis of $R(\gamma\mathbf{n}_C)$ given the angles and vectors for $R(\alpha\mathbf{n}_A)$ and $R(\beta\mathbf{n}_B)$. Before we do this we shall review some well-known results of spherical trigonometry.

3. SPHERICAL TRIGONOMETRY REVISITED

> Perhaps to the student there is no part of elementary mathematics so repulsive as spherical trigonometry.
>
> P. G. Tait (1911). *Encyclopaedia Britannica* XI Edition, vol. xxii, p. 721. Cambridge University Press, Cambridge.

We show in Fig. 1 the spherical triangle *ABC* formed in the Euler construction (Fig. 2.1) by the poles *B* (first rotation) *A* (second rotation) and *C* (resultant rotation). Naturally, all three vectors $\mathbf{l}$ ($\mathbf{n}_A$), $\mathbf{m}$ ($\mathbf{n}_B$), and $\mathbf{n}$ ($\mathbf{n}_C$) are of unit modulus. Notice that, as usual in spherical trigonometry, *A*, *B*, and *C* are also used here to denote the corresponding dihedral angles.

The cosine theorem of spherical trigonometry is well known:

$$\cos a = \cos b \ \cos c + \sin b \ \sin c \ \cos A. \tag{1}$$

This result can be cycled for *a*, *b*, *c*, and *A*, *B*, *C*. Also, one useful trick for deriving new formulae in spherical trigonometry is to construct the

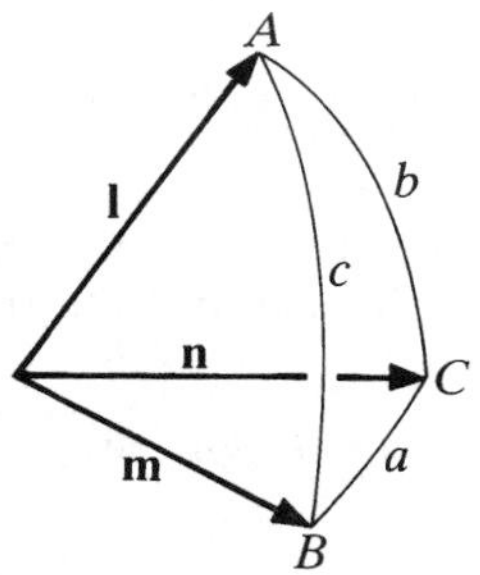

Fig. 9-3.1. The spherical triangle of Fig. 2.1.

supplementary triangle with dihedral angles A', B', C' and sides a', b', c' such that

$$A'+a=a'+A=\pi, \tag{2}$$

and similar relations for the other elements. Any formula written for the supplementary triangle can be transferred to the original triangle by the rules

$$\cos a' \mapsto -\cos A, \qquad \sin a' \mapsto \sin A, \qquad \cos A' \mapsto -\cos a, \tag{3}$$

which follow from (2). Thus, from (1),

$$\cos a' = \cos b' \ \cos c' + \sin b' \ \sin c' \ \cos A', \tag{4}$$

and from (3)

$$\cos A = -\cos B \ \cos C + \sin B \ \sin C \ \cos a. \tag{5}$$

Our task in relation to the Euler construction will be to determine the angle C and the axis $\mathbf{n}$. By cycling (5) and writing $\cos c$ as $\mathbf{l}\,.\,\mathbf{m}$ we can get C in terms of A, B, $\mathbf{l}$, and $\mathbf{m}$:

$$\cos C = -\cos A \ \cos B + \sin A \ \sin B \ (\mathbf{l}\,.\,\mathbf{m}). \tag{6}$$

We shall also need the various steps leading to the sin theorem, starting with the identity

$$(\mathbf{l}\times\mathbf{m})\times(\mathbf{l}\times\mathbf{n}) = (\mathbf{l}\times\mathbf{m}\,.\,\mathbf{n})\mathbf{l}. \tag{7}$$

Thus, since l is of unit length,

$$|(\mathbf{l}\times\mathbf{m})\times(\mathbf{l}\times\mathbf{n})| = \mathbf{l}\times\mathbf{m}\,.\,\mathbf{n}, \tag{8}$$

which gives (see Fig. 1):

$$\sin c \ \sin b \ \sin A = \mathbf{l}\times\mathbf{m}\,.\,\mathbf{n}. \tag{9}$$

Therefore, on cycling eqn (9) and then dividing by $\sin a \ \sin b \ \sin c$, we obtain

$$\frac{\sin A}{\sin a} = \frac{\sin B}{\sin b} = \frac{\sin C}{\sin c}. \tag{10}$$

The next and harder problem is to express $\mathbf{n}$ in terms of the vectors $\mathbf{l}$ and $\mathbf{m}$. We must take for this purpose the triple of vectors, $\mathbf{l}$, $\mathbf{m}$, and $\mathbf{p}=\mathbf{l}\times\mathbf{m}$, so that

$$\mathbf{n} = f\mathbf{l} + g\mathbf{m} + h\mathbf{p}, \tag{11}$$

where

$$f=\mathbf{n}\,.\,\mathbf{l}^*, \qquad g=\mathbf{n}\,.\,\mathbf{m}^*, \qquad h=\mathbf{n}\,.\,\mathbf{p}^*, \tag{12}$$

and $\mathbf{l}^*$, $\mathbf{m}^*$, $\mathbf{p}^*$ are the reciprocal vectors of $\mathbf{l}$, $\mathbf{m}$ and $\mathbf{p}$ constructed in the usual manner. (Problem 1.) When this is done, one obtains

$$\mathbf{n}=\frac{\{\mathbf{l}\,.\,\mathbf{n}-(\mathbf{l}\,.\,\mathbf{m})(\mathbf{m}\,.\,\mathbf{n})\}\mathbf{l}+\{\mathbf{m}\,.\,\mathbf{n}-(\mathbf{l}\,.\,\mathbf{m})(\mathbf{l}\,.\,\mathbf{n})\}\mathbf{m}+(\mathbf{l}\times\mathbf{m}\,.\,\mathbf{n})(\mathbf{l}\times\mathbf{m})}{1-(\mathbf{l}\,.\,\mathbf{m})^2}. \quad (13)$$

We introduce in (13) $\mathbf{l}\,.\,\mathbf{m}=\cos c$, etc., as well as (9), and write the result in the following form:

$$\sin c\ \mathbf{n}=\frac{\cos b-\cos c\ \cos a}{\sin c\ \sin a}\sin a\ \mathbf{l}+$$
$$\frac{\cos a-\cos b\ \cos c}{\sin b\ \sin c}\sin b\ \mathbf{m}+\sin b\ \sin A(\mathbf{l}\times\mathbf{m}). \quad (14)$$

We now introduce (1), as well as its cycled from, and multiply both sides by (10), to find our final formula:

$$\sin C\ \mathbf{n}=\sin A\ \cos B\ \mathbf{l}+\cos A\ \sin B\ \mathbf{m}+\sin A\ \sin B(\mathbf{l}\times\mathbf{m}). \quad (15)$$

Problems 9-3 (p. 289)

1. Given three non-coplanar vectors $\mathbf{l}$, $\mathbf{m}$, $\mathbf{p}$, obtain three vectors $\mathbf{l}^*$, $\mathbf{m}^*$, $\mathbf{p}^*$ (*reciprocal vectors*) that satisfy the relations

$$\mathbf{l}\,.\,\mathbf{l}^*=\mathbf{m}\,.\,\mathbf{m}^*=\mathbf{p}\,.\,\mathbf{p}^*=1, \quad (16)$$
$$\mathbf{l}\,.\,\mathbf{m}^*=\mathbf{l}\,.\,\mathbf{p}^*=\mathbf{m}\,.\,\mathbf{p}^*=\mathbf{m}\,.\,\mathbf{l}^*=\mathbf{p}\,.\,\mathbf{l}^*=\mathbf{p}\,.\,\mathbf{m}^*=0. \quad (17)$$

   Hence verify eqn (12).
2. Obtain $\mathbf{l}^*$, $\mathbf{m}^*$, $\mathbf{p}^*$, as in Problem 1, for $\mathbf{p}=\mathbf{l}\times\mathbf{m}$.

4. THE EULER CONSTRUCTION IN FORMULAE. THE EULER–RODRIGUES PARAMETERS

All that we need in order to apply the results of the preceding section to the Euler construction of Fig. 2.1 is to take the following values for the dihedral angles,

$$A=\tfrac{1}{2}\alpha, \qquad B=\tfrac{1}{2}\beta, \qquad C=\pi-\tfrac{1}{2}\gamma, \quad (1)$$

as results immediately on comparison with Fig. 3.1. On introducing (1) into the fundamental equations (3.6) and (3.15) we obtain the angle and axis of rotation in terms of those of the two original rotations:

$$\cos\tfrac{1}{2}\gamma=\cos\tfrac{1}{2}\alpha\ \cos\tfrac{1}{2}\beta-\sin\tfrac{1}{2}\alpha\ \sin\tfrac{1}{2}\beta\ (\mathbf{l}\,.\,\mathbf{m}), \quad (2)$$
$$\sin\tfrac{1}{2}\gamma\ \mathbf{n}=\sin\tfrac{1}{2}\alpha\ \cos\tfrac{1}{2}\beta\ \mathbf{l}+\cos\tfrac{1}{2}\alpha\ \sin\tfrac{1}{2}\beta\ \mathbf{m}+\sin\tfrac{1}{2}\alpha\ \sin\tfrac{1}{2}\beta\ \mathbf{l}\times\mathbf{m}. \quad (3)$$

Equations (2) and (3) are exactly in the form given by Rodrigues (1840), except, of course, for the use of the then non-existent vector notation. We want to express these results in a more compact fashion. Consider the rotations given in the form $R(\phi_i\mathbf{n}_i)$. We multiply them as follows:

$$R(\phi_1\mathbf{n}_1)R(\phi_2\mathbf{n}_2) = R(\phi_3\mathbf{n}_3), \tag{4}$$

ϕ_1, ϕ_2, ϕ_3, $\mathbf{n}_1$, $\mathbf{n}_2$, and $\mathbf{n}_3$ being α, β, γ, $\mathbf{l}$, $\mathbf{m}$, and $\mathbf{n}$ respectively in (2) and (3). From (2) and (3), however, it is clear that it is most convenient to replace ϕ and $\mathbf{n}$ by the following new parameters

$$\lambda = \cos \tfrac{1}{2}\phi, \qquad \boldsymbol{\Lambda} = \sin \tfrac{1}{2}\phi \ \mathbf{n}. \tag{5}$$

Correspondingly, of course, $R(\phi\mathbf{n})$ will now be replaced by the new symbol $R(\lambda; \boldsymbol{\Lambda})$. If we then re-write (4) as

$$R(\lambda_1; \boldsymbol{\Lambda}_1)R(\lambda_2; \boldsymbol{\Lambda}_2) = R(\lambda_3; \boldsymbol{\Lambda}_3), \tag{6}$$

it follows at once from (2) and (3) that the new parameters combine as follows:

2,3 $$\lambda_3 = \lambda_1\lambda_2 - \boldsymbol{\Lambda}_1 \cdot \boldsymbol{\Lambda}_2, \qquad \boldsymbol{\Lambda}_3 = \lambda_1\boldsymbol{\Lambda}_2 + \lambda_2\boldsymbol{\Lambda}_1 + \boldsymbol{\Lambda}_1 \times \boldsymbol{\Lambda}_2. \tag{7}$$

The rotation parameters (5) are called the *Euler–Rodrigues parameters*. In fact, they are precisely quaternion components, as we shall see later.

Remarks

(1) We know that both $R(\phi\mathbf{n})$ and $R(-\phi, -\mathbf{n})$ denote the same rotation and clearly, from (5), the Euler–Rodrigues parameters are left invariant by this change. The parameters $R(-\phi, -\mathbf{n})$, however, are ruled out by our conventions, so that the relation between the Euler–Rodrigues parameters and the rotation poles is one-to-one, within our conventions. On the other hand, also from (5), adding 2π to ϕ changes $R(\lambda; \boldsymbol{\Lambda})$ into $R(-\lambda; -\boldsymbol{\Lambda})$. Thus,

$$R(\lambda; \boldsymbol{\Lambda}) = R(-\lambda; -\boldsymbol{\Lambda}). \tag{8}$$

Naturally, because of our conventions, each operation of SO(3) will be parametrized with Euler–Rodrigues parameters of the form shown in (L8), but on multiplying two such operations by means of (6), (R6) can appear in the negative form shown in (R8), because the resultant ϕ_3 in (R4) may go out of range and involve an extra turn by 2π. Rather than a disadvantage, this will be most useful, since it means that the Euler–Rodrigues parameters will allow us to keep track of extra turns by 2π which can be introduced when multiplying rotations and this will be most important in determining projective factors.

(2) Notice that the Euler–Rodrigues parameters are four in number (since $\boldsymbol{\Lambda}$ has three components), whereas rotations can always be

parametrized by only three parameters, such as the three Euler angles. The Euler–Rodrigues parameters (5), however, must clearly satisfy the normalization condition

$$\lambda^2 + \mathbf{\Lambda}^2 = 1, \tag{9}$$

which in principle would allow us to eliminate one parameter. There are notable advantages, however, in working with a slightly redundant system.

(3) The Euler–Rodrigues parameters are intimately connected to the Cayley–Klein parameters defined in (**6**-4.9),

$$a = \lambda - \mathrm{i}\Lambda_z, \qquad b = -\Lambda_y - \mathrm{i}\Lambda_x. \tag{10}$$

(4) It is easy to see that for an infinitesimal rotation angle ϕ,

$$\lambda \approx 1 - \tfrac{1}{2}(\tfrac{1}{2}\phi)^2, \qquad \mathbf{\Lambda} \approx \tfrac{1}{2}\phi\ \mathbf{n}. \tag{11}$$

It is thus clear that for $\phi \to 0$, the parameters λ, $\mathbf{\Lambda}$ tend continuously to $\lambda = 1$, $\mathbf{\Lambda} = \mathbf{0}$, which are the parameters pertaining to the group identity, a most important condition, as discussed in §**3**-6. In practice, from (11), infinitesimal rotations become parametrized to first-order by the single parameter $\mathbf{\Lambda}$, which is a vector parallel to the rotation axis with modulus equal to half the rotation angle.

(5) We have now completed the study of the four available parametrizations of SO(3): $R(\phi\mathbf{n})$, where $\phi\mathbf{n}$ is the vector of modulus ϕ and in the direction of the angle of rotation (§**3**-6); $R(\alpha\beta\gamma)$ where α, β, γ are the Euler angles (§**3**-1); $R(ab)$ where ab are the complex Cayley–Klein parameters (**6**-4.9); and $R(\lambda; \mathbf{\Lambda})$. The advantages of the Euler–Rodrigues parametrization are enormous. First, the parameters are as easy to determine geometrically as $R(\phi\mathbf{n})$. Secondly, they are the only parameters for which the group multiplication rule can be given in closed form, as in (6) and (7). Thirdly, they behave correctly near the identity whereas the Euler angles become undetermined for $\beta = 0$, π (§**3**-1). Fourthly, with our convention for $\mathfrak{H}$ they uniquely determine rotation poles, whereas the Euler angles are unable to do this (see §**3**-4). Fifthly, they keep track of 2π turns introduced on multiplying rotations or, what is the same, of changes in phase factors thereby introduced, a property that will be much exploited later on in the book. The relation between the Euler angles and the Euler–Rodrigues parameters will be given in §**12**-11.

Problems 9-4 (p. 289)

1. Verify that λ_3 and $\mathbf{\Lambda}_3$ in (7) satisfy the normalization condition (9).
2. Show that, to the first order, the composition rule of two infinitesimal rotations agrees with their description by the vectors defined at the end of Remark 4 and that it leads to their commutation.

3. Write λ_3 and $\mathbf{\Lambda}_3$ in (7) as λ_{12}, $\mathbf{\Lambda}_{12}$ respectively, in order to indicate the order of the factors in (7). Likewise, λ_{21}, $\mathbf{\Lambda}_{21}$, are defined. Prove that

$$\lambda_{12} = \lambda_{21}, \qquad \mathbf{\Lambda}_{12} - \mathbf{\Lambda}_{21} = 2(\mathbf{\Lambda}_1 \times \mathbf{\Lambda}_2). \tag{12}$$

4. Prove that the rotation angle of the product of two rotations is independent of the order of the factors.
5. Prove that two rotations do not in general commute, but that rotations around the same axis (*coaxial rotations*) always commute.
6. Prove that the necessary and sufficient condition for the commutation of two rotations is, either that they be coaxial, or:

$$\lambda_1 = \lambda_2 = 0, \qquad \mathbf{\Lambda}_1 \perp \mathbf{\Lambda}_2. \tag{13}$$

   What are these rotations called?
7. Prove that the product of an arbitrary rotation $R(\phi\mathbf{n}_1)$ times a binary rotation $R(\pi\mathbf{n}_2)$, with $\mathbf{n}_2 \perp \mathbf{n}_1$, is another binary rotation $R(\pi\mathbf{n})$ around an axis $\mathbf{n}$ normal to $\mathbf{n}_1$ and at an angle $\frac{1}{2}\phi$ to $\mathbf{n}_2$.
8. Show that the rotation matrix (**3**-3.11) takes the following form in terms of the Euler–Rodrigues parameters:

$$\hat{R}^1_{\mathbf{r}}(\lambda;\mathbf{\Lambda}) = \begin{bmatrix} \lambda^2+\Lambda_x^2-\Lambda_y^2-\Lambda_z^2 & 2(\Lambda_x\Lambda_y-\lambda\Lambda_z) & 2(\Lambda_z\Lambda_x+\lambda\Lambda_y) \\ 2(\Lambda_x\Lambda_y+\lambda\Lambda_z) & \lambda^2-\Lambda_x^2+\Lambda_y^2-\Lambda_z^2 & 2(\Lambda_y\Lambda_z-\lambda\Lambda_x) \\ 2(\Lambda_z\Lambda_x-\lambda\Lambda_y) & 2(\Lambda_y\Lambda_z+\lambda\Lambda_x) & \lambda^2-\Lambda_x^2-\Lambda_y^2+\Lambda_z^2 \end{bmatrix}. \tag{14}$$

5. THE CONICAL TRANSFORMATION

In order to complete our geometrical treatment of rotations we re-derive, now geometrically, the conical transformation of a vector $\mathbf{r}$ obtained in (**3**-3.15). This is illustrated in Fig. 1 for a rotation $R(\phi\mathbf{n})$ which trans-

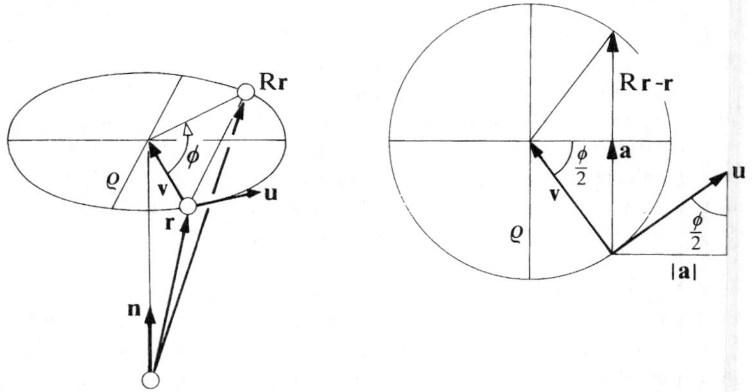

Fig. 9-5.1. Conical rotation of a vector $\mathbf{r}$ by an angle ϕ around the axis $\mathbf{n}$.

forms the vector **r** into $R\mathbf{r}$. In order to obtain $R\mathbf{r}$ it is sufficient to calculate **a**, which is $\frac{1}{2}(R\mathbf{r}-\mathbf{r})$. The vector **a** is written as

$$\mathbf{a}=p\mathbf{u}+q\mathbf{v}, \qquad \mathbf{u}=\mathbf{n}\times\mathbf{r}, \qquad \mathbf{v}=\mathbf{n}\times(\mathbf{n}\times\mathbf{r}), \qquad |\mathbf{u}|=|\mathbf{v}|=\rho. \quad (1)$$

It easily follows that

$$R\mathbf{r}=\mathbf{r}+\sin\phi\ (\mathbf{n}\times\mathbf{r})+2\sin^2\tfrac{1}{2}\phi\{\mathbf{n}\times(\mathbf{n}\times\mathbf{r})\} \quad (2)$$

$$=\mathbf{r}+\sin\phi\ (\mathbf{n}\times\mathbf{r})+(1-\cos\phi)\{(\mathbf{n}\,.\,\mathbf{r})\mathbf{n}-n^2\mathbf{r}\} \quad (3)$$

$$=\cos\phi\ \mathbf{r}+\sin\phi\ (\mathbf{n}\times\mathbf{r})+(1-\cos\phi)(\mathbf{n}\,.\,\mathbf{r})\mathbf{n}. \quad (4)$$

Equation (2) coincides with (**3**-3.15) but in writing (4) we have taken **n** to be of unit length.

Problems 9-5 **(p. 291)**

1. Verify that eqn (4) leads to the matrix $\hat{R}^1_{\mathbf{r}}(\phi\mathbf{n})$ in (**3**-3.11), which transforms $|xyz\rangle$.
2. Prove that

$$R(\lambda\ ;\boldsymbol{\Lambda})\mathbf{r}=(1-2\Lambda^2)\mathbf{r}+2\lambda(\boldsymbol{\Lambda}\times\mathbf{r})+2(\boldsymbol{\Lambda}\,.\,\mathbf{r})\boldsymbol{\Lambda}. \quad (5)$$

3. Given the rotations

$$R_+=R(\lambda;\boldsymbol{\Lambda}+\mathbf{a}), \qquad R_-=R(\lambda;\boldsymbol{\Lambda}-\mathbf{a}), \quad (6)$$

prove that

$$(R_+-R_-)\mathbf{r}=4\{-(\boldsymbol{\Lambda}\,.\,\mathbf{a})\mathbf{r}+\lambda\mathbf{a}\times\mathbf{r}+(\boldsymbol{\Lambda}\,.\,\mathbf{r})\mathbf{a}+(\mathbf{a}\,.\,\mathbf{r})\boldsymbol{\Lambda}\}. \quad (7)$$

4. Prove that, to second order, the commutator of two infinitesimal rotations R_1 and R_2 is

$$[R_1, R_2]=R(1;2\boldsymbol{\Lambda}_1\times\boldsymbol{\Lambda}_2)-E. \quad (8)$$

Verify that (**4**-4.10) is a special case of (8).

10

THE TOPOLOGY OF ROTATIONS

The whole of method consists in the order and disposition of the objects to which the mind's attention is directed, that we may discover some truth. And we shall exactly observe this method, if we reduce involved and obscure propositions step by step to simpler ones, and then, from an intuition of the simplest ones of all, try to ascend through the same steps to the knowledge of all the others.

René Descartes (*ca.* 1628–9). *Regulae ad Directionem Ingenii. Oeuvres* (eds C. Adam and P. Tannery). Cerf, Paris (1908), vol. X, p. 379.

The argument so far runs as follows. SU(2) matrices, although two-dimensional, 'represent' three-dimensional rotations. Nevertheless, they do not form a decent representation: we have seen that although the angles add up correctly, there is an uncertainty as to sign, which spoils the game. For instance, for rotations $R(\phi)$ around the **z** axis,

$$\hat{R}(\phi)\hat{R}(\phi') = \pm\hat{R}(\phi+\phi'), \qquad (1)$$

as follows from (**6**-5.2) and (**6**-5.3).

It is clear from (1) that the SU(2) matrices cannot form a vector representation of SO(3). So, we must aim at constructing a projective representation from them. If, however, the factor ± 1 on (R1) is to be used for this purpose as a projective factor, it must be determined uniquely by the operations $R(\phi)$ and $R(\phi')$ in their given order.

The reader may think that we have already solved this problem. In Table **8**-1.1, e.g., we produced matrices $\boldsymbol{I}, \boldsymbol{I}_x, \boldsymbol{I}_y, \boldsymbol{I}_z$ that map the operations of $\mathbf{D}_2$ and which, as we showed, formed a projective representation of $\mathbf{D}_2$ with a factor system in Table **8**-1.3. In doing this, however, we cheated, because in Table **8**-1.2 we wrote the multiplication rules of the matrices in a rather arbitrary way. We assert in that table, e.g., that

$$\boldsymbol{I}_x\boldsymbol{I}_y = \boldsymbol{I}_z, \qquad (2)$$

although (R2), from the fundamental property of the SU(2) matrices, could just as well have been written as $-\boldsymbol{I}_z$. The factor system that we constructed in Table **8**-1.3, therefore, is entirely *ad hoc* and it has no valid status at this stage. What we have done, however, is not rubbish, since, with foreknowledge, we arranged things behind the scenes in such a way that the factor system for $\mathbf{D}_2$, and thus the projective representation, agree with the definitive rules to be formulated in this chapter. These rules must determine uniquely the projective factor corresponding to a product of rotations $R(\phi_1\mathbf{n}_1)$, $R(\phi_2\mathbf{n}_2)$. It is for this reason, in fact, that we carefully determined in Chapter **9** the operation $R(\phi_3\mathbf{n}_3)$ that is the result of this product.

The way in which this work has to be done was discovered by Hermann Weyl (1925*b*). He was entirely concerned with SO(3) which, of course, is a continuous group, and was able to show that continuity conditions in this group determine uniquely the projective factors. We are concerned mainly with point groups, but these must be embedded in SO(3) – that is, studied as subgroups of this group – whereupon the rules that result from Weyl's theory can be applied to them. Although in order to do this work we must study continuity, that is the topology of SO(3), we shall do this at first in an entirely geometrical fashion that follows on from our geometrical study of the composition of rotations in Chapter **9**. Towards the end of this chapter we shall show that the geometrical work involved can be entirely bypassed by using the composition rules (**9**-4.6) and (**9**-4.7) of rotations in the Euler–Rodrigues parametrization. Because these rules have such an admirable application, we shall develop in Chapter **12** a whole new algebra from them, the *quaternion algebra.*

In order to study the continuity problems in SO(3) we must define a parametric space for this group and discuss in it some simple topological concepts, such as paths and homotopy. This will be done in the next few sections.

1. THE PARAMETRIC BALL

The parametric ball should not be confused with the unit sphere. Whereas the latter is where the action takes place, the parametric ball is used to *record* the results of this action. I emphasize this word because the reader will remember that whereas we always stressed that rotations have no history, we found it necessary to introduce in (**8**-2.3) a symbol g_i, g_j that establishes a difference between g_k, identical with g_i, g_j, and the succession of operations (g_j first, g_i second) that leads to it. In order to establish this distinction between g_k performed in one go, say, and the *same operation* done as by the product g_i, g_j, we saw in §**8**-2 that an element of history has to be introduced, and this will be provided by the parametric ball, which will allow us to establish the distinction mentioned in a clear and rigorous way.

We saw in §**3**-6 that, when parametrizing a rotation in the form $R(\phi\mathbf{n})$ the symbol $\phi\mathbf{n}$ had to be understood as a single symbol, namely a vector parallel to the axis of rotation and of modulus ϕ. This prescription was refined in §**9**-1. The angle ϕ is now always taken as positive, in the range $0 \leqslant \phi \leqslant \pi$, (**9**-1.2), but the direction of the vector $\phi\mathbf{n}$ is chosen so that this vector, when prolonged, cuts the sphere at a point belonging to $\mathfrak{H}$ for positive rotations or to $\bar{\mathfrak{H}}$ for negative rotations. The parametrization of binary rotations is made unique by agreeing that their poles are always in $\mathfrak{H}$. The *parametric ball* is a ball where the vector parameters $\phi\mathbf{n}$ of all rotations $R(\phi\mathbf{n})$ in SO(3) are plotted. It is a ball of radius π which we take to be oriented precisely parallel to the unit sphere in its reference (i.e. unrotated) position. The surface of the parametric ball is divided into a positive and a negative hemisphere by precisely the same rules as used for the unit sphere, so that it is always possible to orient $\phi\mathbf{n}$ correctly.

We must make sure that there is a one-to-one mapping of the unit ball onto SO(3), a question which we shall now discuss. Let us first consider rotations $R(\phi\mathbf{n})$ for $\phi < \pi$. Each of these rotations will uniquely be represented by one and only one point in the body of the parametric ball, at a distance ϕ from the origin and oriented towards $\mathfrak{H}$ or $\bar{\mathfrak{H}}$ in accordance to whether the rotation is positive or negative. Vice versa, each point inside the ball will map one and only one such rotation. Consider now binary rotations $R(\pi\mathbf{n})$: they will all be mapped to points on the surface of the parametric ball belonging to $\mathfrak{H}$, which we shall call *podal points*. However, for the mapping to be one-to-one we must agree that the corresponding *antipodal points* (which are on $\bar{\mathfrak{H}}$) must be removed from the parametric ball. (Clearly, the antipodal point corresponds to the same binary rotation as its podal point.) This means that the negative hemisphere $\bar{\mathfrak{H}}$ must be removed from the parametric ball.

With this definition of the parametric ball, there is a one-to-one mapping both ways between SO(3) and the points of the parametric ball. The point of the parametric ball that maps $R(\phi\mathbf{n})$ of SO(3) will be called the *parametric point* of $R(\phi\mathbf{n})$.

Naturally, it is inconvenient to work with the parametric ball with $\bar{\mathfrak{H}}$ peeled off and we shall agree to keep $\bar{\mathfrak{H}}$ in the parametric ball in the understanding that every point of it will be deemed to be identical with its corresponding podal point on $\mathfrak{H}$. When we say identical, we mean that a podal point and its antipodal point are now to be regarded as one and the same object. Thus, the topology of our parametric ball is profoundly different from that of a closed ball (i.e. a ball plus its sphere). In visualizing this topology, it is convenient to imagine that each antipodal point is linked or joined in some manner to its podal point in such a way that when a parametric point reaches $\bar{\mathfrak{H}}$ it *jumps back* from it to the corresponding podal point.

It should be noticed that the identity of SO(3) is parametrized by the vector $0\mathbf{n}$ of zero length, i.e. by the centre of the parametric ball. It is clear that the identity and its neighbourhood (the infinitesimal operations) which are the vital elements of SO(3) as a continuous group, are safely away from the surface of the parametric ball with its eccentric topology. The wisdom of not using the range of 0 to 2π for the rotation angles, that is so popular (see eqn 1 of Chapter **3** and its following sentence) will now be appreciated.

2. PATHS

We shall now begin to see more clearly the *raison d'être* of the parametric ball. We know that all the properties of a group are contained in its multiplication table. For a continuous group such an object, alas, cannot be constructed. The nearest thing to it is to form a parametric space for the group and to describe *paths* in this space. These paths will take the place, in a way, of the multiplication rules.

Let us consider as an example successive rotations around the **z** axis of the unit sphere (remember that this is where rotations take place: the parametric ball never changes!). We first perform a rotation by ϕ_1 and then follow it with another rotation by ϕ_2 (notice that we are 'multiplying' rotations). Suppose that ϕ_2 is such that $\phi_1+\phi_2=\phi_3<\pi$. We plot this in the parametric ball (see Fig. 1*a*). The first rotation gives a point along the **z** axis at a distance ϕ_1 from the origin. When the second rotation is performed, the parametric point is displaced by ϕ_2 and reaches ϕ_3. Likewise, if we perform a sequence of infinitesimal rotations around **z** until ϕ_3 is reached, the parametric point will move continuously along the **z** axis from O to ϕ_3. This is an example of a path (see *b* in the figure).

In general, a path is a line in the parametric ball that describes the change of the parametric point as a succession of rotations is effected. The same parametric point can be reached from the identity by more than one path. This can be done, in fact, in an infinite number of ways but we illustrate two such paths in Fig. 2.

We perform in Fig. 2*c* a sequence of infinitesimal positive rotations around **z** that add up to a total angle of $+\frac{1}{2}\pi$. The parametric point moves from the identity along the **z** axis until it reaches p, as it is illustrated in

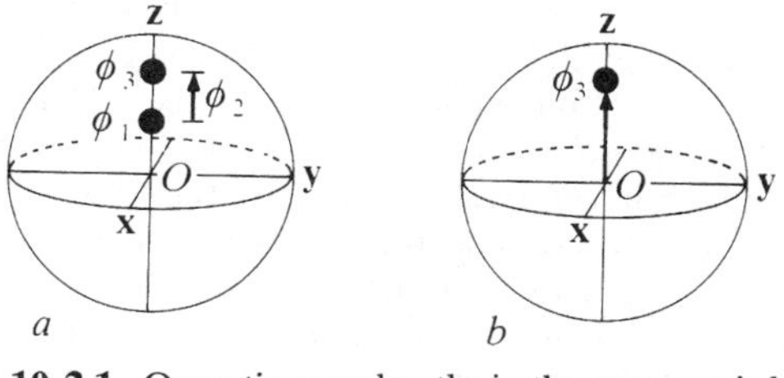

Fig. 10-2.1. Operations and paths in the parametric ball.

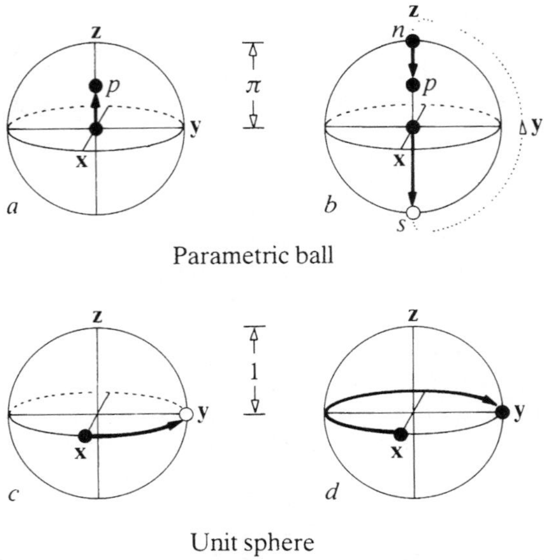

Fig. 10-2.2. The same rotation can be effected through two different paths.

Fig. 2*a*. The identical parametric point can also be reached by performing negative rotations, as in Fig. 2*d*. When a total angle of $-\pi$ is attained, the parametric point (see *b* in the figure) reaches the antipodal point *s*. However, this is topologically identical to *n*, so that we jump to *n*. Negative rotations continue and the angle $+\pi$ at *n* diminishes gradually until we reach *p* again.

It must be emphatically re-stated that, envisaged as single rotations, the two rotations depicted in Fig. 2*c* and *d* are one and the same rotation, as it is obvious because both of them are mapped to the same parametric point *p*. On the other hand, if Fig. 2*c* and *d* are envisaged as products of successive rotations, these are truly different as witnessed by the fact that they are mapped by different paths.

Paths of rotations are easily obtained by means of the Euler construction, as illustrated in Fig. 3. We find in Fig. 3*a* the product $C_{\alpha x}C_{2z}$, where $C_{\alpha x}$ is a rotation by an infinitesimal angle α. In the Euler construction, we have to join the two poles by a great circle. This must be swept around **z** by $\pm\frac{1}{2}\pi$ and around **x** by $\pm\frac{1}{2}\alpha$. It follows, as in the Euler construction, that *P* is the pole of the product and that half the angle of rotation γ is the angle $\frac{1}{2}\gamma$ shown in the figure. The parametric point of *P*, *p* is thus easily marked in Fig. 3*b*. If we now assume that we start from the identity *O* with infinitesimal rotations around **z** until we reach an angle π (point *m*)

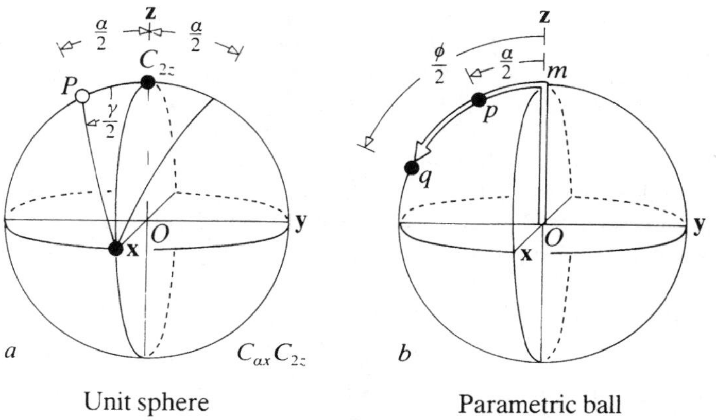

Fig. 10-2.3. Path of a rotation from the Euler construction.

and that we then follow with infinitesimal rotations around **x** until we reach some finite angle ϕ, the path in parametric space that describes this process is the one shown with the double arrow, *Ompq*.

It should be noticed that paths in parameter space can be deformed continuously, as illustrated in Fig. 4. The point p in the parametric ball denotes the rotation $C_{2x}C_{2z}$ and the double arrow leading to it is a path obtained as in Fig. 3 (with ϕ equal to π, i.e. a binary rotation around **x**). This path can be deformed as follows. Once C_{2z} is reached from infinitesimal rotations around **z** (parametric point m) we start rotating around **y** until an angle α is reached. Comparison with Fig. 3*a* (allowance

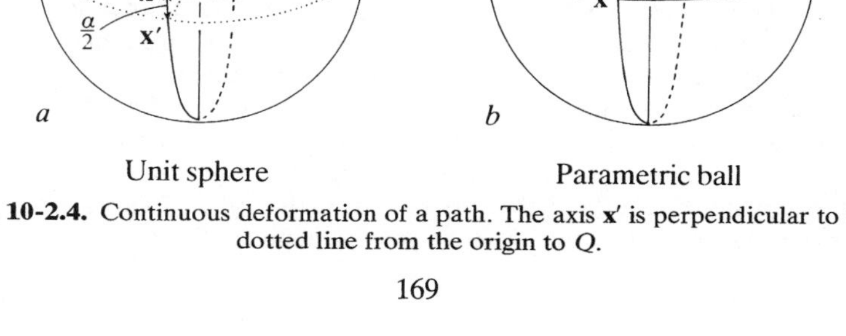

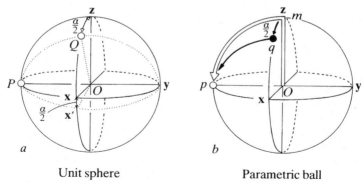

Fig. 10-2.4. Continuous deformation of a path. The axis **x**′ is perpendicular to the dotted line from the origin to Q.

being made for the change in orientation) should show that the pole must move from $\mathbf{z}$ to Q in Fig. 4a. A binary rotation around $\mathbf{x}'$ performed infinitesimally, moves Q to P along the great circle shown, as can be seen by comparison with Fig. 3a. As a result, the path *Omp* has been deformed into *Omqp*. On taking α as small as we please the difference between the two paths is infinitesimal. Therefore, the finite deformation just described can be effected continuously, by increasing α infinitesimally. This idea of the continuous deformation of paths will play a very important part in the theory.

3. PROGRAMME: CONTINUITY

Let us denote the full rotation group SO(3) with G and its rotations with g. Consider now rotations that multiply as follows:

$$g_i g_j = g_k, \tag{1}$$

$$g'_i g'_j = g_k. \tag{2}$$

We shall assume here that the paths that lead from the origin to g_k in either case are continuously deformable one into the other, just as in the example in §2. Imagine now that the g's are mapped onto some matrices $\check{G}(g)$ in such a way that

$$\check{G}(g_i)\check{G}(g_j) = +\check{G}(g_k). \tag{3}$$

Then, for the product in (2) we must require that

$$\check{G}(g'_i)\check{G}(g'_j) = +\check{G}(g_k). \tag{4}$$

In other words, we must reject any possible factor of -1 on (R4): ghastly discontinuities would otherwise appear in going from the group to its mapping. It is this idea that we shall exploit in order to eliminate the uncertainty as to the sign of the rotation matrices.

4. HOMOTOPY

As follows from the last section, the concept of the continuous deformation of paths is essential in studying continuity conditions in SO(3). It is given a systematic expression by means of the concept of *homotopy*, a good review of which can be found in Wallace (1957) or Roman (1975). Two coterminous paths in parameter space are said to be *homotopic* if they can be continuously deformed one into the other. Clearly, the two paths illustrated in Fig. 2.4b are homotopic. On the other hand, the paths a and b of Fig. 2.2 are not so, although they are coterminous. That they cannot be continuously deformed one into the other follows from the following fact. Imagine we stop short of reaching s by an infinitesimal:

this prevents us from ever reaching n. We can of course retrace our steps and go back to the origin and even reach p again, but we cannot possibly do this by keeping always at an infinitesimal distance of the original path through the whole of its length. It must be clearly understood that although by repeated continuous deformation we can distort a path by a great deal, we must do this in such a way that at each stage the two paths that are deformed differ infinitesimally through the whole of their lengths. Such a process is represented in Fig. 1, where the black and white arrows denote two homotopic paths.

It can be shown that homotopy is an equivalence relation, i.e. if two paths are homotopic to a third one, they are homotopic amongst themselves. This allows the classification of all coterminous paths in sets that contain only paths that are mutually homotopic. Such sets are called *classes of homotopy*. The parametric space of SO(3) has the remarkably simple property that there are only two classes of homotopy. As we shall see, it is possible to describe one class as the class of paths with no jump and we shall call this the *class* 0. The other will be called *class* 1 and corresponds to paths with one jump only.

The reason for this situation is the following result. It is always possible to deform a path continuously (i.e. keeping within the same class of homotopy) in such a way as to remove two jumps, but it is not possible to do so removing only one jump. Assuming this result to be true for the moment, it is clear that all paths with an even number of jumps are homotopic to a path with no jumps and therefore belong to the same class (class 0). On the other hand, all paths with an odd number of jumps are homotopic to a path with one jump only and belong to the same class (class 1).

We shall now demonstrate by two different methods that two jumps can be continuously removed but that this cannot be done for only one jump. The first method is based on a construction given by Wigner (1959)

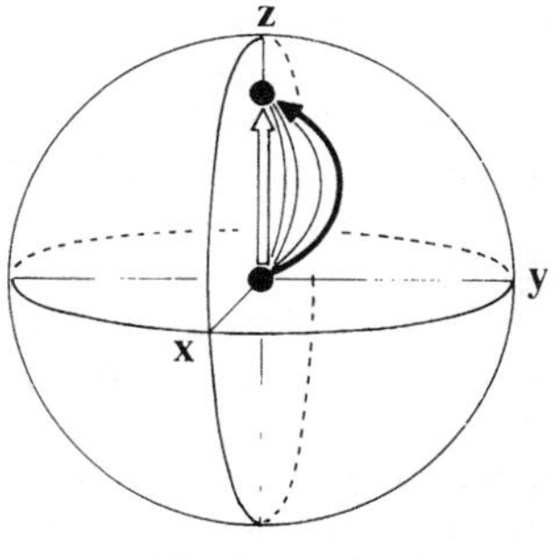

Fig. 10-4.1. Homotopy. The figure depicts the parametric ball.

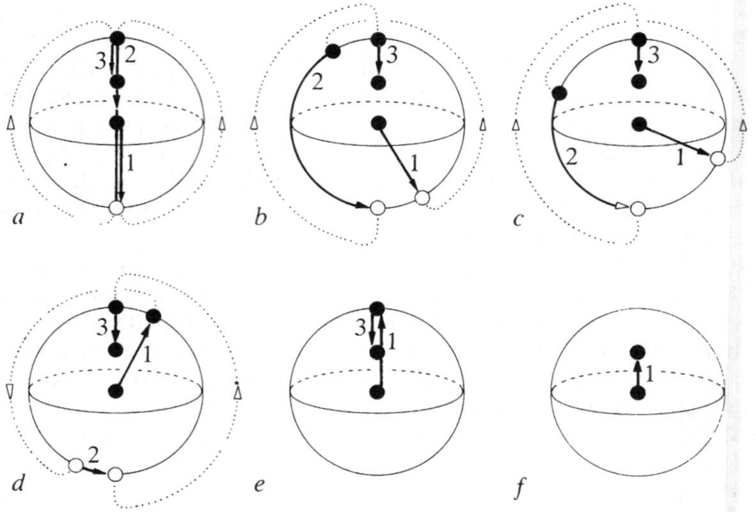

Fig. 10-4.2. Removal of two jumps. All six figures depict the parametric ball.

and it is illustrated in Fig. 2, in which we start with a path with two jumps in Fig. 2*a* which is then deformed continuously until, in Fig. 2*f*, no jump remains. In all cases the path starts at the origin and ends at the rotation $+\frac{1}{2}\pi$. In Fig. 2*a*, however, we proceed from the origin negatively (leg 1) until we reach the rotation $-\pi$. We jump from there to the podal point $+\pi$ and continue to rotate negatively, in leg 2, this time by -2π, until $-\pi$ is reached again. We then jump once more and start leg 3, which we follow until $+\frac{1}{2}\pi$ is reached. In Fig. 2*b* we deform infinitesimally leg 1, and we then jump to the podal point to start leg 2, which is arranged to end at $-\pi$ as before. Leg 2 can be effected as in Fig. 2.3*b*, by repeated rotation around the $\mathbf{x}$ axis until the point $-\pi$ is reached. (Strictly speaking, this leg should lie infinitesimally near the leg 2 shown in Fig. 2*a* but, having started this way, we could deform leg 2 in Fig. 2*b* until it becomes the chord of the curved arrow in that figure, in which case it will only differ from the previous leg 2 by an infinitesimal.) We repeat in Fig. 2*c* the same process and the leg 2 becomes smaller, until in Fig. 2*d* it is infinitesimal and finally disappears altogether in Fig. 2*e*. We then cut short the leg 1 until it reaches $\pi-\delta$ instead of π, and start leg 3 at $\pi-\delta$ so that, by repeated diminution by δ the leg 3 disappears and we reach Fig. 2*d*, having removed two jumps from the coterminous path in Fig. 2*a*.

We show in Fig. 3 that a single jump cannot be removed. All three paths in the figure are coterminous with the path in Fig. 2, but entail a

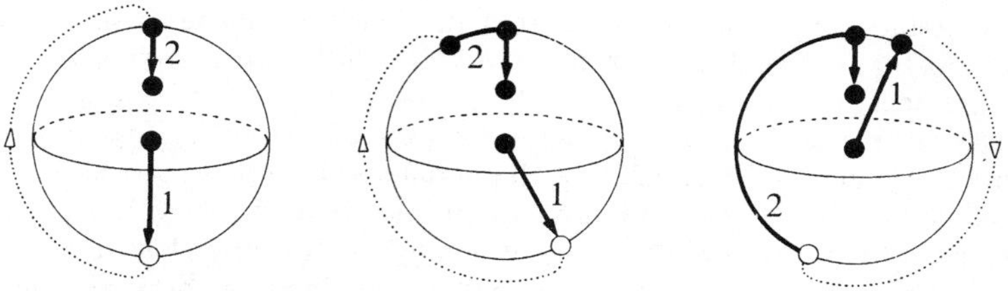

Fig. 10-4.3. Impossibility of removal of a single jump. All three figures depict the parametric ball.

single jump. It can be seen in this case that there is no way in which leg 2 can be removed: on the contrary, the more we deform leg 1, the longer leg 2 becomes.

The same result will now be demonstrated in a somewhat more topological way. Figure 4 has been obtained from the parametric ball by considering the definition of its positive hemisphere given in Fig. **9**-1.1, and by using the following prescription. Cut across the surface of the sphere along the line *ABC* and peel off the surface of the top (positive) hemisphere without upsetting the line *AB′C*. As the top hemisphere is peeled off, it is inverted like a glove and it is then folded back until it covers the lower (negative) hemisphere. A podal point, in general, will not cover its antipodal point, so that the jump between the two is done as follows. The path must jump from the antipodal point on the lower inner surface (negative hemisphere) to the outer sheet, and then move on it until it reaches the podal point. This latter part of the path is irrelevant when ascertaining the homotopy class of the path, since it can continuously be reduced to zero. Thus, all we have to consider from the point of

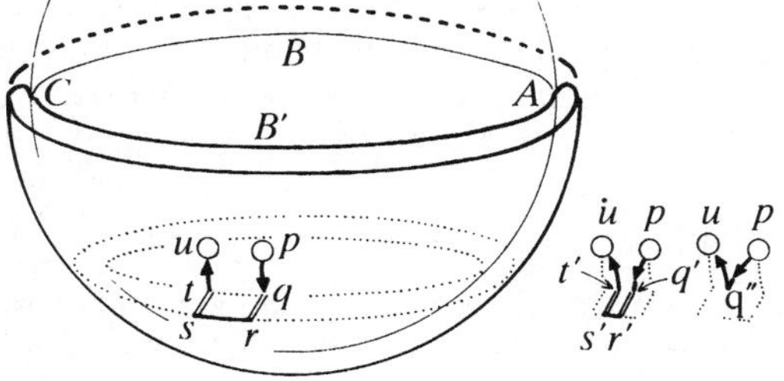

Fig. 10-4.4. Removal of two jumps in the parametric ball.

view of the jump is the passage from the inner to the outer sheet. We show in Fig. 4 a path from p to u with two jumps qr and st, shown in double lines. In the two insets we show how q and t can be continuously moved nearer to q' and t' and, finally, how the two jumps disappear when these two points collapse. Clearly, the initial and final paths are coterminous and they have been continuously deformed one into the other, so that they must belong to the same class of homotopy. It is also clear that there is no way in which a single jump could be removed. A third, and more rigorous, proof of this result can be found in Pontryagin (1966), p. 370.

5. THE PROJECTIVE FACTORS

We are now in a position to fix uniquely the sign of the rotation matrices by imposing the very important condition of continuity in the parameter space. We now consider again in more detail the programme that we sketched in §3. First of all, we shall see how it is that we must make a distinction between a product of two operations g_i, g_j and their result $g_i g_j = g_k$. This is so because the path entailed when obtaining the transformation g_k will, in general, be different from the path that arises when g_j is first effected and it is then followed by g_i. Thus, the symbol g_i, g_j which we introduced in **8**-2.3 as distinct from $g_i g_j$ acquires a clear and precise meaning: the symbol g_i, g_j is used to denote the path from the identity to $g_i g_j$ when g_i is performed after g_j. We shall construct a projective representation of SO(3),

$$\check{G}(g_i)\check{G}(g_j) = [g_i, g_j]\check{G}(g_i g_j), \tag{1}$$

with the following rule to choose the projective factor:

$$[g_i, g_j] = +1, \qquad \text{if path } g_i, g_j \text{ is of class 0,} \tag{2}$$

$$= -1, \qquad \text{if path } g_i, g_j \text{ is of class 1.} \tag{3}$$

This rule ensures continuity in the matrix representation. Consider a path g_i', g_j' coterminous with g_i, g_j (this means that the product $g_i' g_j'$, just as $g_i g_j$, must be g_k). Assume also that g_i', g_j' belongs to the same homotopy class as g_i, g_j. It is clear that, on continuous deformation of g_i', g_j' until it coincides with g_i, g_j the sign of the matrix on (R1), as given by the projective factor, never changes.

6. OPERATIONS, TURNS, AND CONNECTIVITY

We must refine our discussion a little. Given a group of rotations g_i, we define some path of class 0 associated with each rotation g_i, (the *standard path*). Thus, if $g_i g_j$ equals g_k, a path of class 0 is always associated with g_k and therefore with its identical symbol $g_i g_j$. When g_i, however, is performed after g_j, there is no guarantee that the final path, which we now denote with g_i, g_j, be of class 0, as illustrated, e.g., in Fig. 2.2. We can now understand the distinction we made in §**8**-2 between $g_i g_j \varphi(\mathbf{r})$ and $g_i, g_j \varphi(\mathbf{r})$. (See eqn **8**-2.4.) If we call g_k the product $g_i g_j$, the first symbol means the transform of $\varphi(\mathbf{r})$ by g_k through a path of class 0, whereas the second symbol denotes the transform of $\varphi(\mathbf{r})$ by g_k through a defined path which need not necessarily be class 0.

We can also see that a rotation g_k, as a linear transformation, entails no history (i.e. it is fully specified by the initial and final positions of the unit sphere), but when it is realized as a product $g_i g_j = g_k$, the precise details of how this product is effected, that is, of its path g_i, g_j are relevant insofar as they determine a projective factor. Whereas the operations themselves entail no history, their paths in parameter space, which record the multiplication rules of the group, do so and, as one would expect, this is reflected through the phase factors that affect the transformed functions and which appear as projective factors.

One important property of SO(3), to be denoted for simplicity as G, with elements g, now emerges. Every operation g must be mapped to a unique parametric point p_g in the parameter space of the group, chosen so that the path from the identity to p_g be of class 0. When the operation g, however, is obtained as a result of products of other operations of G, the same point p_g must be reached from the identity, but this might now be done through a path of class 1. Thus, if we consider a closed path (loop) that starts and ends at p_g, it is clear that such loops can also be classified as class 0 or class 1. In the first case, the loop can be continuously deformed to nothing, i.e. it can continuously be contracted to the single point p_g itself, whereas this is not so in the second case. A (topological) space in which all loops can be continuously deformed into a single point of the space is called a *singly connected space*. In this case, it is clear that all loops passing through a given point must belong to the same class of homotopy. If, instead, they belong to n district classes of homotopy, the space is said to be *n-fold connected*. (See Roman 1975, vol I, p. 242 for a simple discussion.) Thus, SO(3) is a *doubly connected group*.

Finally, it is worthwhile formalizing the terminology about the so-called rotations by 2π. As an operation – that is, as a linear transformation – a rotation by 2π is indistinguishable from the identity. On the other hand, when multiplying two or more rotations (such as two rotations by π around the same axis), the identity may be reached through a path in which the accumulated angle of rotation adds up to 2π. This path is a path of class 1 because it must entail one jump (see Fig. **11**-5.1), and thus entails a negative phase factor associated with the identity. The expression 'rotation by 2π', often used to describe this situation, should be referred to in our system as the identity realized through a path of class 1, and, as a shorthand, I shall call this a *turn by* 2π. This will help to dispel confusion as regards the meaning of the word 'rotation'.

The idea of a turn by 2π is handy in order to provide an intuitive geometrical meaning to the different classes of homotopy. Thus, the two non-homotopic paths illustrated for $R(\frac{1}{2}\pi\mathbf{z})$ in Fig. 2.2*a*,*b*, can be said to differ by a 2π turn. In fact, if we consider the rotations of the unit sphere, depicted in *c* and *d* of that figure, as given by products of infinitesimal rotations, it is clear that, in order to go from *d* to *c* one must, starting from the point **y**, rotate positively until the accumulated angles of rotation add up to 2π. In the same manner, all rotations of SO(3) can be realized in two ways, one through a class 0 path from the identity (the *standard path*), and the other through the standard path to which a 2π turn is added (thus producing a class 1 path from the identity). A turn by 4π, on the other hand, entails two jumps – that is a path of class 0 – and is topologically identical to the identity.

Our major task is now completed and all that remains is to provide a scheme in order to establish in practical cases the class of homotopy of a path: this will be done in the next chapter.

11

THE SPINOR REPRESENTATIONS

Nella vita del signor Palomar c'è stata un'epoca in cui la sua regola era questa: primo, costruire nella sua mente un modello, il piú perfetto, logico, geometrico possibile; . . .

Italo Calvino (1983). *Palomar.* Einaudi, Torino, p. 110.

There was a time in Mr Palomar's life when his guiding rule was this, first to build in his mind a model, the most perfect, logical geometrical model possible; . . .

The first problem which we shall consider in this chapter is the determination of the class of homotopy of a given path in terms of the product of two finite rotations. The projective representations corresponding to the factor system so defined will be called *spinor representations* for reasons which will be discussed in §8. Before we deal with these representations for SO(3) and O(3) we shall be concerned with point groups and we shall investigate some of the properties of their spinor representations. We have seen, in particular, that characters are not class functions for projective representations but we shall be able to show that, with the factor system chosen, this very useful property is regained for the spinor representations. In doing this work we shall at first continue to use the geometrical procedures introduced in Chapters **9** and **10**, but we shall later on in this chapter replace them by an algebraic method based on the Euler–Rodrigues parametrization of rotations.

1. DETERMINATION OF THE PROJECTIVE FACTORS

The determination of the class of homotopy of a path, which we need in order to determine the projective factors from (**10**-5.2) and (**10**-5.3), is very simply done by means of the Euler construction. We illustrate this work in Fig. 1, where we consider the products $C_{2z}C_{2x}$ and $C_{2x}C_{2z}$ and their paths. We know from Table **8**-1.3 that these operations commute and that the result of both products is C_{2y}. We first fix the pole of C_{2x} in Fig. 1a making sure that it is in $\mathfrak{H}$ as defined in Fig. **9**-1.1, after which we trace a path in parameter space (Fig. 1c) along the $\mathbf{x}$ axis from the origin until π is reached (parametric point labelled C_{2x}). We then return to Fig. 1a and start performing infinitesimal rotations around $\mathbf{z}$, of which the first

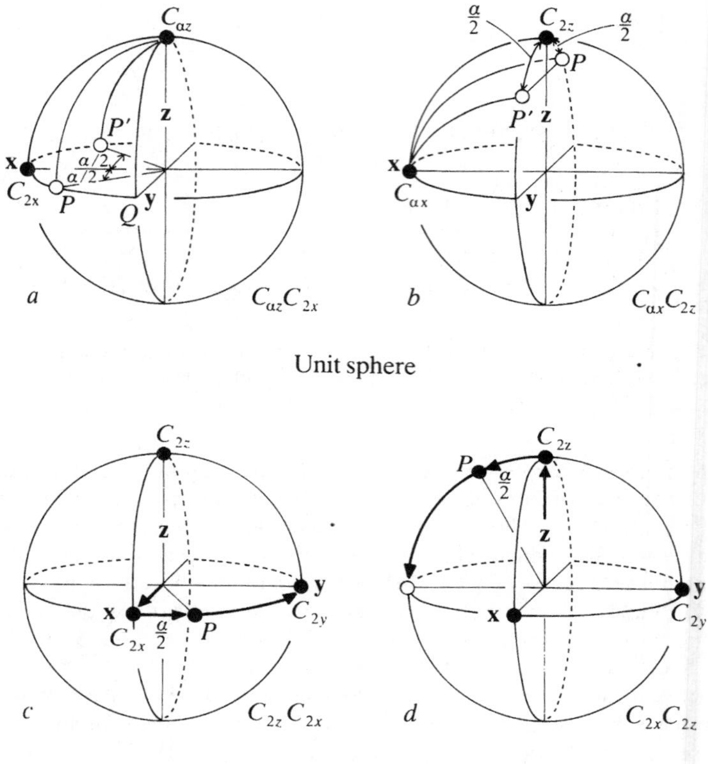

Parametric ball

Fig. 11-1.1. Determination of projective factors from the Euler construction. Notice the change of orientation of the **x**, **y** axes in the unit sphere with respect to the parametric ball. This choice has been made in order to facilitate comparison with Fig. 2.1. Verify that P in (a) is the pole of the product rotation because it is left invariant by it. The choice of $\mathfrak{H}$ in the parametric ball is that depicted in Fig. 9-1.1. Parametric points that belong to $\mathfrak{H}$ and $\bar{\mathfrak{H}}$ are shown with full and open circles, respectively.

one is labelled $C_{\alpha z}$, by an infinitesimal angle α. We follow for this the Euler prescription, sweeping the great circle through $C_{\alpha z}$ and C_{2x} by $\pm\frac{1}{2}\alpha$ around **z** and then by $\pm\frac{1}{2}\pi$ around **x**. The pole of the resulting rotation is P and the rotation angle is twice the angle $C_{\alpha z}PQ$, that is, π. From Fig. 1a we can now fix on the parametric ball in Fig. 1c the parametric point P of the rotation $C_{\alpha z}C_{2x}$. As more rotations around **z** are performed until the angle π is reached, the parametric point moves away from C_{2x} by $\frac{1}{2}\pi$ and reaches the point marked, which is C_{2y} and still in $\mathfrak{H}$ (see Fig. **9**-1.1). Thus, the path from the identity to the parametric point of $C_{2z}C_{2x}$ (path C_{2z}, C_{2x}) has no jumps and it is therefore of class 0, so that from (**10**-5.2),

$$[C_{2z}, C_{2x}] = +1. \tag{1}$$

The same procedure is followed exactly in Fig. 1*b* and *d* for the product $C_{2x}C_{2z}$ (path C_{2x}, C_{2z}) and although the result is still C_{2y} as we must expect, the parametric point is now the antipodal point of C_{2y} and it therefore belongs to $\bar{\mathfrak{H}}$; hence the path entailed is of class 1. Thus,

$$[C_{2x}, C_{2z}] = -1. \tag{2}$$

The reader should notice that (1) and (2) coincide with their values in Table **8**-1.3 – which is, of course, the result of foresight in the choice of consistent conventions. Although the Euler construction is very simple, it can become somewhat tedious for a large point group and we shall later on give an algebraic method to get the projective factors. The geometry of the method described here, nevertheless, must remain as the fundamental model from which the correct conventions can be formulated in order to make the algebraic methods consistent.

Problems 11-1 (p. 292)

1. Show that for any binary rotation C_2,

$$[C_2, C_2] = -1. \tag{3}$$

(Compare this result with the corresponding factors shown for C_{2x}, C_{2y} and C_{2z} in Table **8**-1.3.)

2. Determine $[C_{2y}, C_{2x}]$ by the method of Fig. 1 and compare with Table **8**-1.3.

2. THE INTERTWINING THEOREM

We shall discuss here properties of some projective factors that will be important in §3 in the study of the characters of the spinor representations. For simplicity of notation we shall denote SO(3) with g and the operations g, g_i and g_j will all belong to it. We shall then say that g_i and g_j are *intertwined* by g if

$$g g_i = g_j g. \tag{1}$$

If g_i and g_j are identical, (1) is a commutation relation. If this is not so, then g_j is the conjugate of g_i by g:

1

$$g_j = g g_i g^{-1} =_{\text{def}} g_i^g. \tag{2}$$

Thus, the advantage of the concept of intertwining is that it encompasses both commutation and conjugation. From (1) and (2), the intertwining relation can be written as

$$gg_i = g_i^g g, \tag{3}$$

and it entails commutation when g_i^g equals g_i. We shall now see that the projective factors of the products on either side of (3) have remarkably simple relations. We must consider for this purpose the only three cases by which two rotations g and g_i can be related: they can be either general, in which case they do not commute, or coaxial, which commute, or bilateral binary (henceforth called BB), which also commute. (See Problems **2**-1.8, **2**-2.3, **9**-4.6.)

Theorem 1. *Intertwining*

For two general (non-commuting) rotations g_i and g,

$$[g, g_i] = [g_i^g, g], \qquad \text{(general).} \tag{4}$$

The proof is done by means of Fig. 1, in which g is a rotation by α and g_i a rotation by β. In Fig. 1*a* we use the Euler construction to find that the pole of gg_i is *P*, and we deal in the same manner in Fig. 1*b* with the product $g_i^g g$. The first thing we must do in this figure is to fix the pole of g_i^g which is done as in (**9**-1.5) by rotating the pole of g_i by the rotation g (by α). Because the pole of g_i^g is the mirror image of g_i with respect to the great circle through g and *P*, it follows at once that the pole of $g_i^g g$ is identical with *P*. In a way, this is not surprising, since both sides of (3) are identical, but what is important is that in no way Fig. 1*b* could lead to the

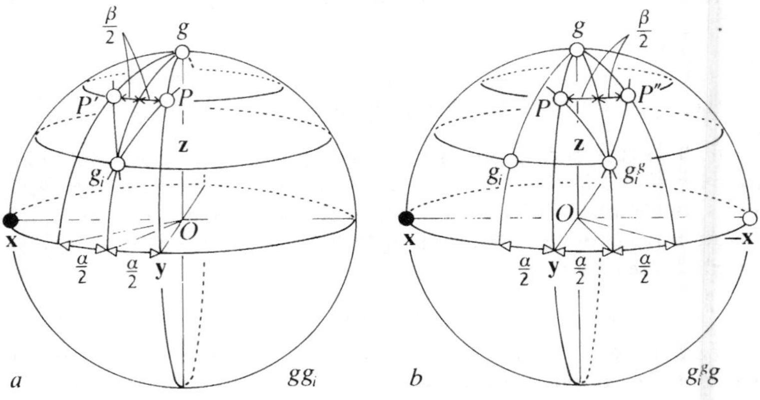

Fig. 11-2.1. The intertwining theorem. Compare (*a*) with Fig. **9**-2.1.

antipole of P, which verifies that the two projective factors in (4) must have the same sign.

We shall now consider the two cases when g_i and g commute, i.e. when g_i^g coincides with g_i:

3 $$gg_i = g_i g. \tag{5}$$

When g and g_i are coaxial it is pretty obvious from Fig. **10**-2.1 that the two paths entailed on both sides of (5) must be identical, whence

$$[g, g_i] = [g_i, g], \qquad \text{(coaxial)}. \tag{6}$$

That is, (4) is still valid in this case. When g and g_i are BB, however, a pathological event takes place. We shall leave g in Fig. 1 on the $\mathbf{z}$ axis and, in order to have g_i normal to g, we place it at the point marked $\mathbf{x}$. That is, we take g and g_i as C_{2z} and C_{2x} respectively. We now go to Fig. 1*b* and fix first of all the pole of g_i^g. Clearly, since g is now a rotation by π, it takes the pole of g_i at $\underline{\mathbf{x}}$ in Fig. 1*a* to the point marked $-\mathbf{x}$ in Fig. 1*b*. This point, however, is in $\bar{\mathfrak{H}}$ and we know that this is a binary rotation which cannot have a pole on the negative sphere. Thus, we must take back this pole of g_i^g (which is C_{2x}) to its podal point at $\mathbf{x}$. (It would have been sufficient, of course, to recognize that g_i^g is identical with g_i, that is, with C_{2x}, to place its pole correctly at $\mathbf{x}$. Remember that g_i^g stands for the single operation that is its result.) If we look back at Fig. 1.1, parts *a* and *b* of that figure coincide exactly with the situation now obtaining in Fig. 1, and we saw in that figure that the two projective factors differ in sign:

$$[g, g_i] = -[g_i, g], \qquad \text{(BB)}. \tag{7}$$

It is now clear that BB's stand out as exceptional with respect to all other rotations (i.e. general non-commuting, and coaxial). For this reason general (non-commuting) and coaxial rotations are called *regular rotations* and BB rotations are called *irregular*. The results obtained can thus be summarized in the following

Theorem 2. *Intertwining for regular and irregular operations*

For all regular operations

$$[g, g_i] = [g_i^g, g] \tag{8}$$

whereas for all irregular operations

$$[g, g_i] = -[g_i^g, g] = -[g_i, g]. \tag{9}$$

(We use here the fact that for BB's, which commute, g_i^g equals g_i.)

A more detailed discussion of these theorems can be found in Altmann (1979) and in §7.

3. THE CHARACTER THEOREM

The characters of projective representations pose awkward problems in general, since, not being class functions, they are not of much use. The situation, however, is far more satisfactory for spinor representations for which, as we shall see, characters behave very much like those of vector representations. We saw in (**8**-4.2) that, for projective representations, the following relation between the characters of two operations in the same class obtains:

$$\chi(g_i^g \mid \check{G}) = [g_i^g, g][g, g_i]^{-1}\chi(g_i \mid \check{G}), \qquad \forall g \in G. \tag{1}$$

From Theorem 2 of §2, the product of the projective factors on (R1) is unity for all regular operations. Thus, for all regular operations, the characters are class functions:

$$\chi(g_i^g \mid \check{G}) = \chi(g_i \mid \check{G}). \tag{2}$$

For irregular operations (BBs) we know two things. First, g_i^g equals g_i and the product of the two projective factors in (1), from (2.9), is -1. Therefore, from (1),

$$\chi(g_i \mid \check{G}) = -\chi(g_i \mid \check{G}) \qquad \Rightarrow \qquad \chi(g_i \mid \check{G}) = 0, \tag{3}$$

which means that even for the irregular operations the characters are still class functions. We can summarize these results in the following theorem.

Theorem 1. The characters of spinor representations are class functions which vanish over the irregular classes (classes of BB operations). See §7 for necessary and sufficient conditions for this result.

4. THE IRREDUCIBLE REPRESENTATIONS

As stated in §**8**-4, there is no general rule to give the total number of irreducible projective representations for any given factor system. For spinor representations, however, there is a very simple theorem, which was first given in a slightly different form by Brown (1970). (See Altmann 1979.)

Theorem 1. The number of irreducible spinor representations of a point group equals the number of regular classes in the group (that is, classes of regular elements or, which is the same, all classes that do not contain BBs).

The proof of this result is based on the fact that irreducible unitary projective representations satisfy precisely the same orthogonality relations for the characters as the vector representations. These are (Jansen and Boon 1967, p. 111):

$$\sum_{g} \chi(g \mid^{i} \check{G})^{*} \chi(g \mid^{j} \check{G}) = |G| \, \delta_{ij}. \tag{1}$$

$$\sum_{i}^{|I|} \chi(g_m \mid^{i} \check{G})^{*} \chi(g_n \mid^{i} \check{G}) = |G| \, |C(g_m)|^{-1} \, \delta_{mn}. \tag{2}$$

The first summation here is over all elements g of G and entails products of characters of the same element g in two different irreducible representations $^{i}\check{G}$ and $^{j}\check{G}$. $|G|$ on (R1) is the order of the group. The second summation is over all irreducible spinor representations of the group, $|I|$ in number, and entails products of characters of two different group elements g_m and g_n in the same irreducible representation $^{i}\check{G}$. $|C(g_m)|$ on (R2) is the total number of elements in the class $C(g_m)$ to which g_m belongs.

The summation over g in (1) can be taken over elements belonging to the regular classes only, since all other characters vanish. Also, the character of all elements in the same regular class is the same. Thus, (L1) defines orthogonality relations for vectors of dimension $|R|$, equal to the number of regular classes, and the number of such vectors is the same as that of the irreducible representations, $|I|$. Thus, $|I| \leqslant |R|$. In (L2), likewise, g_m and g_n need be taken over elements of the regular classes only, whence (L2) defines orthogonality relations for $|R|$ vectors of dimension $|I|$, so that $|R| \leqslant |I|$. It follows at once from the two inequalities obtained that $|I| = |R|$, as stated by the theorem.

We know from §**8**-4 that the sum of the square of the dimensionality of all $|I|$ irreducible spinor representations of the group must equal its order, a result which combined with Theorem 1 permits the calculation of the dimensionalities of the representations. It should also be noticed that, just as for vector representations, eqn (1) provides a very handy criterion for the irreducibility of spinor representations. The results of this section, combined with the character theorem of §3, make the work with the spinor representations just as easy as with vector representations.

We shall now move on from the geometrical and topological definition of the projective factors to one in which they are entirely determined algebraically from the Euler–Rodrigues parametrization of rotations.

5. THE PROJECTIVE FACTORS FROM THE EULER–RODRIGUES PARAMETERS

We shall be concerned here with a proper point group G. The first thing we must do is to choose a positive hemisphere $\mathfrak{H}$ (more about this choice in §6 and §**15**-1), which shall be used both for the unit sphere and for the parametric ball. We shall say that a vector $\mathbf{n}$ in either object belongs to $\mathfrak{H}$

whenever it belongs to that region of the ball of which $\mathfrak{H}$ is the surface. We then choose all the rotation angles ϕ in the range $-\pi<\phi\leqslant\pi$ and take $\mathbf{n}\in\mathfrak{H}$ for all positive rotations and $\mathbf{n}\in\bar{\mathfrak{H}}$ for all negative rotations. Each operation $g\in G$ will have a unique parametric point and the parametric points so defined shall be called *standard parametric points.* Because the sign of the rotation is now given by whether $\mathbf{n}$ belongs to $\mathfrak{H}$ or $\bar{\mathfrak{H}}$, ϕ for the standard parametric points is always in the range $0\leqslant\phi\leqslant\pi$. All standard parametric points are either in the parametric ball or on $\mathfrak{H}$. This is so because the only points on the surface of the parametric ball (radius π) are binary rotations for which ϕ, by our definition of its range, $-\pi<\phi\leqslant\pi$, must always be positive, $-\pi$ not being admitted in this range. It is clear that all standard parametric points are reached from the identity by a path of class 0.

Once the standard parametric points are defined, a corresponding set of *standard Euler–Rodrigues parameters* is constructed. Because ϕ for them is in the range $0\leqslant\phi\leqslant\pi$, $\lambda=\cos\frac{1}{2}\phi$ is such that $\lambda\geqslant 0$ and $\sin\frac{1}{2}\phi$ is always positive, whence $\mathbf{\Lambda}=\sin\frac{1}{2}\phi\ \mathbf{n}$ will belong to $\mathfrak{H}$ or $\bar{\mathfrak{H}}$ for positive and negative rotations respectively. When $\lambda=0$ (binary rotations), $\mathbf{\Lambda}\in\mathfrak{H}$ for the reasons explained above. Thus the standard Euler–Rodrigues parameters of an operation g have the following properties:

$$\textbf{either}\quad \lambda_g>0 \qquad\qquad \textbf{or} \qquad\qquad \lambda_g=0 \quad\text{and}\quad \mathbf{\Lambda}_g\in\mathfrak{H}. \tag{1}$$

We must now realize that, given a rotation g of SO(3) with standard parameters $R(\lambda_g;\mathbf{\Lambda}_g)$ and standard parametric point p_g, then $R(-\lambda_g;-\mathbf{\Lambda}_g)$ corresponds to the identical parametric point p_g, but p_g is now reached from the identity through a path of class 1. The proof of this assertion follows at once from the fact that $R(-\lambda_g;-\mathbf{\Lambda}_g)$ corresponds to the same values of ϕ and $\mathbf{n}$ as $R(\lambda_g;\mathbf{\Lambda}_g)$ for $\lambda=\cos\frac{1}{2}\phi$, $\mathbf{\Lambda}=\sin\frac{1}{2}\phi\ \mathbf{n}$, (see eqn **9**-4.8), except that an angle 2π must be added to ϕ. This entails a turn by 2π which changes the class of the path. For binary rotations, $R(0;-\mathbf{\Lambda}_g)$ will correspond to the antipodal point of p_g but the two parametric points involved are identical, except that the antipodal point entails a path of class 1 from the identity, again in agreement with the previous assertion.

When multiplying two operations belonging to g we use the multiplication rule (**9**-4.7):

$$R(\lambda_1;\mathbf{\Lambda}_1)R(\lambda_2;\mathbf{\Lambda}_2)=R(\lambda_1\lambda_2-\mathbf{\Lambda}_1\cdot\mathbf{\Lambda}_2;\lambda_1\mathbf{\Lambda}_2+\lambda_2\mathbf{\Lambda}_1+\mathbf{\Lambda}_1\times\mathbf{\Lambda}_2). \tag{2}$$

In this equation, the parameters of the two operations on the left must satisfy (1) since each $g\in G$ is uniquely given by a pair λ, $\mathbf{\Lambda}$ in the standard set. The result on (R2), however, may not belong to the standard set. If $R(\lambda;\mathbf{\Lambda})$ on (R2) satisfies (1), the path entailed by (L2) is class 0, but if it does not, this operation must appear parametrized as $R(-\lambda;-\mathbf{\Lambda})$ and the path must be class 1. Thus, if we write (2) as

$$R(\lambda_1;\mathbf{\Lambda}_1)R(\lambda_2;\mathbf{\Lambda}_2)=R(\lambda_3;\mathbf{\Lambda}_3), \tag{3}$$

then,

$$[R(\lambda_1;\mathbf{\Lambda}_1), R(\lambda_2;\mathbf{\Lambda}_2)]=1, \tag{4}$$

if

$$\textbf{either}\quad \lambda_3>0 \qquad \textbf{or} \qquad \lambda_3=0 \text{ and } \mathbf{\Lambda}_3\in\mathfrak{H}. \tag{5}$$

On the other hand,

$$[R(\lambda_1;\mathbf{\Lambda}_1)R(\lambda_2;\mathbf{\Lambda}_2)]=-1, \tag{6}$$

if

$$\textbf{either}\quad \lambda_3<0 \qquad \textbf{or} \qquad \lambda_3=0 \text{ and } \mathbf{\Lambda}_3\in\bar{\mathfrak{H}}. \tag{7}$$

This ability of the Euler–Rodrigues parameters to recognize the homotopy class of a path is unique to them and it makes them immensely superior to the Euler angles. We shall now give a few examples of how this works.

We shall first show that the choice of the set of standard parameters for the operations of G entails that the corresponding set of projective factors is standardized, i.e. that $[e, g]$ is unity for all $g\in G$. (See § **8**-3). This is so because the standard parameters for the identity are $R(1;\mathbf{0})$ whence, if the standard parameters of g are $R(\lambda, \mathbf{\Lambda})$, then, from (2),

$$R(1;\mathbf{0})R(\lambda,\mathbf{\Lambda})=R(\lambda;\mathbf{\Lambda}). \tag{8}$$

Since the parameters on (R8) are standardized, they satisfy (1) and therefore (5), so that the projective factor is unity, as asserted.

Consider now the projective factor for any binary rotation C_2 with itself, $[C_2, C_2]$. C_2 is given by the standard parameters $R(0,\mathbf{\Lambda})$ with $|\mathbf{\Lambda}|=1$ and $\mathbf{\Lambda}\in\mathfrak{H}$. Thus, (R2) is $R(-1;\mathbf{0})$ which is the negative identity $R(1;\mathbf{0})$ and clearly satisfies (7). Hence,

$$[C_2, C_2]=-1, \forall C_2. \tag{9}$$

(See Fig. 1a and Table **8**-1.3.) Consider any C_3 which for convenience we shall take along the **z** axis,

$$C_3^+\colon \lambda=\tfrac{1}{2},\ \mathbf{\Lambda}=\tfrac{\sqrt{3}}{2}(001); \qquad C_3^-\colon \lambda'=\tfrac{1}{2},\ \mathbf{\Lambda}'=\tfrac{\sqrt{3}}{2}(00\bar{1}). \tag{10}$$

We know, of course, that $C_3^+C_3^+$ equals C_3^-. On writing this product as in (3), we have from (2),

$$\lambda_3=\tfrac{1}{4}-\tfrac{3}{4}=-\tfrac{1}{2}, \qquad \mathbf{\Lambda}_3=\tfrac{1}{2}\mathbf{\Lambda}+\tfrac{1}{2}\mathbf{\Lambda}=\mathbf{\Lambda}, \tag{11}$$

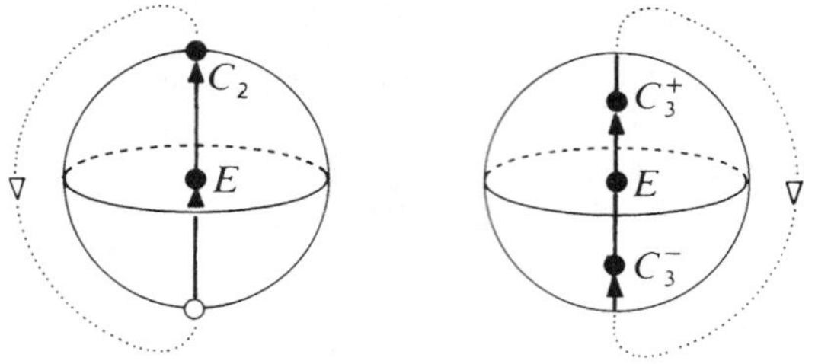

a $[C_2, C_2] = -1$

b $[C_3^+, C_3^+] = -1$

Fig. 11-5.1. Projective factors for C_2, C_2, and C_3^+, C_3^+.

which are the non-standard parameters of C_3^-. Thus:

$$[C_3^+, C_3^+] = -1, \tag{12}$$

as illustrated in Fig. 1*b*. Because the parameters of C_3^- in (11) are the negative of the standard ones (10) for that operation, an extra turn by 2π is added to the path.

Let us now try this method to get $[C_{2z}, C_{2x}]$ and $[C_{2x}, C_{2z}]$ which are $+1$ and -1 respectively from eqns (1.1) and (1.2) derived geometrically from Fig. 1.1. The standard parameters for these operations are

$$C_{2z} = R(0; \mathbf{z}), \qquad C_{2x} = R(0; \mathbf{x}), \tag{13}$$

whence

13|2 $$C_{2z}C_{2x} = R(0; \mathbf{z})R(0; \mathbf{x}) = R(0; \mathbf{z}\times\mathbf{x}) = R(0; \mathbf{y}), \tag{14}$$

13|2 $$C_{2x}C_{2z} = R(0; \mathbf{x})R(0; \mathbf{z}) = R(0; \mathbf{x}\times\mathbf{z}) = R(0; -\mathbf{y}). \tag{15}$$

If one had a rather classical view of the rotations, one would argue that the difference in the sign of $\mathbf{y}$ in (14) and (15) does not matter and that both sets of parameters give the same operation C_{2y}. This is so, but whereas in (14) we obtain a pole on $\mathfrak{H}$, in (15) the pole is on $\bar{\mathfrak{H}}$; that is, we do get the same operation but effected through paths of different homotopy. Equations (14) and (15), with (5) and (7), give $+1$ and -1 for the corresponding projective factors, just as obtained geometrically before.

Problem 11-5 (p. 292)

1. Determine the factor system for $\mathbf{D}_2$ in Fig. **2**-5.1 by the methods given in this section, on using $\mathfrak{H}$ as defined in Fig. **9**-1.1, and compare with Table **8**-1.3.

6. INVERSES AND CONJUGATES IN THE EULER–RODRIGUES PARAMETRIZATION

The inverse g^{-1} of g is defined by the condition $gg^{-1}=E$, which will allow us to get the parameters of g^{-1} in terms of those of g:

5.2 $$R(\lambda_g;\mathbf{\Lambda}_g)R(\lambda_{g^{-1}};\mathbf{\Lambda}_{g^{-1}})=$$

$$=R(\lambda_g\lambda_{g^{-1}}-\mathbf{\Lambda}_g\cdot\mathbf{\Lambda}_{g^{-1}};\lambda_g\mathbf{\Lambda}_{g^{-1}}+\lambda_{g^{-1}}\mathbf{\Lambda}_g+\mathbf{\Lambda}_g\times\mathbf{\Lambda}_{g^{-1}}) \quad (1)$$

$$=R(1;\mathbf{0}). \quad (2)$$

In order to satisfy (2) we must take, on using the condition $\lambda^2+\mathbf{\Lambda}^2=1$,

$$\lambda_{g^{-1}}=\lambda_g, \qquad \mathbf{\Lambda}_{g^{-1}}=-\mathbf{\Lambda}_g. \quad (3)$$

If g is not binary, $\lambda_g>0$; so $\lambda_{g^{-1}}>0$ and it is automatically standardized by condition (5.1). For g binary, however, λ_g equals zero, and $\mathbf{\Lambda}_g\in\mathfrak{H}$ (since the parameters of g must be standard), so that $\mathbf{\Lambda}_{g^{-1}}\in\bar{\mathfrak{H}}$. In this case, on the other hand, we know that g coincides with g^{-1} and there is no difficulty in ensuring that there is a correct operation g^{-1} in the group with standard parameters. It is, of course, desirable, when using g^{-1} in forming a conjugate such as gg_ig^{-1}, to have a unique definition of g^{-1}, and fortunately, as well as most importantly, we shall be able to prove that, when (3) is adopted for this purpose for all g (binary or otherwise) then, and only then, the results are consistent with our fundamental geometrical picture of conjugation.

In order to obtain the parameters for the conjugate of g_i by g, g_i^g, we shall use the following notation:

$$g_i^g=gg_ig^{-1}=_{\text{def}}gg'=_{\text{def}}g''; \; g'=_{\text{def}}g_ig^{-1}. \quad (4)$$

3|5.2 $$\lambda_{g'}=\lambda_{g_i}\lambda_g+\mathbf{\Lambda}_{g_i}\cdot\mathbf{\Lambda}_g, \qquad \mathbf{\Lambda}_{g'}=-\lambda_{g_i}\mathbf{\Lambda}_g+\lambda_g\mathbf{\Lambda}_{g_i}-\mathbf{\Lambda}_{g_i}\times\mathbf{\Lambda}_g. \quad (5)$$

Proceeding in the same way for the product gg' one easily finds, after a little simplification,

$$\lambda_{(g_i)^g}=\lambda_{g_i}, \quad (6)$$

$$\mathbf{\Lambda}_{(g_i)^g}=\mathbf{\Lambda}_{g_i}-2\Lambda_g^2\mathbf{\Lambda}_{g_i}+2\lambda_g\mathbf{\Lambda}_g\times\mathbf{\Lambda}_{g_i}+2(\mathbf{\Lambda}_g\cdot\mathbf{\Lambda}_{g_i})\mathbf{\Lambda}_g. \quad (7)$$

The notation here is delicate. Notice that, whereas g_i^g and g'' are one and the same operation, we have used neither symbol as a suffix on (L6) and (L7). The reason is that, since $g''\in G$, $\lambda_{g''}$ and $\mathbf{\Lambda}_{g''}$ (or the same symbols with suffix g_i^g) must be used to denote the *standard* parameters of g'', whereas we do not yet know whether (R6) and (R7) will give precisely these standard parameters. To remind ourselves of this fact, we use the modified suffices shown in (L6) and (L7). Before we discuss this question of the standardization of (6) and (7), the important geometrical meaning of these expressions must be clearly understood. First, we know that conjugation leaves invariant the rotation angle, so that (6) is as expected. Secondly, as regards (7), it can easily be proved (see Problem 1) that it agrees with the parameters of the conjugate pole of g_i by g, which, as was shown below eqn (**9**-1.4), is obtained by acting with g on the pole of g_i: that is, by subjecting the pole of g_i to the conical transformation under the rotation g. We must again refine our notation in order to make this point sufficiently clear. Let us call $\Pi(g_i)$ and $\Pi(g_i^g)$ the poles of g_i and g_i^g (or g''). Because g_i and g'' belong to G, these poles must be standard. Let us call $\Pi(g_i)^g$ the *conjugate pole* of $\Pi(g_i)$ by g; that is, the pole obtained by acting with g on $\Pi(g_i)$. Although we made no distinction in §**9**-1 between $\Pi(g_i^g)$ and $\Pi(g_i)^g$ we must do so now, because whereas the former must be a standard pole, the latter may not be so. It can, in fact, be the antipole of $\Pi(g_i^g)$ as will be later on exemplified.

Looking back at eqns (6) and (7), they must be regarded to pertain, as their notation shows, to $\Pi(g_i)^g$ rather than $\Pi(g_i^g)$ until the standardization of their right-hand sides is established, as we shall now do. Three cases must be considered: (i) g_i is not binary, in which case $\lambda_{g_i} > 0$; hence, from (6), $\lambda_{(g_i)^g} > 0$, and the conjugate pole satisfies (5.1) so that it is standard; (ii) g is binary but g_i and g are not BB – in this case the conjugate pole, from (6), is also binary ($\lambda_{g_i} = 0$) but $\mathbf{\Lambda}$ as given by (7) may or may not belong to $\mathfrak{H}$, as will be illustrated in an example below; (iii) g_i and g are BB – in this case, the conjugate pole of g_i by g can never be standard, as we shall now prove. For BBs, $\lambda_{g_i} = \lambda_g = 0$, $\mathbf{\Lambda}_{g_i}^2 = \mathbf{\Lambda}_g^2 = 1$, and the conditions $\mathbf{\Lambda}_g \in \mathfrak{H}$, $\mathbf{\Lambda}_{g_i} \in \mathfrak{H}$ must be satisfied. Also, because the rotation axes must be

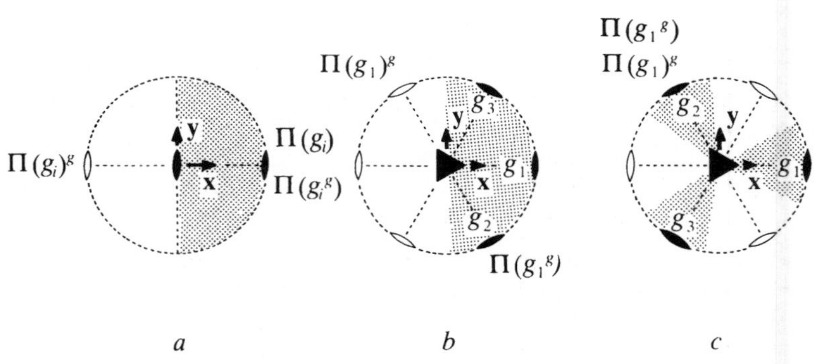

Fig. 11-6.1. Conjugation and the choice of the positive hemisphere. The **z** axis is above the figure. The range of the positive hemisphere on the **x**, **y** plane is shaded. Poles, on $\mathfrak{H}$, and antipoles, on $\bar{\mathfrak{H}}$, are depicted with full and open digons respectively. The operation g is C_{2z} in (*a*) and C_3^+ in (*b*) and (*c*). Its pole is at the head of the positive **z** axis.

perpendicular, $\boldsymbol{\Lambda}_g \cdot \boldsymbol{\Lambda}_{g_i}$ vanishes. Therefore

7 $$\boldsymbol{\Lambda}_{(g_i)^g} = \boldsymbol{\Lambda}_{g_i} - 2\boldsymbol{\Lambda}_{g_i} = -\boldsymbol{\Lambda}_{g_i} \quad \Rightarrow \quad \boldsymbol{\Lambda}_{(g_i)^g} \in \bar{\mathfrak{H}}. \tag{8}$$

This case is illustrated in Fig. 1*a*. Here, g_i and g are C_{2x} and C_{2z} respectively and $g_i^g \equiv g_i$. The operation C_{2z} takes the pole $\Pi(g_i)$ into its antipole, as given by (8). We use in the figure the same choice of $\mathfrak{H}$ as in Fig. **9**-1.1 but it is clear that no choice of $\mathfrak{H}$ could possibly permit both a pole and its antipole to belong to the same positive hemisphere. Notice also that the agreement between (8) and Fig. 1*a* would disappear if the use of (3) for the inverse element of g had not been extended also for the case when g is a binary rotation: we have ensured that the algebraic definition of the conjugate pole in (6) and (7) agrees with its geometrical construction, as further shown in Problem 1.

Conjugation and the choice of the positive hemisphere

We shall now illustrate the case (ii) above, in which the conjugate of a binary rotation g_i is formed by an operation g such that g and g_i are not BB. We shall discuss for this purpose the group $\mathbf{D}_3$ depicted in Fig. 1*b*,*c*, where g_1, g_2, g_3 are binary axes, perpendicular to C_3^+ and C_3^-. Consider the operation g_1^g for g equal to C_3^+. Since the product $C_3^+ \, g_1 \, C_3^-$ equals g_2 (see Problem 3) the pole of g_1^g, $\Pi(g_1^g)$, must coincide with $\Pi(g_2)$. The position of the latter in Fig. 1*b* is dictated by the choice of $\mathfrak{H}$, which was taken in this case to be the connected surface of Fig. **9**-1.1. We have thus fixed the pole of the conjugate of g_1. The conjugate pole of $\Pi(g_1)$, $\Pi(g_1)^g$, on the other hand, is simply obtained by rotating $\Pi(g_1)$ under C_3^+, as shown in the figure. It is clear that $\Pi(g_1)^g$ and $\Pi(g_1^g)$ do not coincide in this case. The reason for this is that although g_2 is the conjugate of g_1 by C_3^+, the choice of $\mathfrak{H}$ precludes putting the standard pole $\Pi(g_2)$ in the site to which $\Pi(g_1)$ is taken under C_3^+. This can be remedied by taking $\mathfrak{H}$ as the disconnected surface illustrated in Fig. 1*c*. (We still assume that $\mathfrak{H}$ is given for $z > 0$, but when z vanishes the curve that defines $\mathfrak{H}$ on the **x**, **y** plane is disconnected, as shown in the figure.) In this case, it is clear that the standard set of poles for g_1, g_2, and g_3, is such that the conjugate poles (i.e. the poles obtained by rotation) coincide with the poles of the conjugate operations. Further discussion of this question can be found in §**15**-1.

Problems 11-6 (p. 292)

1. Obtain eqn (7) by the conical transformation of the pole of g_i by the rotation g.
2. Prove that

$$[g, g^{-1}] = 1, \qquad \forall g : g \neq C_2. \tag{9}$$

What happens when $g = C_2$?

3. Prove geometrically that for g_1 and g defined as in Fig. 1*b*, $g_1^g = g_2$.
4. Find, from (6) and (7), the pole $\Pi(g_1)^g$ for g_1 and g defined in Fig. 1*b*. Compare it with $\Pi(g_1^g)$.

7. THE CHARACTER THEOREM PROVED IN THE EULER–RODRIGUES PARAMETRIZATION

We know from (3.1) that the character of g_i^g, $\chi(g_i^g \mid \check{G})$, equals the character of g_i times a factor $[g_i^g, g][g, g_i]^{-1}$, and we showed geometrically in §3 that either this factor is unity or the character over the class of g_i vanishes. We shall now obtain algebraically this result and at the same time we shall tighten the theorem by providing necessary and sufficient conditions for its validity. In proving the theorem, it must first be stressed that the symbol g_i^g both in the character symbol and in the projective factor stands for the single operation (called g'' in §6) that is the result of the product gg_ig^{-1}, no consideration being taken of its path. (This can be verified in eqns **8**-4.3 and **8**-4.4, where the path appears, implicitly, in the latter equation only through a projective factor.) This means that whenever the Euler–Rodrigues parameters of g_i^g are used one must ensure that they are standardized, as is the case for any group operation. It follows from the discussion in §6 that the (standardized) values of λ and $\boldsymbol{\Lambda}$ for g_i^g are related to the values λ and $\boldsymbol{\Lambda}$ of the conjugate pole $\Pi(g_i)^g$, given in (6.6) and (6.7), by the conditions

$$\lambda_{g_i^g} = \pm\lambda_{(g_i)^g}, \qquad \boldsymbol{\Lambda}_{g_i^g} = \pm\boldsymbol{\Lambda}_{(g_i)^g}, \tag{1}$$

where the plus sign obtains either when g_i is not binary or when g_i is binary and $\mathfrak{H}$ has been chosen in accordance to the rules of §6 (to contain all conjugate binary poles). The negative sign in (1) obtains when $\mathfrak{H}$ has not been so chosen (thus causing $\Pi(g_i)^g$ to belong to $\bar{\mathfrak{H}}$) or when g_i and g are BB. With this identification, we can write at once the values of λ and $\boldsymbol{\Lambda}$ for the product that appears in $[g_i^g, g]$:

6.6, 6.7|5.2
$$\lambda_{g_i^g g} = \pm\{\lambda_{g_i}\lambda_g - (\boldsymbol{\Lambda}_{g_i}\cdot\boldsymbol{\Lambda}_g - 2\Lambda_g^2\boldsymbol{\Lambda}_{g_i}\cdot\boldsymbol{\Lambda}_g + 2\boldsymbol{\Lambda}_g\cdot\boldsymbol{\Lambda}_{g_i}\Lambda_g^2)\} \tag{2}$$
$$= \pm(\lambda_{g_i}\lambda_g - \boldsymbol{\Lambda}_{g_i}\cdot\boldsymbol{\Lambda}_g). \tag{3}$$

6.6, 6.7|5.2
$$\boldsymbol{\Lambda}_{g_i^g g} =$$
$$= \pm[\lambda_{g_i}\boldsymbol{\Lambda}_g + \lambda_g\{\boldsymbol{\Lambda}_{g_i} - 2\Lambda_g^2\boldsymbol{\Lambda}_{g_i} + 2\lambda_g\boldsymbol{\Lambda}_g\times\boldsymbol{\Lambda}_{g_i} + 2(\boldsymbol{\Lambda}_g\cdot\boldsymbol{\Lambda}_{g_i})\boldsymbol{\Lambda}_g\}$$
$$+ \boldsymbol{\Lambda}_{g_i}\times\boldsymbol{\Lambda}_g - 2\Lambda_g^2\boldsymbol{\Lambda}_{g_i}\times\boldsymbol{\Lambda}_g + 2\lambda_g(\boldsymbol{\Lambda}_g\times\boldsymbol{\Lambda}_{g_i})\times\boldsymbol{\Lambda}_g] \tag{4}$$
$$= \pm(\lambda_g\boldsymbol{\Lambda}_{g_i} + \lambda_{g_i}\boldsymbol{\Lambda}_g + \boldsymbol{\Lambda}_g\times\boldsymbol{\Lambda}_{g_i}). \tag{5}$$

In going from (4) to (5) we have used some simple algebra. The values of λ and $\boldsymbol{\Lambda}$ for the product that appears in $[g, g_i]$ are easy to obtain:

5.2
$$\lambda_{gg_i} = \lambda_g\lambda_{g_i} - \boldsymbol{\Lambda}_g\cdot\boldsymbol{\Lambda}_{g_i}, \tag{6}$$

5.2
$$\boldsymbol{\Lambda}_{gg_i} = \lambda_g\boldsymbol{\Lambda}_{g_i} + \lambda_{g_i}\boldsymbol{\Lambda}_g + \boldsymbol{\Lambda}_g\times\boldsymbol{\Lambda}_{g_i}. \tag{7}$$

It is necessary and sufficient for the factors $[g_i^g, g]$ and $[g, g_i]$ to be equal that the parameters λ in (3) and (6) and $\boldsymbol{\Lambda}$ in (5) and (7) be respectively equal. On comparing these equations, it is clear that

$$\chi(g_i^g \mid \check{G}) = \pm\chi(g_i \mid \check{G}). \tag{8}$$

The negative sign, however, will only appear when $\mathfrak{H}$ has not properly been chosen to contain all conjugate binary poles or when g_i and g are BB. In the latter case, however, g_i^g is identical with g_i, whence (8) requires that the character vanishes.

Problem 11-7 **(p. 293)**

1. Verify the character theorem for $g_i = g_1$ and $g = g$ as defined in Fig. 6.1*b* and 6.1*c*.

8. THE SU(2) REPRESENTATION OF SO(3)

It is easy at this stage to write down the SU(2) matrices in terms of the Euler–Rodrigues parameters, but it will be useful to recollect how these matrices appeared, since they have been turning up along the book in a variety of ways. We obtained in Chapter **4** the important relation between a rotation $R(\phi\mathbf{n})$ and its infinitesimal generator **I**:

4-2.15
$$R(\phi\mathbf{n}) = \exp(-\mathrm{i}\phi\mathbf{n}\,.\,\mathbf{I}). \tag{1}$$

Later on, we introduced matrices in this equation, with the matrix of the generator written as $\frac{1}{2}\boldsymbol{\sigma}$:

4-9.10
$$\hat{R}^{1/2}(\phi\mathbf{n}) = \exp\,(-\tfrac{1}{2}\mathrm{i}\phi\mathbf{n}\,.\,\boldsymbol{\sigma}). \tag{2}$$

The generator matrices $\boldsymbol{\sigma}$ (we use the plural because the vector notation in $\boldsymbol{\sigma}$ entails three generators $\boldsymbol{\sigma}_x$, $\boldsymbol{\sigma}_y$, $\boldsymbol{\sigma}_z$) were obtained as follows. When the bilinear transformation was discussed in Chapter **6**, we were able to construct three SU(2) matrices $\boldsymbol{I}_x$, $\boldsymbol{I}_y$, $\boldsymbol{I}_z$ (eqn **6**-4.1) corresponding to the three fundamental BBs around the **x**, **y**, **z** axes respectively, and we found that, on being multiplied by i, they give the infinitesimal generators $\boldsymbol{\sigma}_x$, $\boldsymbol{\sigma}_y$, $\boldsymbol{\sigma}_z$ in (2):

$$\begin{matrix} \boldsymbol{\sigma}_x & \boldsymbol{\sigma}_y & \boldsymbol{\sigma}_z \\ \begin{bmatrix} & 1 \\ 1 & \end{bmatrix}, & \begin{bmatrix} & -\mathrm{i} \\ \mathrm{i} & \end{bmatrix}, & \begin{bmatrix} 1 & \\ & -1 \end{bmatrix}, \end{matrix} \tag{3}$$

which coincide with the Pauli matrices (**4**-9.9). (It is important to ensure that this is the case, so as to satisfy the Condon and Shortley convention.) The matrix $\mathbf{n}\,.\,\boldsymbol{\sigma}$ in (2) is now fully defined and from its properties it is easy to prove, as it was done in Problem **4**-9.4, that (2) can be rewritten in the following form:

$$R^{1/2}(\phi\mathbf{n}) = (\cos\tfrac{1}{2}\phi)\mathbf{1}_2 - \mathrm{i}(\sin\tfrac{1}{2}\phi)\mathbf{n}\,.\,\boldsymbol{\sigma}. \tag{4}$$

Although this expression is a matrix, we do not embellish the R with a caret (normal or inverted) since it is not clear what type of representation, if any, is entailed by these matrices, a point which we shall now work out. For this purpose, we introduce in (4) the Euler–Rodrigues parameters (**9**-4.5); that is, λ equal to $\cos\frac{1}{2}\phi$ and the vector $\boldsymbol{\Lambda}$ equal to $\sin\frac{1}{2}\phi$ $\mathbf{n}$. We shall call the corresponding matrix $\check{R}^{1/2}(\lambda;\boldsymbol{\Lambda})$, because we shall claim that it forms a projective representation of SO(3) and because we use the superscript $\frac{1}{2}$ (equal to j) to denote the dimension $2j+1$ of the representation:

9-4.5|4
$$\check{R}^{1/2}(\lambda\,;\boldsymbol{\Lambda}) = \lambda\mathbf{1}_2 - \mathrm{i}\boldsymbol{\Lambda}\,.\,\boldsymbol{\sigma}. \tag{5}$$

On introducing here the matrices (3), it follows at once that

$$\check{R}^{1/2}(\lambda\,;\boldsymbol{\Lambda}) = +\begin{bmatrix}\lambda - \mathrm{i}\Lambda_z & -\Lambda_y - \mathrm{i}\Lambda_x \\ \Lambda_y - \mathrm{i}\Lambda_x & \lambda + \mathrm{i}\Lambda_z\end{bmatrix} = \begin{bmatrix} a & b \\ -b^* & a^*\end{bmatrix}. \tag{6}$$

Several points have to be noticed here. First, this is clearly a SU(2) matrix, as we stress by putting it in terms of the a, b (Cayley–Klein) parameters, whereupon it takes the usual SU(2) form (**6**-2.5), with precisely the same values of a, b as in (**9**-4.10). Secondly, when the relation between λ, $\boldsymbol{\Lambda}$ and ϕ, $\mathbf{n}$ is introduced into (6), this matrix coincides with (**4**-9.12). Thirdly, the positive sign in (6) is inserted in order to stress that in the (alleged) projective representation each rotation $R(\lambda;\boldsymbol{\Lambda})$ is represented by the *positive* SU(2) matrix in (6), a point that we shall now discuss. We have seen in §5, in fact, that each rotation $g \in$ SO(3) is mapped by a unique pair of Euler–Rodrigues parameters λ, $\boldsymbol{\Lambda}$, the so-called *standard parameters* which must satisfy condition (5.1). The parameters $-\lambda$, $-\boldsymbol{\Lambda}$ that map the same rotation are neither allowed in the standard set, nor in (6), and the positive sign in it is used as a reminder of this fact.

We shall now prove that, as stated above, the matrices (6) form a projective representation of SO(3) which corresponds precisely to the factor system constructed in §5, the definition of which will now be summarized for greater clarity in a slightly different notation. The product of two rotations of SO(3) is

$$R(\lambda_1;\boldsymbol{\Lambda}_1)R(\lambda_2;\boldsymbol{\Lambda}_2) = R(\lambda_3';\boldsymbol{\Lambda}_3'), \tag{7}$$

with

5.2
$$\lambda_3' = \lambda_1\lambda_2 - \boldsymbol{\Lambda}_1\,.\,\boldsymbol{\Lambda}_2, \qquad \boldsymbol{\Lambda}_3' = \lambda_1\boldsymbol{\Lambda}_2 + \lambda_2\boldsymbol{\Lambda}_1 + \boldsymbol{\Lambda}_1 \times \boldsymbol{\Lambda}_2. \tag{8}$$

We use the primes on (R7) to stress the fact that, whereas the parameters in (L7) must be standard, (because they pertain to the operations of SO(3) chosen as factors), those in (8) may or may not be so. The resulting operation on (R7) will have standard parameters λ_3, $\mathbf{\Lambda}_3$ which will be either identical with λ'_3, $\mathbf{\Lambda}'_3$ in (8) or with their negatives. In the first case, the projective factor $[R(\lambda_1, \mathbf{\Lambda}_1), R(\lambda_2, \mathbf{\Lambda}_2)]$ is $+1$ and in the second case it is -1. We must prove that, with the matrices (6)

$$\check{R}^{1/2}(\lambda_1; \mathbf{\Lambda}_1)\check{R}^{1/2}(\lambda_2; \mathbf{\Lambda}_2) = [R(\lambda_1; \mathbf{\Lambda}_1), R(\lambda_2; \mathbf{\Lambda}_2)]\check{R}^{1/2}(\lambda_3; \mathbf{\Lambda}_3). \quad (9)$$

(Notice, in comparison with eqn 7, that the parameters on R9 are now unprimed since the matrices given by eqn 6 must always entail standard parameters.) We now multiply the matrices on (L9):

6|L9 $$\{R^{1/2}(\lambda_1; \mathbf{\Lambda}_1)R^{1/2}(\lambda_2; \mathbf{\Lambda}_2)\}_{11} =$$

$$= (\lambda_1 - \mathrm{i}\Lambda_{1z})(\lambda_2 - \mathrm{i}\Lambda_{2z}) + (-\Lambda_{1y} - \mathrm{i}\Lambda_{1x})(\Lambda_{2y} - \mathrm{i}\Lambda_{2x}) \quad (10)$$

$$= \lambda_1\lambda_2 - \mathbf{\Lambda}_1 \cdot \mathbf{\Lambda}_2 - \mathrm{i}(\lambda_1\mathbf{\Lambda}_2 + \lambda_2\mathbf{\Lambda}_1 + \mathbf{\Lambda}_1 \times \mathbf{\Lambda}_2)_z \quad (11)$$

8 $$= \lambda'_3 - \mathrm{i}\Lambda'_{3z}. \quad (12)$$

In the same way,

$$\{R^{1/2}(\lambda_1; \mathbf{\Lambda}_1)R^{1/2}(\lambda_2; \mathbf{\Lambda}_2)\}_{12} = -\Lambda'_{3y} - \mathrm{i}\Lambda'_{3x}. \quad (13)$$

Thus, the result of the product of the matrices on (L9) is a matrix of the form (6) with the parameters λ'_3, $\mathbf{\Lambda}'_3$ in (8). If these parameters are standard, this matrix is $\check{R}^{1/2}(\lambda_3; \mathbf{\Lambda}_3)$ and the factor in (R9) is $+1$. If the parameters λ'_3, $\mathbf{\Lambda}'_3$ are not standard, the matrix product of the matrices on (L9) must be multiplied by (-1) in order to write it as $\check{R}^{1/2}(\lambda_3; \mathbf{\Lambda}_3)$ and, in order to cancel this factor, another factor of (-1) must be introduced, which is the projective factor on (R9). This agrees exactly with the prescription given.

It is easy to see that the projective representation of SO(3) formed by the SU(2) matrices in the manner defined is spanned by the basis $\langle u^{1/2}_{1/2} u^{1/2}_{-1/2}|$ of the eigenfunctions of the angular momentum for $j = \frac{1}{2}$, defined in §**4**-7. For $\lambda = \cos\frac{1}{2}\phi$, $\mathbf{\Lambda} = \sin\frac{1}{2}\phi\ \mathbf{z}$, i.e. for rotations by ϕ around the $\mathbf{z}$ axis, it follows at once that the matrix (6) is diagonal with diagonal elements $\exp(-\frac{1}{2}\mathrm{i}\phi)$ and $\exp(\frac{1}{2}\mathrm{i}\phi)$, thus coinciding with the matrix (**4**-7.9) for the angular momentum spin eigenfunctions. It is for this reason that all the projective representations for the same factor system that corresponds to the SU(2) matrices (6) are called *spinor representations.* A fundamental property of this factor system is that, for all C_2, $[C_2, C_2]$ equals -1. This means that when two successive rotations by π are applied to the bases of the spinor representations, they are multiplied by a phase factor of -1, as was discussed for spinors in §**6**-7.

For SO(3), the dimension of the spinor representations will be an even integer, $2j+1$ with j half-integral, but these irreducible representations of

SO(3) may reduce into irreducible representations of smaller dimension for the point groups. These will be projective representations for the stated factor system and thus will still be called *spinor representations*. The spinor representation formed by the (positive) SU(2) matrices themselves for the standard values of the Euler–Rodrigues parameters will be called the *standard spinor representation* (abbreviated when convenient to *standard representation*).

It should be noticed that the number of matrices of SU(2) in the standard representation is identical with the number of matrices of SO(3). It should also be clear that the whole of SU(2) forms a vector representation which, when subduced to the standard matrices, coincides with the standard projective representation. Thus, SU(2) is the covering group of SO(3). (See §**8**-6.)

9. $\mathbf{C}_i$ AND THE IRREDUCIBLE REPRESENTATIONS OF O(3). THE SU′(2) REPRESENTATION OF O(3)

The group O(3) is the group of all proper and improper rotations in $\mathbb{R}^3$ and it is given by the direct product

$$O(3) = SO(3) \otimes \mathbf{C}_i, \qquad \mathbf{C}_i = E \oplus \boldsymbol{i}. \tag{1}$$

The group $\mathbf{C}_i$ is thus the key to the study of O(3) and improper rotations and we shall now discuss its properties.

The representations of $\mathbf{C}_i$

The first thing we must do is to construct a factor system suitable for the spinor bases, for which all that we need is the projective factor $[\boldsymbol{i}, \boldsymbol{i}]$, which is obtained as follows. Matrices for the inversion on spinor bases were obtained in (**6**-7.10) and (**6**-7.11) and we now give them after their bases, transcribed in dual form:

$$\langle u^{1/2}_{1/2} u^{1/2}_{-1/2}|, \begin{bmatrix} -\mathrm{i} & \\ & -\mathrm{i} \end{bmatrix}; \qquad \langle u^{1/2}_{1/2} u^{1/2}_{-1/2}|^*, \begin{bmatrix} \mathrm{i} & \\ & \mathrm{i} \end{bmatrix}. \tag{2}$$

Since $\boldsymbol{i}^2$ equals E, represented by $\mathbf{1}_2$, the square of each of the matrices in (2), which is in both cases $-\mathbf{1}_2$, leads to

$$[\boldsymbol{i}, \boldsymbol{i}] = -1. \tag{3}$$

We can now write the multiplication table and factor system for $\mathbf{C}_i$ in Table 1.

It is clear that the matrices in (2) form two projective representations of $\mathbf{C}_i$:

$$\begin{array}{ccc} & \check{E} & \check{\boldsymbol{i}} \\ \text{(a)} & \begin{bmatrix} 1 & \\ & 1 \end{bmatrix} & \begin{bmatrix} -\mathrm{i} & \\ & -\mathrm{i} \end{bmatrix}; \end{array} \qquad \begin{array}{ccc} & \check{E} & \check{\boldsymbol{i}} \\ \text{(b)} & \begin{bmatrix} 1 & \\ & 1 \end{bmatrix} & \begin{bmatrix} \mathrm{i} & \\ & \mathrm{i} \end{bmatrix}. \end{array} \tag{4}$$

Table **11**-9.1. $\mathbf{C}_i$ multiplication and factor tables.
$g_i g_j$ and $[g_i, g_j]$ appear in the intersection of the g_i row with the g_j column.

	$g_i g_j$: E	$g_i g_j$: $\boldsymbol{i}$		$[g_i, g_j]$: E	$[g_i, g_j]$: $\boldsymbol{i}$
E	E	$\boldsymbol{i}$	E	1	1
$\boldsymbol{i}$	$\boldsymbol{i}$	E	$\boldsymbol{i}$	1	−1

These representations are inequivalent (since the $\check{\boldsymbol{i}}$ matrices have different characters) but they are both reducible. In Table 2 we display the two projective representations thus obtained, labelled $\bar{A}_g$ and $\bar{A}_u$, together with the two well-known vector representations. We also list bases for the representations, those for the projective representations arising from (2), except that we add a suffix c the meaning of which will later be explained. The basis u_0^0 for A_g is the spherical harmonic of degree zero, a *scalar.* The basis $\bar{u}_0^0$ for A_u is a *pseudoscalar* (invariant under all rotations but ungerade with respect to inversion). We have already met it in (**R5**-3.8), since the u^0 function there, like the whole of that basis, transforms under inversion by (**R5**-3.9) so that it is ungerade. Thus, the pseudoscalar is the $j=0$ component of the direct product of two spinors on (L2).

We shall now verify that $\bar{A}_g$ and $\bar{A}_u$ are gauge-equivalent to the vector representations A_g and A_u respectively. (This is a particular example of a general result for all cyclic groups, namely that their projective representations are always gauge-equivalent to vector representations. See, e.g., Altmann 1977, p. 76.) We recall that a gauge equivalence (see eqn **8**-2.17) arises when each of the matrices of a projective representation is multiplied by an individual phase factor. Thus, when the matrices in either $\bar{A}_g$ or $\bar{A}_u$ are multiplied by 1 for E and by i for $\boldsymbol{i}$, then A_g and A_u appear. Clearly, at the same time, the factor system in Table 1 changes into the trivial factor system (in which all projective factors are the unity). The factor system in Table 1 will be called the *Cartan gauge* whereas the trivial factor system will be called the *Pauli gauge*. (See the end of §**5**-3.)

Table **11**-9.2. $\mathbf{C}_i$ irreducible representations.

	E	$\boldsymbol{i}$	Bases
A_g	1	1	u_0^0
A_u	1	−1	$\bar{u}_0^0$
$\bar{A}_g$	1	−i	$(u_{1/2}^{1/2})_c$
$\bar{A}_u$	1	i	$(u_{1/2}^{1/2})_c^*$

Table **11**-9.3. $\mathbf{C}_i$ irreducible representations. Pauli gauge.

	E	i	Integral bases	Half-integral bases
A_g	1	1	u_0^0	$u_{1/2}^{1/2}$
A_u	1	−1	$\bar{u}_0^0$	$(u_{1/2}^{1/2})^*$

A change of gauge cannot affect the energy levels pertaining to the representations (see §**8**-2) so that, from this point of view, it is only the simpler Pauli gauge that need be used, the representations $\bar{A}_g$ and $\bar{A}_u$ in Table 2 now being replaced by A_g and A_u respectively. The bases of $\bar{A}_g$ and $\bar{A}_u$ in Table 2 were given a suffix c to indicate the gauge; that is, the fact that the corresponding spinors transform under inversion by the matrices in (2). When these matrices, in going to the Pauli gauge, are multiplied by i, $u_{1/2}^{1/2}$ and $(u_{1/2}^{1/2})^*$ become gerade (A_g) and ungerade (A_u) respectively, and they are listed accordingly in Table 3, without a suffix, it being understood that spinor bases are given, unless indication to the contrary, always in the Pauli gauge. Table 3 contains the only two representations of $\mathbf{C}_i$ that need in practice be used.

Although, as it is traditional, we have changed from the Cartan to the Pauli gauge, it must be emphasized that the gauge used is significant when symmetrizing functions. We have seen in §**5**-3, e.g., that a representation transforming like the spherical vectors (or spherical harmonics for $l=1$) cannot be obtained by a tensor product of spinor representations unless the latter are used in the Cartan gauge.

The irreducible representations of O(3)

From (1), each irreducible representation of SO(3), whether vector or projective, will split into two representations of O(3), gerade and ungerade respectively in accordance with the representation of $\mathbf{C}_i$ from Table 3 used in forming the direct product. We shall first consider the vector representations, for which the bases of SO(3) are the bases (**4**-7.1), $\langle u_j^j \ldots u_{-j}^j|$, for integral j, which shall now be denoted with the shortened symbol $\langle u^j|$. From (**4**-7.5) we know that these bases are g (gerade) and u (ungerade) for j even and odd respectively. The bases of the vector representations of O(3) will be direct products of $\langle u^j|$ (j integral) with u_0^0 and $\bar{u}_0^0$ respectively in Table 3 and each $\langle u^j|$ will lead to one g and one u basis as follows:

$$j = \text{even} \qquad \langle u^j|\, u_0^0, \quad (g); \qquad \langle u^j|\, \bar{u}_0^0, \quad (u). \tag{5}$$

$$j = \text{odd} \qquad \langle u^j|\, u_0^0, \quad (u); \qquad \langle u^j|\, \bar{u}_0^0, \quad (g). \tag{6}$$

The projective (spinor) representations of O(3) are obtained by forming the bases $\langle u^j|$ and $\langle u^j|^*$ (j half-integral for the representations A_g and A_u, respectively, of Table 3.

The SU′(2) representation of O(3)

The operations of O(3) belong to two cosets in one of which all the operations are of the form $g \in SO(3)$, that is they all are proper rotations. The operations of the other coset are then the improper rotations $\boldsymbol{i}g$, with $g \in \mathrm{SO}(3)$. Assume that the standard representation of SO(3) is formed (§8) in which every element g is mapped by the positive SU(2) matrix $A(g)$ expressed in terms of the standard Euler–Rodrigues parameters of g. Then, the direct product $\mathrm{O}(3) = \mathrm{SO}(3) \otimes \mathbf{C}_i$ and the two representations of $\mathbf{C}_i$ in (4) provide two representations of O(3), by the mappings

$$g \mapsto A(g), \qquad \boldsymbol{i}g \mapsto -\mathrm{i}A(g), \tag{7}$$

$$g \mapsto A(g), \qquad \boldsymbol{i}g \mapsto +\mathrm{i}A(g). \tag{8}$$

From (2), the bases of (7) and (8) are the $j = \frac{1}{2}$ spinors and conjugate spinors respectively. The representations themselves are projective representations in the Cartan gauge in which $[\boldsymbol{i}, \boldsymbol{i}] = -1$, and they are clearly inequivalent. If, however, in (8), all the matrices for all g and all $\boldsymbol{i}g$ are multiplied by $+1$ and -1 respectively, then (8) becomes identical with (7), although the transformed gauge is still the original Cartan gauge. Thus, (8) is merely gauge-equivalent to (7) which is the only SU′(2) representation of interest in the Cartan gauge. Representation (7) itself can still undertake a change of gauge, by multiplying the matrices of g and $\boldsymbol{i}g$ by $+1$ and i respectively. The mapping now becomes

$$g \mapsto A(g), \qquad \boldsymbol{i}g \mapsto A(g), \tag{9}$$

and the gauge is now the Pauli gauge, since the matrix representative for $\boldsymbol{i}$ becomes $\mathbf{1}_2$, so that $[\boldsymbol{i}, \boldsymbol{i}] = 1$. Whereas (7) is a faithful representation, (9) is not so. This latter representation is universally used as the standard representation for O(3), in the sense that the multiplication rules for O(3) are taken from the homomorphism (9). This is, of course, very simple and convenient but can cause some confusion because, in this gauge, no distinction appears to be made between SO(3) and O(3), the inversion in the latter being taken as the identity – a practice which would be fatally wrong in dealing with the vector representations of O(3). Since all improper point groups are subgroups of O(3) it must be remembered that the same situation obtains for them.

The factor system for O(3)

Since we already have the representations of O(3), we have all the information we need for this group. It will be useful, however, to discuss its factor system in more detail, since all improper point groups are subgroups of O(3), so that their factor systems must be constructed as subsets of the factor system of O(3). It will be convenient for this purpose to use the structure of O(3) given by (1). In order to simplify the notation, let us call M, G, H respectively in this equation the groups O(3), SO(3), and $\mathbf{C}_i$. The elements of M are of the form gh, and $M = G \otimes H$. From (**8**-5.2) and (**8**-5.3) we know that the matrix representative of gh, $\check{M}(gh)$ is given by

$$\check{M}(gh) = [g, h]^{-1} \check{G}(g) \otimes \check{H}(h), \tag{10}$$

and that the factor system of the corresponding representation is

$$[gh, g'h'] = [g, g'][h, h'][gg', hh'][g, h]^{-1}[g', h']^{-1}. \tag{11}$$

Here, h can take only two values, E or $\boldsymbol{i}$. Thus, in (R11), besides the projective factors $[g, g']$ that, coming from SO(3), are fully defined by the rules given in §5, we need projective factors of the form $[g, E]$, $[g, \boldsymbol{i}]$, $[\boldsymbol{i}, E]$ and $[\boldsymbol{i}, \boldsymbol{i}]$. As always, in order to keep to the standardization of the factor system, we take

$$[g, E] = [E, g] = [\boldsymbol{i}, E] = [E, \boldsymbol{i}] = 1. \tag{12}$$

It is easy to see, in fact, that this agrees with the projective factors in Table 1, which arose from the multiplication of the matrices in (2). Likewise, from Table 1, for the same reasons,

$$[\boldsymbol{i}, \boldsymbol{i}] = -1. \tag{13}$$

All that remains, therefore, is to consider projective factors of the form $[g, \boldsymbol{i}]$ or $[\boldsymbol{i}, g]$, which we shall now obtain. We have seen in Fig. **9**-1.2 that the inversion leaves the pole of any rotation $g \in \mathrm{SO}(3)$ invariant. Thus, to each operation of O(3) of the form $g\boldsymbol{i}$ or $\boldsymbol{i}g$ a pole can be assigned which is identical to the pole of g. Now, as we know, all the operations of SO(3) are given standard poles so that the path from the identity to them is always class 0. Likewise, because the pole of $g\boldsymbol{i}$ or $\boldsymbol{i}g$ is the same as the pole of g, the path from the identity to this pole is class 0 and this homotopy cannot be altered by $\boldsymbol{i}$, since it leaves this pole invariant. Thus,

$$[g, \boldsymbol{i}] = [\boldsymbol{i}, g] = 1, \qquad \forall g \in \mathrm{SO}(3). \tag{14}$$

It follows from (12) and (14) that

$$[g, h] = 1, \qquad \forall g \in \mathrm{SO}(3), \qquad \forall h \in \mathbf{C}_i. \tag{15}$$

On introducing this result into (10) and (11), we have:

$$\check{M}(gh) = \check{G}(g) \otimes \check{H}(h), \tag{16}$$

and

$$[gh, g'h'] = [g, g'][h, h'], \qquad \text{(Cartan)}. \tag{17}$$

where $[h, h']$ will always be unity unless both h and h' are the inversion, in which case it is -1. This is so in the Cartan gauge. In the Pauli gauge even this factor is unity and

$$[gh, g'h'] = [g, g'], \qquad \text{(Pauli)}. \qquad (18)$$

Problem 11-9 (p. 293)

1. Prove that the *orthogonal mirrors* (OM) σ_x, σ_y, σ_z, respectively normal to the axes **x**, **y**, **z**, have the products and projective factors displayed in Table 4.

Table **11**-9.4. Products and projective factors for orthogonal mirrors.
A factor $g_i g_j$ or $[g_i, g_j]$ appears in the intersection of the g_i row with the g_j column. The Cartan gauge is used.

	$g_i g_j$				$[g_i, g_j]$		
	σ_x	σ_y	σ_z		σ_x	σ_y	σ_z
σ_x	E	C_{2z}	C_{2y}	σ_x	1	-1	1
σ_y	C_{2z}	E	C_{2x}	σ_y	1	1	-1
σ_z	C_{2y}	C_{2x}	E	σ_z	-1	1	1

10. IMPROPER POINT GROUPS

The basic work for improper groups is all done. An improper point group M will always be embedded in O(3), i.e. treated as a subgroup of O(3) with a factor system that is the appropriate subset of the factor system defined in §9. Every operation m of the improper group can be written in the form gh (or hg, since h, being either e or $\boldsymbol{i}$ always commutes with g), with $g \in SO(3)$ and $h \in \mathbf{C}_i$. Since, as follows from §9, the pole of gh is always the same as the pole of g, each operation $m = gh$ of the improper group can be assigned a set of standard Euler–Rodrigues parameters $R_g(\lambda; \boldsymbol{\Lambda})$, which are identical with those of g. Then, the factor system, in either the Cartan or the Pauli gauges, is fully defined by (9.17) and (9.18). Notice that in the Pauli gauge both operations of $\mathbf{C}_i$ are treated as the identity.

We must now consider the character theorem for the group M:

8-4.2 $$\chi(m_i^m \mid \check{M}) = [m_i^m, m][m, m_i]^{-1}\chi(m_i \mid \check{M}). \qquad (1)$$

On writing, in an obvious notation,

$$m_i = g_i h_i, \qquad m = gh = hg, \qquad (2)$$

we shall prove that

$$\chi(m_i^{gh} \mid \check{M}) = [g_i^g, g][g, g_i]^{-1}\chi(m_i \mid \check{M}). \tag{3}$$

In order to do this, we first remark that

2.2,2
$$m_i^{gh} = ghm_ih^{-1}g^{-1} = ghh^{-1}m_ig^{-1} = m_i^g. \tag{4}$$

Also, for $m_i = g_ih_i$,

,2
$$m_i^g = gg_ig^{-1}gh_ig^{-1} = g_i^gh_igg^{-1} = g_i^gh_i, \tag{5}$$

so that, with (5) and (4) into (1), we have:

$$\chi(m_i^{gh} \mid \check{M}) = [g_i^gh_i, gh][gh, g_ih_i]^{-1}\chi(m_i \mid \check{M}). \tag{6}$$

On using (9.17), we obtain

$$[g_i^gh_i, gh][gh, g_ih_i]^{-1} = [g_i^g, g][h_i, h][g, g_i]^{-1}[h, h_i]^{-1} \tag{7}$$

$$= [g_i^g, g][g, g_i]^{-1}, \tag{8}$$

since, as it is evident from the factor table for $\mathbf{C_i}$ in Table 9.1, $[h_i, h]$ equals $[h, h_i]$ for all h and h_i in $\mathbf{C}_i$. On introducing (8) into (6), eqn (3) follows. It is evident from this proof that this result is true also in the Pauli gauge, in which case (9.18) would be used instead of (9.17) on (L7), so that the factors pertaining to H on (R7) would not even appear in (7).

The implications of (3) are pretty obvious. We know from §7 that, for proper point groups, the product of the projective factors on (R3) is unity unless g and g_i are BBs. The only irregular case, thus, arises when g and g_i are BBs but this case can now, for improper groups, arise either when m_i and m in (2) are BBs (h and h_i being the identity) or when they are *orthogonal mirrors*, $m_i = g_i\boldsymbol{i}$, $m = g\boldsymbol{i}$, with g_i and g BB's. In this case, because m_i and m commute, m_i^m equals m_i; hence, in (3),

$$\chi(m_i \mid \check{M}) = [g_i, g][g, g_i]^{-1}\chi(m_i \mid \check{M}) = -\chi(m_i \mid \check{M}), \tag{9}$$

since, as discussed in §7, the product of the projective factors in (9) is -1 when g_i and g are BB. It follows from (9) that the character $\chi(m_i \mid \check{M})$ must vanish. Thus, the character theorem for improper groups takes the following form: the characters are class functions that vanish over the irregular classes, these being the classes formed either by BBs or by orthogonal mirrors. Theorem 1 of §4 is still entirely valid; that is, the number of irreducible spinor representations is equal to the number of regular classes. We are assuming here, of course, that $\mathfrak{H}$ is properly chosen as specified in §§6 and 7.

12

THE ALGEBRA OF ROTATIONS: QUATERNIONS

> The invention of the calculus of quaternions is a step towards the knowledge of quantities related to space which can only be compared for its importance, with the invention of triple coordinates by Descartes. The ideas of this calculus, as distinguished from its operations and symbols, are fitted to be of the greatest use in all parts of science.
>
> J. C. Maxwell (1869). *Proceedings of the London Mathematical Society* **3**, p. 226.

The Euler–Rodrigues parametrization of rotations has allowed us, in the last six sections of the previous chapter, to work out rotation products and projective factors in an entirely algebraic way. We have thus completed all the necessary work in order to obtain consistently and unambiguously the spinor representations of O(3) and of any of its finite subgroups. We shall now, nevertheless, take a wider view of the subject, partly in order to make contact with some ideas used in other branches of mathematics, and partly because, in generalizing our outlook, we shall obtain a deeper insight into the all-important relation between SU(2) and SO(3). Before we start serious work, we shall play about a little with binary rotations.

1. AN ENTERTAINMENT ON BINARY ROTATIONS

When we met the equation

$$\mathrm{i}^2 = -1 \tag{1}$$

for the first time in our lives, we were probably worried because we could not conceive of a God-made object with this property. A Spanish priest, indeed, is said to have raised this matter with the Inquisition. We have seen in Chapter **1** that a realization of such an object was provided by Argand by means of a 90° rotation which, repeated twice, changes the sign of any vector. This was the picture used by Hamilton when he invented quaternions and some confusion has thereby arisen. Sadly, because the topology of rotations was not then known, Hamilton was in no position to understand that it is the binary rotations C_2 that provide an immediate realization of the imaginary unit. Take a binary rotation around the **z** axis, say, given in terms of the Euler–Rodrigues parameters as $R(0; \mathbf{k})$.† On using the multiplication rule for rotations in these parameters, eqns (**11**-8.7) and (**11**-8.8), we have,

$$C_2C_2 = R(0\,;\mathbf{k})R(0\,;\mathbf{k}) = R(-\mathbf{k}\cdot\mathbf{k}\,;\mathbf{0}) = R(-1\,;\mathbf{0}). \tag{2}$$

What we deduce from (2) is that $[C_2, C_2]$ equals -1, in agreement of course with the fact that the path C_2, C_2 in Fig. **11**-5.1 is of class 1. We shall, however, force the notation a little. First, because $R(1, \mathbf{0})$ is the identity, we can write (R2) as $-E$, that is, (2) as $C_2^2 = -E$. Secondly, because the identity is always represented by the unity, we can just as well write

$$C_2^2 = -1. \tag{3}$$

Strictly speaking, of course, the number one here should be the unit matrix **1**, because what we are implicitly doing in going from (2) to (3) is to pass from the group operations to their representatives. We have written (3) in that form, however, because we see at once, on comparison with (1), that binary rotations provide a realization of the imaginary unit. We shall presently see that the three BB rotations around **i, j, k** will provide us not with one but with three imaginary units, and this idea will be at the very core of the concept of quaternions.

One more reference to Hamilton before we start our formal work. It was not only the lack of topology that led Hamilton to his view of the imaginary unit as realized by a 90° rotation. Rather, his primary interest in quaternions arose from their algebra, and when quaternions are regarded in this way it is not only possible but also natural to use Hamilton's interpretation. It must be emphasized, however, that, as will be shown in §9, this leads to a parametrization of rotations that disagrees with that of Rodrigues and that is entirely unacceptable from the point of view of the rotation group.

2. THE DEFINITION OF QUATERNIONS

We shall start our work by streamlining a little the symbol for a rotation in terms of its Euler–Rodrigues parameters, $R(\lambda\,;\boldsymbol{\Lambda})$, which shall now be replaced by $[\![\lambda, \boldsymbol{\Lambda}]\!]$. With this notation, the multiplication rule for rotations, (**11**-8.7), (**11**-8.8), takes the following form

$$[\![\lambda_1, \boldsymbol{\Lambda}_1]\!][\![\lambda_2, \boldsymbol{\Lambda}_2]\!] = [\![\lambda_1\lambda_2 - \boldsymbol{\Lambda}_1\cdot\boldsymbol{\Lambda}_2, \lambda_1\boldsymbol{\Lambda}_2 + \lambda_2\boldsymbol{\Lambda}_1 + \boldsymbol{\Lambda}_1\times\boldsymbol{\Lambda}_2]\!]. \tag{1}$$

Within the context of rotation products, the remark that the object on

† *Warning.* In order to coincide with the usual notation in this field, **i, j, k** in this chapter are unit vectors along the orthogonal directions **x, y, z** *fixed in space*. That is, **i, j, k** in this chapter coincide with the unit vectors **x, y, z** used elsewhere in the book.

(R1) is of the same nature as the factors on the left (being given by a scalar and a vector) is obvious. In a more general context, however, this remark is significant. Let us call a quaternion $\mathbb{A}$ an object $\llbracket a, \mathbf{A}\rrbracket$, given by a real number (scalar) a and a vector $\mathbf{A}$, that obeys the non-commutative multiplication rule (1). That is, if $\mathbb{B} = \llbracket b, \mathbf{B}\rrbracket$, then

$$\mathbb{A}\mathbb{B} = \llbracket a, \mathbf{A}\rrbracket\llbracket b, \mathbf{B}\rrbracket = \llbracket ab - \mathbf{A}\,.\,\mathbf{B},\, a\mathbf{B} + b\mathbf{A} + \mathbf{A}\times\mathbf{B}\rrbracket. \tag{2}$$

Again, the object on (R2) is given by a scalar and a vector and it is therefore a quaternion. This point, however, requires further discussion (see §3). It is easy to prove (Problem 1) that the product is *associative*, that is,

$$(\mathbb{A}\mathbb{B})\mathbb{C} = \mathbb{A}(\mathbb{B}\mathbb{C}). \tag{3}$$

A quaternion of the form $\llbracket a, \mathbf{0}\rrbracket$ is called a *real quaternion*. Because they multiply precisely like real numbers,

2 $$\llbracket a, \mathbf{0}\rrbracket\llbracket b, \mathbf{0}\rrbracket = \llbracket ab, \mathbf{0}\rrbracket, \tag{4}$$

they can actually be identified with real numbers,

$$\llbracket a, \mathbf{0}\rrbracket \equiv a, \tag{5}$$

very much like the complex number $a + ib$ for $b = 0$ is identified with the real number a. We can, in this way, define the product of a real number times a quaternion:

5,2 $$a\llbracket b, \mathbf{B}\rrbracket = \llbracket a, \mathbf{0}\rrbracket\llbracket b, \mathbf{B}\rrbracket = \llbracket ab, a\mathbf{B}\rrbracket. \tag{6}$$

A quaternion of the form $\llbracket 0, \mathbf{A}\rrbracket$ is called a *pure quaternion*. If $\mathbb{A}$ and $\mathbb{B}$ are pure quaternions, their product is as follows:

2 $$\mathbb{A}\mathbb{B} = \llbracket 0, \mathbf{A}\rrbracket\llbracket 0, \mathbf{B}\rrbracket = \llbracket -\mathbf{A}\,.\,\mathbf{B}, \mathbf{A}\times\mathbf{B}\rrbracket. \tag{7}$$

Notice that it entails the two types of products between vectors. The vector $\mathbf{A}$ in the pure quaternion $\llbracket 0, \mathbf{A}\rrbracket$ can be written as $A\mathbf{n}$ where A is $|\mathbf{A}|$ and $\mathbf{n}$ is a unit vector. Thus

,6 $$\llbracket 0, \mathbf{A}\rrbracket = \llbracket 0, A\mathbf{n}\rrbracket = A\llbracket 0, n\rrbracket =_{\text{def}} A\mathbb{n}. \tag{8}$$

A pure quaternion such as $\mathbb{n}$, of the form $\llbracket 0, \mathbf{n}\rrbracket$, where $\mathbf{n}$ is a unit vector ($\mathbf{n}^2 = 1$), will be called a *unit quaternion*. They will always be denoted in this chapter with lower-case letters.

We shall now establish an additive form for quaternions, in which a quaternion $\llbracket a, \mathbf{A}\rrbracket$ is written as a real quaternion $\llbracket a, \mathbf{0}\rrbracket$ plus a pure quaternion $\llbracket 0, \mathbf{A}\rrbracket$. For this notation to be valid it must be possible to multiply two quaternions so written,

$$\llbracket a, \mathbf{A}\rrbracket = \llbracket a, \mathbf{0}\rrbracket + \llbracket 0, \mathbf{A}\rrbracket, \tag{9}$$

$$\llbracket b, \mathbf{B}\rrbracket = \llbracket b, \mathbf{0}\rrbracket + \llbracket 0, \mathbf{B}\rrbracket, \tag{10}$$

by the usual rules of algebra

$$\llbracket a, \mathbf{A}\rrbracket\llbracket b, \mathbf{B}\rrbracket = \llbracket a, \mathbf{0}\rrbracket\llbracket b, \mathbf{0}\rrbracket + \llbracket a, \mathbf{0}\rrbracket\llbracket 0, \mathbf{B}\rrbracket + \llbracket 0, \mathbf{A}\rrbracket\llbracket b, \mathbf{0}\rrbracket + \llbracket 0, \mathbf{A}\rrbracket\llbracket 0, \mathbf{B}\rrbracket, \quad (11)$$

and to obtain the correct multiplication rule (2). In fact, on performing the four quaternion multiplications on (R11) by (2) and then collecting scalar and vector parts, it is easy to prove that (11) coincides with (2).

Once (9) is established, we can replace the real and pure quaternions in it by their forms (5) and (8) respectively,

$$\llbracket a, \mathbf{A}\rrbracket = a + \mathbb{n}A, \quad (12)$$

whereupon the quaternion is written as the sum of a real number plus the product of another real number times a unit quaternion, very much like a complex number is written in the form $a + ib$. The similarity is even greater, since the square of a unit quaternion is indeed -1, so that $\mathbb{n}$ behaves like an imaginary unit:

,2,5 $$\mathbb{n}^2 = \llbracket 0, \mathbf{n}\rrbracket\llbracket 0, \mathbf{n}\rrbracket = \llbracket -\mathbf{n}^2, \mathbf{n}\times\mathbf{n}\rrbracket = \llbracket -1, \mathbf{0}\rrbracket = -1. \quad (13)$$

Equation (12) is called the *binary form* of the quaternion.

We must establish the geometrical meaning of a unit quaternion. Let us, for this purpose, write the quaternion that represents the rotation $R(\lambda; \mathbf{\Lambda})$:

9-4.5 $$\llbracket \lambda, \mathbf{\Lambda}\rrbracket = \llbracket \cos\tfrac{1}{2}\phi, \sin\tfrac{1}{2}\phi\ \mathbf{n}\rrbracket, \qquad \mathbf{n}^2 = 1. \quad (14)$$

For this quaternion to be pure, ϕ must equal π, in which case the quaternion takes the form $\llbracket 0, \mathbf{n}\rrbracket$ and is identical with the unit quaternion $\mathbb{n}$ in (13). Thus the unit quaternion $\mathbb{n}$ is a binary rotation around $\mathbf{n}$, which agrees with our realization, in §1, of an imaginary unit by a binary rotation. A pure quaternion $\llbracket 0, \mathbf{A}\rrbracket$ with $\mathbf{A} = A\mathbf{n}$, $|\mathbf{n}| = 1$, is $A\mathbb{n}$; that is, a scalar multiple of a binary rotation around $\mathbf{n}$. Historically, however, the pure quaternion $\llbracket 0, \mathbf{A}\rrbracket$ has often been identified with the vector $\mathbf{A}$. This identification has serious limitations, which will be discussed in the next section, and it is not sufficiently general to be usable without the greatest care. (See §§8,9.)

Problems 12-2 (p. 294)

1. Prove the associative property (3) of the quaternion product.
2. Prove that the product of two pure quaternions is a pure quaternion if and only if their corresponding vectors are orthogonal. Identify the pure quaternions with binary rotations and compare with Problem **2**-1.14.
3. Prove that the product of two quaternions of the binary form (12), $\mathbb{A} = a + A\mathbb{n}, \mathbb{B} = b + B\mathbb{n}$, where $\mathbb{n}^2 = -1$, is a quaternion of the same

binary form, but that this result fails if the unit quaternions in $\mathbb{A}$ and $\mathbb{B}$ are not identical.

4. Prove that the product of a general quaternion $[\![a, \mathbf{A}]\!]$ with a pure quaternion $[\![0, \mathbf{B}]\!]$ is a pure quaternion if and only if $\mathbf{A}\,.\,\mathbf{B}=0$.

3. INVERSION OF QUATERNIONS. CHARACTERIZATION OF THEIR SCALAR AND VECTOR PARTS

In the active interpretation the inversion $\boldsymbol{i}$ leaves fixed the $\mathbf{i}, \mathbf{j}, \mathbf{k}$ vectors (now the fixed space vectors) and inverts all points in the configuration space. It allows us to classify vectors into *polar* and *axial vectors*, henceforth called *vectors* and *pseudovectors* respectively:

$$\boldsymbol{i}\mathbf{r}=-\mathbf{r} \quad \Rightarrow \quad \mathbf{r} \text{ vector}; \qquad \boldsymbol{i}\mathbf{r}=\mathbf{r} \quad \Rightarrow \quad \mathbf{r} \text{ pseudovector}. \tag{1}$$

The cross product $\mathbf{r}\times\mathbf{s}$, e.g., is a pseudovector in accordance with (1) whenever $\mathbf{r}$ and $\mathbf{s}$ are of the same type and, contrariwise, it can be a vector only when $\mathbf{r}$ and $\mathbf{s}$ are of different types. The unit vectors $\mathbf{i}, \mathbf{j}, \mathbf{k}$ are also pseudovectors, because by definition they do not change sign under inversion. This classification agrees with the fact that $\mathbf{i}=\mathbf{j}\times\mathbf{k}$ must always be a pseudovector, since it would be absurd to use a basis $\mathbf{i}, \mathbf{j}, \mathbf{k}$ containing mixed types.

If real numbers A are defined in configuration space that are functions, $A(\mathbf{r})$, of the position vector $\mathbf{r}$ in this space such that

$$A(\mathbf{r})=\pm A(\boldsymbol{i}\mathbf{r})=\pm A(-\mathbf{r}). \tag{2}$$

then A is a *scalar* or a *pseudoscalar* when the + and − signs obtain, respectively.

Note that $\mathbf{r}$ in (1) means the vector *not* the triple of its components. The components of a vector and a pseudovector are pseudoscalars and scalars respectively. Thus $\mathbf{r}$, given by $r_x\mathbf{i}+r_y\mathbf{j}+r_z\mathbf{k}$, is a vector because each term in this sum is a pseudoscalar times a pseudovector. A most important but often forgotten rule in vector algebra, is that $\mathbf{A}+\mathbf{B}$ has a meaning only when $\mathbf{A}$ and $\mathbf{B}$ are both vectors or both pseudovectors. When this is not the case, $\mathbf{A}+\mathbf{B}$ is neither even nor odd with respect to inversion and it is a nonsense. Likewise, to add up a scalar with a pseudoscalar in this type of work is a step towards perdition.

Having thus brushed up on the facts of life, we can now see what the nature is of the objects a, and $\mathbf{A}$ that appear in a quaternion $[\![a, \mathbf{A}]\!]$, for which purpose we look at the product of two quaternions:

2.2 $$[\![a, \mathbf{A}]\!][\![b, \mathbf{B}]\!]=[\![ab-\mathbf{A}\,.\,\mathbf{B},\, a\mathbf{B}+b\mathbf{A}+\mathbf{A}\times\mathbf{B}]\!]. \tag{3}$$

We can take a to be scalar or pseudoscalar and $\mathbf{A}$ to be vector or pseudovector, so that we could in principle have four quaternion types.

We shall accept for the time being the restriction that all quaternions be of the same type. Under this restriction, the term on the left of the comma on (R3) must always be a true scalar whence a and b on (L3) must also be scalars. Likewise, $\mathbf{A}\times\mathbf{B}$ is always, under the restriction stated, a pseudovector, so that for the vector sum on (R3) to make sense $\mathbf{A}$ and $\mathbf{B}$ must also be pseudovectors (since a and b must be scalars). We conclude that in any quaternion $[\![a, \mathbf{A}]\!]$, a must be a scalar and $\mathbf{A}$ a pseudovector. Thus:

$$\boldsymbol{i}[\![a, \mathbf{A}]\!] = [\![a, \mathbf{A}]\!]. \tag{4}$$

This is precisely the behaviour we expect of the rotation quaternion in (2.14). Inversion changes ϕ into $-\phi$ and $\mathbf{n}$ into $-\mathbf{n}$ (see Fig. 9-1.2). Thus $\cos\frac{1}{2}\phi$ is a scalar, $\sin\frac{1}{2}\phi$ a pseudoscalar, and the product of $\sin\frac{1}{2}\phi$ with the vector $\mathbf{n}$ a pseudovector.

It is clear that the restriction that we have stated cannot be given up without much care, because (R3) becomes otherwise nonsensical. If it is given up with care, however, it is found that the only way in which $\mathbf{A}$ in (4) can be a vector is by taking a as a pseudoscalar. The product of two such objects, however, does not close, being a quaternion of the form given in (4). Such objects, thus, cannot be used in representations of the rotation group. There is a more serious difficulty with them, however. When the pseudovector $\mathbf{A}$ in (4) is written as $A_x\mathbf{i}+A_y\mathbf{j}+A_z\mathbf{k}$, the coefficients here are scalars, like a is. It is thus possible to construct an algebra from them, since an algebra requires the formation of linear combinations of a basis with scalar coefficients (see §5). The components of a vector, instead, are pseudoscalars and it is not possible to construct a proper algebra.

Since we must accept that all quaternions $[\![a, \mathbf{A}]\!]$ are given by a scalar and a pseudovector, it is clear that the identification of a pure quaternion $[\![0, \mathbf{A}]\!]$ with a vector $\mathbf{A}$ cannot be general. It is indeed most unfortunate that vectors were historically introduced this way (as invented by Hamilton), since confusion can easily arise. (See §9.)

4. CONJUGATE AND NORMALIZED QUATERNIONS. INVERSE QUATERNIONS

The conjugate $\mathbb{A}^*$ of a quaternion $[\![a, \mathbf{A}]\!]$ is the quaternion

$$\mathbb{A}^* =_{\text{def}} [\![a, -\mathbf{A}]\!]. \tag{1}$$

When $\mathbb{A}$ is written in the binary form (2.12), $a + A\mathbb{n}$, $\mathbb{A}^*$ is $a - A\mathbb{n}$, which shows the similarity between the quaternion conjugate and the conjugate of a complex number. It is easy to prove (Problem 1) that

$$(\mathbb{A}\mathbb{B})^* = \mathbb{B}^*\mathbb{A}^*. \tag{2}$$

The *norm* $|\mathbb{A}|$ of a quaternion $\mathbb{A}$, as that of a complex number, is defined as the square root of $\mathbb{A}\mathbb{A}^*$:

2.2,2.5 $$\mathbb{A}\mathbb{A}^* = \llbracket a, \mathbf{A}\rrbracket\llbracket a, -\mathbf{A}\rrbracket = \llbracket a^2 + A^2, \mathbf{0}\rrbracket = a^2 + A^2 =_{\text{def}} |\mathbb{A}|^2. \tag{3}$$

A quaternion of unit norm is called a *normalized quaternion,* not to be confused with a unit quaternion $\llbracket 0, \mathbf{n}\rrbracket$, with $n^2 = 1$, which is a pure normalized quaternion:

$$|\mathbb{n}|^2 = \llbracket 0, \mathbf{n}\rrbracket\llbracket 0, -\mathbf{n}\rrbracket = \llbracket \mathbf{n}\cdot\mathbf{n}, \mathbf{0}\rrbracket = 1. \tag{4}$$

Given a quaternion $\llbracket a, \mathbf{A}\rrbracket$ written in the forms $\llbracket a, A\mathbf{n}\rrbracket$ or $a + A\mathbb{n}$, ($|\mathbf{n}| = 1$), the simplest way in which the normalization condition,

3 $$a^2 + \mathbf{A}^2 = 1, \tag{5}$$

can be satisfied is by writing $a = \cos\alpha$, $A = \sin\alpha$, that is

$$\mathbb{A} = \llbracket \cos\alpha, \sin\alpha\ \mathbf{n}\rrbracket = \cos\alpha + \sin\alpha\ \mathbb{n}, \qquad |\mathbf{n}| = 1. \tag{6}$$

Equally well, the rotation quaternion in the Euler–Rodrigues parameters (2.14) is also normalized:

$$\mathbb{A} = \llbracket \cos\tfrac{1}{2}\phi, \sin\tfrac{1}{2}\phi\ \mathbf{n}\rrbracket = \cos\tfrac{1}{2}\phi + \sin\tfrac{1}{2}\phi\ \mathbb{n}, \qquad |\mathbf{n}| = 1. \tag{7}$$

The advantage of the normalized form (7) is, of course, that ϕ is the rotation angle, whereas α in (6) is double that angle. The difference between the two forms, however, is rather more subtle than that, since, under certain conditions, α in (6) can be claimed to be a 'rotation angle', obviously in a different sense from that in (7). Since this parametrization, the use of which in the rotation group is entirely unacceptable, was employed by Hamilton (see §9), a normalized quaternion in the form (6), shall be called a *Hamilton quaternion.* Such forms were called *versors* by Hamilton himself. The normalized form (7) was first used by Rodrigues, before Hamilton's invention of quaternions, and shall be called a *Rodrigues quaternion.* The conjugate $\mathbb{A}^*$ of the quaternion (7) is given by $\llbracket \cos\tfrac{1}{2}\phi, -\sin\tfrac{1}{2}\phi\ \mathbf{n}\rrbracket$ and is obtained by reversing the sign of ϕ, thus denoting the inverse rotation. We shall in fact see that the conjugate quaternion is deeply connected to the inverse quaternion, now to be defined.

The *inverse* $\mathbb{A}^{-1}$ of $\mathbb{A}$ is defined by the condition

,2.5 $$\mathbb{A}\mathbb{A}^{-1} = \llbracket 1, \mathbf{0}\rrbracket = 1. \tag{8}$$

On the other hand, from (3),

$$\mathbb{A}\mathbb{A}^*\,|\mathbb{A}|^{-2} = 1, \tag{9}$$

whence

9|8 $$\mathbb{A}^{-1} = \mathbb{A}^*\,|\mathbb{A}|^{-2}. \tag{10}$$

Clearly, in (3),

$$\mathbb{A}^*\mathbb{A} = |\mathbb{A}|^2, \tag{11}$$

whence (10) is also the left inverse, i.e. it satisfies the relation $\mathbb{A}^{-1}\mathbb{A}$ equals unity. Since the only quaternion of zero norm is the null quaternion $[\![0, \mathbf{0}]\!]$, it is clear from (10) that the inverse of a non-vanishing quaternion always exists. This means, that, as long as the denominator does not vanish, two quaternions can always be divided, because

$$\mathbb{A}/\mathbb{B} = \mathbb{C} \qquad \Rightarrow \qquad \mathbb{C} = \mathbb{A}\mathbb{B}^{-1} = \mathbb{A}\mathbb{B}^* \, |\mathbb{B}|^{-2}, \tag{12}$$

an operation which can always be performed if $|\mathbb{B}|$ is not zero, that is, if $\mathbb{B}$ is not null.

It is clear from (10) that for a normalized quaternion,

$$\mathbb{A}^{-1} = \mathbb{A}^*, \tag{13}$$

whence the inverse of a normalized quaternion is itself normalized. We have already remarked that the conjugate of (7) denotes the inverse rotation.

It can readily be proved (Problem 2) that the product of two normalized quaternions is itself a normalized quaternion. This is a most important property: complex numbers represent rotations in the plane because the product of two complex numbers of unit modulus, $\exp(\mathrm{i}\alpha)\exp(\mathrm{i}\beta)$ is itself a complex number of unit modulus, $\exp\{\mathrm{i}(\alpha + \beta)\}$. As regards the Rodrigues quaternions, which are, of course, normalized, we have seen in Problem **9**-4.1 that their product is also normalized and it is clear that if this were not so this product could not parametrize a rotation.

Problems 12-4 (p. 294)

1. Prove that $(\mathbb{A}\mathbb{B})^* = \mathbb{B}^*\mathbb{A}^*$.
2. Prove that the product of two normalized quaternions is a normalized quaternion.
3. Prove that $\mathbb{A}$ is a pure quaternion if and only if

$$\mathbb{A}^* = -\mathbb{A}. \tag{14}$$

4. Prove that the norm of the product of two quaternions is equal to the product of their norms:

$$|\mathbb{A}\mathbb{B}| = |\mathbb{A}| \, |\mathbb{B}|. \tag{15}$$

5. Given a pure quaternion $\mathbb{r}$ and a quaternion $\mathbb{R}$, prove that

$$\bar{\mathbb{r}} = \mathbb{R}\mathbb{r}\mathbb{R}^* \tag{16}$$

is a pure quaternion of norm

$$|\bar{\mathbb{r}}| = |\mathbb{R}|^2 \, |\mathbb{r}|. \tag{17}$$

5. THE QUATERNION UNITS

Given a quaternion $[\![a, \mathbf{A}]\!]$, it is natural to express the vector $\mathbf{A}$ in components

$$\mathbf{A} = A_x\mathbf{i} + A_y\mathbf{j} + A_z\mathbf{k}. \tag{1}$$

Consider first a vector $\mathbf{A}$ of the form $A_x\mathbf{i}$. Then, as in (2.9) and (2.12), we write

$$[\![a, A]\!] = [\![a, \mathbf{0}]\!] + [\![0, A_x\mathbf{i}]\!] = a + A_x[\![0, \mathbf{i}]\!] =_{\text{def}} a + A_x\mathbb{i}, \tag{2}$$

in an obvious notation. It is clear that in dealing with this problem, it will be useful to define unit quaternions of the type given in (R2) for all three vectors $\mathbf{i}$, $\mathbf{j}$, and $\mathbf{k}$. These quaternions will be called the *quaternion units*:

$$\mathbb{i} = [\![0, \mathbf{i}]\!], \qquad \mathbb{j} = [\![0, \mathbf{j}]\!], \qquad \mathbb{k} = [\![0, \mathbf{k}]\!]. \tag{3}$$

It is traditional in the quaternion literature not to make a notational distinction between a pure quaternion $[\![0, \mathbf{n}]\!]$ and the vector $\mathbf{n}$ so that the symbols $\mathbf{i}, \mathbf{j}, \mathbf{k}$ are normally used for the quaternion units in (3). As explained in §2, however, pure quaternions are really binary rotations and their identification with their corresponding vectors is somewhat precarious, so that, as in (3), we shall continue making a notational difference. Notice that the quaternion units in (3) are the three unit quaternions corresponding to the orthogonal coordinate axes. Because

$$\mathbf{i}^2 = \mathbf{j}^2 = \mathbf{k}^2 = 1, \qquad \mathbf{i}\times\mathbf{j} = \mathbf{k}, \qquad \mathbf{j}\times\mathbf{k} = \mathbf{i}, \qquad \mathbf{k}\times\mathbf{i} = \mathbf{j}, \tag{4}$$

it is very easy to multiply the quaternions in (3):

$$\mathbb{i}^2 = \mathbb{j}^2 = \mathbb{k}^2 = [\![-1, \mathbf{0}]\!] = -1, \tag{5}$$

$$\mathbb{i}\mathbb{j} = [\![-\mathbf{i}\,.\,\mathbf{j}, \mathbf{i}\times\mathbf{j}]\!] = [\![0, \mathbf{k}]\!] = \mathbb{k}, \tag{6}$$

$$\mathbb{j}\mathbb{i} = [\![-\mathbf{j}\,.\,\mathbf{i}, \mathbf{j}\times\mathbf{i}]\!] = [\![0, -\mathbf{k}]\!] = -\mathbb{k}. \tag{7}$$

The other products can be similarly obtained and are displayed in Table 1 where, for completeness, we have also added the real quaternion 1 equal to $[\![1, 0]\!]$, which is the rotation identity, $\phi = 0$ in (4.7). It is clear that $\mathbb{i}$, $\mathbb{j}$, and $\mathbb{k}$ are binary rotations around $\mathbf{x}, \mathbf{y}, \mathbf{z}$ and thus respectively identical to the rotations C_{2x}, C_{2y}, C_{2x}, as displayed in the table. In fact, if we consider the multiplication table of these rotations in Table **8**-1.3, and incorporate in the products the corresponding projective factors therein given, the agreement between the two tables is clear. Of course, the multiplication rules just described, with the explicit phase factors, are those that must be obeyed by the SU(2) matrices $\boldsymbol{I}, \boldsymbol{I}_x, \boldsymbol{I}_y, \boldsymbol{I}_z$ that represent the identity and the three BB rotations of $\mathbf{D}_2$. We also identify this mapping in Table 1, which can then be seen to coincide exactly with Table **8**-1.2 for $\mathbf{D}_2$.

Table **12**-5.1. Multiplication of quaternion units. The product ab appears on the intersection of the a row with the b column.

			1 E $\boldsymbol{I}$	$\mathbb{i}$ C_{2x} $\boldsymbol{I}_x$	$\mathbb{j}$ C_{2y} $\boldsymbol{I}_y$	$\mathbb{k}$ C_{2z} $\boldsymbol{I}_z$
1	E	$\boldsymbol{I}$	1	$\mathbb{i}$	$\mathbb{j}$	$\mathbb{k}$
$\mathbb{i}$	C_{2x}	$\boldsymbol{I}_x$	$\mathbb{i}$	-1	$\mathbb{k}$	$-\mathbb{j}$
$\mathbb{j}$	C_{2y}	$\boldsymbol{I}_y$	$\mathbb{j}$	$-\mathbb{k}$	-1	$\mathbb{i}$
$\mathbb{k}$	C_{2z}	$\boldsymbol{I}_z$	$\mathbb{k}$	$\mathbb{j}$	$-\mathbb{i}$	-1

We have already remarked in (1.3) that a binary rotation squares exactly like an imaginary unit and we see in Table 1 that $\mathbb{i}$, $\mathbb{j}$, and $\mathbb{k}$ behave in this sense like three imaginary units. We can now complete the work started in eqn (2):

$$[\![a, \mathbf{A}]\!] = [\![a, \mathbf{0}]\!] + [\![0, \mathbf{A}]\!] = a + [\![0, A_x\mathbf{i} + A_y\mathbf{j} + A_z\mathbf{k}]\!] \tag{8}$$

$$= a + A_x[\![0, \mathbf{i}]\!] + A_y[\![0, \mathbf{j}]\!] + A_z[\![0, \mathbf{k}]\!] \tag{9}$$

$$= a + A_x\mathbb{i} + A_y\mathbb{j} + A_z\mathbb{k}. \tag{10}$$

It should be noticed that we are extending in eqn (9) the additive notation introduced in (2.9). As in that case, the justification for doing so is that, on using (10) with the multiplication rules for the quaternion units in Table 1, the multiplication rule for quaternions is satisfied (Problem 1).

Since $\mathbb{i}$, $\mathbb{j}$, and $\mathbb{k}$ behave like imaginary units, eqn (10) shows up the meaning of quaternions as an extension of the complex numbers. This was indeed Hamilton's motivation for his invention: it was his remarkable intuition first, to realize that the complex field cannot be extended by adding just one more imaginary unit and, secondly, to guess that the three units that had to be introduced had to be multiplied as in Table 1. The set of elements $1, \mathbb{i}, \mathbb{j}, \mathbb{k}$ in Table 1 is said to be the *basis* of an *algebra* over the field of the reals, the elements of the algebra being linear combinations of the basis with real coefficients. That is, the elements of the algebra are the quaternions themselves in the form given by (10). The quaternion algebra obeys all the usual arithmetical laws except that the multiplication is non-commutative. As shown in (4.12) this algebra has the property that the quotient of any two elements of the algebra, $\mathbb{A}/\mathbb{B}$, always exists, except when $\mathbb{B}$ is the zero quaternion. An algebra with this property is called a *division algebra* and it considerably contributed to the fame of quaternions as an algebraic structure when Frobenius proved in 1878 that there are only three such algebras: the real numbers themselves, the complex numbers and the quaternions. (See Pontryagin 1966, p. 160, or Littlewood 1950, p. 233.)

It should be noticed that all coefficients of the basis in (10) are scalars. A pure quaternion in (10) takes the form $A_x\mathbb{i}+A_y\mathbb{j}+A_z\mathbb{k}$, which Hamilton, making no distinction between a quaternion unit such as $\mathbb{i}$ and the corresponding unit vector $\mathbf{i}$, wrote as $A_x\mathbf{i}+A_y\mathbf{j}+A_z\mathbf{k}$. He then coined the word *vector* for such an object. It must be stressed, however, that 1, $\mathbb{i}$, $\mathbb{j}$ and $\mathbb{k}$ are the identity (rotation by zero in eqn 4.7) and three BB rotations. They are, however, not isomorphic to the group $\mathbf{D}_2$ defined by these operations, since it is clear from Table 1 that they do not form a group at all. Comparison of this table with Table **8**-1.3 for $\mathbf{D}_2$ shows that these four elements form a projective representation of $\mathbf{D}_2$ with the projective factor system given in that table. That is, the quaternion units (including the identity 1) form a spinor representation of $\mathbf{D}_2$, the covering or double group of which is the quaternion group discussed in §**8**-6. This description of the quaternion units is, of course, lost in the Hamilton parametrization (4.6) where α must take the value $\frac{1}{2}\pi$ for a quaternion unit.

Although eqn (10) is so easy to derive, it has a deep significance in the rotation group. The quaternion on (L10) is an arbitrary rotation of SO(3) and the three quaternions on (R10) are the three BB rotations around the coordinate axes. Equation (10) thus states that any arbitrary rotation of SO(3) can be expressed in terms of the three fundamental BB rotations. This does not mean, of course, that an arbitrary rotation of SO(3) can be expressed as a product of these three BB's, which we know from §**3**-5 is not possible. We know on the other hand from §**6**-4 that the SU(2) matrices $\boldsymbol{I}_x$, $\boldsymbol{I}_y$, $\boldsymbol{I}_z$ that represent the BB's are the infinitesimal generators of SO(3) and thus capable of generating all rotations of SO(3).

Problems 12-5 (p. 295)

1. Given

$$[\![a, \mathrm{A}]\!] = a + A_x\mathbb{i} + A_y\mathbb{j} + A_z\mathbb{k}; \qquad [\![b, B]\!] = b + B_x\mathbb{i} + B_y\mathbb{j} + B_z\mathbb{k}, \quad (11)$$

prove that they multiply by the rule in eqn (2.2).

2. Prove that

$$\mathbb{i}\mathbb{j}\mathbb{k} = -1. \quad (12)$$

6. SO(3), SU(2), AND QUATERNIONS

The connection between SO(3) and quaternions hardly needs stressing. From our point of view normalized quaternions $[\![a, \mathbf{A}]\!]$, with $a + A^2 = 1$, can always be identified with the Rodrigues $[\![\lambda, \mathbf{\Lambda}]\!]$ quaternions and thus with rotations of SO(3). Clearly, the set of all *normalized* quaternions forms a group, closure being provided by Problem 4.2, and the inverse of $\mathbb{A}$ by $\mathbb{A}^*$. This group unfortunately, cannot be called the quaternion group, a name reserved for the double group of $\mathbf{D}_2$. We shall call it the *group of quaternions*, Q, not to be confused with the quaternion group of Table **8**-6.2, which is a subgroup of Q. SO(3) maps onto Q as follows:

$$g \in \mathrm{SO}(3) \qquad g \mapsto \pm \mathbb{A}, \qquad \mathbb{A} \in Q. \tag{1}$$

This precisely parallels the homomorphism between SU(2) and SO(3) given in §**7**-3:

$$g \in \mathrm{SO}(3), \qquad g \mapsto \pm A, \qquad A \in \mathrm{SU}(2). \tag{2}$$

In fact, SU(2) and Q are isomorphic, as discussed below.

Before we do this, it is useful to note that the mapping (1) can be made one-to-one by imposing the standardization condition (**11**-5.1) on the Rodrigues quaternion thus ruling out the negative sign in (1). The set of all normalized quaternions that satisfy this standardization form a faithful projective representation of SO(3) because: (i) then satisfy the multiplication law (**11**-5.2) for rotations; (ii) this multiplication law is associative (Problem 2.1). The identity is $[\![1, \mathbf{0}]\!]$.

In order to establish more precisely the isomorphism between Q and SU(2) we re-write slightly the additive expression (5.10) for a quaternion $\mathbb{A}$:

$$\mathbb{A} = [\![a, \mathbf{A}]\!] = a1 + A_x \mathbb{i} + A_y \mathbb{j} + A_z \mathbb{k}. \tag{3}$$

It follows from Table 5.1 that the quaternions 1, $\mathbb{i}$, $\mathbb{j}$, $\mathbb{k}$ multiply precisely as the SU(2) matrices for the identity, $\boldsymbol{I}$, and the three BB rotations of $\mathbf{D}_2$, $\boldsymbol{I}_x, \boldsymbol{I}_y, \boldsymbol{I}_z$. We copy these matrices from Table **8**-1.1:

$$\begin{matrix} \boldsymbol{I} & \boldsymbol{I}_x & \boldsymbol{I}_y & \boldsymbol{I}_z \\ \begin{bmatrix} 1 & \\ & 1 \end{bmatrix} & \begin{bmatrix} & -i \\ -i & \end{bmatrix} & \begin{bmatrix} & -1 \\ 1 & \end{bmatrix} & \begin{bmatrix} -i & \\ & i \end{bmatrix}. \end{matrix} \tag{4}$$

It is thus clear that the quaternion $\mathbb{A}$ in (3) can be represented by a matrix $\check{A}$ obtained by replacing 1, $\mathbb{i}$, $\mathbb{j}$ and $\mathbb{k}$ in that equation by $\boldsymbol{I}, \boldsymbol{I}_x, \boldsymbol{I}_y, \boldsymbol{I}_z$ respectively:

$$\check{A} = \begin{bmatrix} a - iA_z & -A_y - iA_x \\ A_y - iA_x & a + iA_z \end{bmatrix}. \tag{5}$$

Thus, a rotation $R(\lambda; \mathbf{\Lambda})$ with quaternion $[\![\lambda, \mathbf{\Lambda}]\!]$ will be represented by the matrix

$$\check{R}^{1/2}(\lambda; \mathbf{\Lambda}) = \begin{bmatrix} \lambda - i\Lambda_z & -\Lambda_y - i\Lambda_x \\ \Lambda_y - i\Lambda_x & \lambda + i\Lambda_z \end{bmatrix}. \tag{6}$$

As we must expect, this matrix is identical with the SU(2) matrix previously obtained in eqn (**11**-8.6). The quaternion $[\![-\lambda, -\mathbf{\Lambda}]\!]$, which corresponds to the same rotation just considered, is clearly mapped by the negative of the matrix in (6). The isomorphism between Q and SU(2) is thus established. Since SU(2) is the covering group of SO(3) (see §**11**-8), it follows that Q is also the covering group of SO(3). This is perhaps the most important property of quaternions. It might be worthwhile reminding the reader that the reason why there is a two-to-one homomorphism from Q onto SO(3) is that, in the Rodrigues parametrization, the quaternions $[\![\cos\frac{1}{2}\phi, \sin\frac{1}{2}\phi\ \mathbf{n}]\!]$ and $[\![\cos\frac{1}{2}(\phi+2\pi), \sin\frac{1}{2}(\phi+2\pi)\ \mathbf{n}]\!]$ must map to the same rotation $R(\phi\mathbf{n})$. (Clearly, the second of these quaternions is the negative of the first.) It is also opportune to reflect how the very elementary geometry that has led us to the Rodrigues parametrization in Chapter **9** provides such a deep-seated insight into the structure of the rotation group.

It is useful to notice that the trace of the matrix in (6) is 2λ, that is $2\cos\frac{1}{2}\phi$, so that binary rotations ($\phi=\pi$) must be represented by traceless matrices as $\boldsymbol{I}_x, \boldsymbol{I}_y, \boldsymbol{I}_z$ in (4) indeed are. A binary rotation about an arbitrary axis $\mathbf{r}=(x, y, z)$, with $x^2+y^2+z^2=1$, will be represented by the traceless matrix

6
$$\check{R}^{1/2}(\pi;\mathbf{n}) =_{\text{def}} \check{\mathbb{r}} = \begin{bmatrix} -iz & -y-ix \\ y-ix & iz \end{bmatrix}. \tag{7}$$

We write this matrix as $\check{\mathbb{r}}$ since it represents the unit quaternion $\mathbb{r}=[\![0,\mathbf{r}]\!]$. Since (7) is a SU(2) matrix, we must expect its determinant to be unity. In fact,

$$\det\check{\mathbb{r}} = x^2+y^2+z^2=1. \tag{8}$$

Problems 12-6 (p. 295)

1. Verify that the matrix $\check{\mathbb{r}}$ in (7) satisfies the relation $\check{\mathbb{r}}^2=-\mathbf{1}_2$, in agreement with (2.13).
2. Call $\check{R}^{1/2}[\![\lambda,\boldsymbol{\Lambda}]\!]$ the matrix $\check{R}^{1/2}(\lambda;\boldsymbol{\Lambda})$ in (6), since it is the representative of $[\![\lambda,\boldsymbol{\Lambda}]\!]$. Prove with this notation that $\check{R}^{1/2}[\![\lambda,\boldsymbol{\Lambda}]\!]^*$ is $\{\check{R}^{1/2}[\![\lambda;\boldsymbol{\Lambda}]\!]\}^{-1}$.

Warning: notice that $\check{R}^{1/2}[\![\lambda;\boldsymbol{\Lambda}]\!]^*$ is not $\{\check{R}^{1/2}[\![\lambda;\boldsymbol{\Lambda}]\!]\}^*$.

7. EXPONENTIAL FORM OF QUATERNIONS

Consider two normalized quaternions in the Hamilton parametrization (4.6), written in their binary forms (2.12):

$$\mathbb{A}=\cos\alpha+\mathbb{n}\sin\alpha, \qquad \mathbb{B}=\cos\beta+\mathbb{n}\sin\beta. \tag{1}$$

(Notice that we assume a common rotation axis.) It follows at once, on remembering that $\mathbb{n}^2$ equals -1, that their product is a quaternion of the same form:

$$\mathbb{A}\mathbb{B} = \cos(\alpha+\beta) + \mathbb{n}\sin(\alpha+\beta). \tag{2}$$

It follows, in precisely the same way as for complex numbers, that it is convenient to introduce a symbolic exponential notation in (1):

$$\mathbb{A} = \cos\alpha + \mathbb{n}\sin\alpha =_{\text{def}} \exp(\mathbb{n}\alpha); \qquad \mathbb{B} = \exp(\mathbb{n}\beta), \tag{3}$$

since then the product (2) can be written very quickly:

3 $$\mathbb{A}\mathbb{B} = \exp\{\mathbb{n}(\alpha+\beta)\} =_{\text{def}} \cos(\alpha+\beta) + \mathbb{n}\sin(\alpha+\beta). \tag{4}$$

As a difference with complex numbers, this notation suffers the severe limitation that the unit quaternions in (1) or (3) must be identical.

An unnormalized quaternion can easily be expressed in exponential form, for which purpose it must be written as follows:

2.12,4.3 $$\mathbb{A} = |\mathbb{A}|\,(a\,|\mathbb{A}|^{-1} + A\,|\mathbb{A}|^{-1}\mathbb{n}). \tag{5}$$

The quaternion in the bracket is now normalized and it can be written as in (3) as

$$\mathbb{A} = |\mathbb{A}|\,(\cos\alpha + \mathbb{n}\sin\alpha) = |\mathbb{A}|\exp(\mathbb{n}\alpha). \tag{6}$$

8. THE CONICAL TRANSFORMATION

The problem we want to solve in this section is this: to find the transform of an arbitrary vector **r** under an arbitrary rotation $R(\phi\mathbf{n})$. This is the conical transformation of a vector depicted in Fig. **9**-5.1. We appear to have here a difficult problem, because we do not know how to express a vector **r** in the quaternion language, the pure quaternion $[\![0, \mathbf{r}]\!]$ being, strictly speaking, a binary rotation rather than a vector.

In order to keep within the (normalized) quaternion language, the only objects that we can talk about are rotations. So, we start with the observation that the only way in which we can denote points (or vectors) in the configuration space is by using the poles of rotations on the sphere. This means that in order to fix the point **r** (that we want to transform) we mark on the sphere the pole of a rotation $R(\phi'\mathbf{r})$, for any arbitrary rotation angle ϕ'. The quaternion for this rotation is $[\![\cos\frac{1}{2}\phi', \sin\frac{1}{2}\phi'\,\mathbf{r}]\!]$ and, obviously, it will simplify matters if we take ϕ' equal to π, not so much because the quaternion becomes pure but, rather, because its vector part becomes precisely **r**. As before, we denote this quaternion with the symbol $\mathbb{r} = [\![0, \mathbf{r}]\!]$. So, we have now fixed the point **r** on the sphere as the pole of $R(\pi\mathbf{r})$. The next thing we do is to rotate this pole **r** under the arbitrary rotation $R(\phi\mathbf{n})$ specified for the work. The pole **r** goes into a new pole $\bar{\mathbf{r}}$ which, from (**9**-1.5), must be the pole of the rotation conjugate of $R(\pi\mathbf{r})$, of which $\mathbf{r}$ is the pole, by $R(\phi\mathbf{n})$:

$$\bar{\mathbf{r}} = R(\phi\mathbf{n})\mathbf{r} = \text{pole of } \{R(\phi\mathbf{n})R(\pi\mathbf{r})R(\phi\mathbf{n})^{-1}\}. \tag{1}$$

We also know, (see eqn **11**-6.6), that conjugation leaves invariant the rotation angle. Thus the rotation in braces in (1) has pole $\bar{\mathbf{r}}$ and angle π and it is $R(\pi\bar{\mathbf{r}})$: all that we need in order to find $\bar{\mathbf{r}}$ is to form the product of the three rotations on (R1) and thus obtain its rotation axis. Of course, we need nothing more for this purpose than to multiply the three corresponding quaternions. The quaternion corresponding to $R(\phi\mathbf{n})$ is the standard Rodrigues quaternion

$$\mathbb{R} = [\![\cos \tfrac{1}{2}\phi, \mathbf{n} \sin \tfrac{1}{2}\phi]\!]. \tag{2}$$

The quaternion corresponding to $R(\pi\mathbf{r})$ is $[\![0, \mathbf{r}]\!]$, which we have called $\mathbb{r}$. The quaternion corresponding to $R(\phi\mathbf{n})^{-1}$ is the inverse of $\mathbb{R}$ in (2), i.e. from (4.13),

$$\mathbb{R}^* = [\![\cos \tfrac{1}{2}\phi, -\mathbf{n} \sin \tfrac{1}{2}\phi]\!]. \tag{3}$$

As we have said, the product of the three rotations in (1) will be the binary rotation around $\bar{\mathbf{r}}$, $R(\pi\bar{\mathbf{r}})$ with quaternion $[\![0, \bar{\mathbf{r}}]\!]$ which, in analogy with (1) will be written as

$$\bar{\mathbb{r}} = [\![0, \bar{\mathbf{r}}]\!]. \tag{4}$$

All that we now have to do is to write

$$\bar{\mathbb{r}} = \mathbb{R}\mathbb{r}\mathbb{R}^*, \tag{5}$$

introduce (2) and (3) into this equation, and turn the handle of the quaternion multiplication rule:

$$\bar{\mathbb{r}} = [\![\cos \tfrac{1}{2}\phi, \mathbf{n} \sin \tfrac{1}{2}\phi]\!][\![0, \mathbf{r}]\!][\![\cos \tfrac{1}{2}\phi, -\mathbf{n} \sin \tfrac{1}{2}\phi]\!] \tag{6}$$

$$= [\![\cos \tfrac{1}{2}\phi, \mathbf{n} \sin \tfrac{1}{2}\phi]\!][\![\mathbf{r} \cdot \mathbf{n} \sin \tfrac{1}{2}\phi, \mathbf{r} \cos \tfrac{1}{2}\phi + (\mathbf{n} \times \mathbf{r}) \sin \tfrac{1}{2}\phi]\!] \tag{7}$$

$$= [\![0, \cos \phi \;\; \mathbf{r} + \sin \phi (\mathbf{n} \times \mathbf{r}) + (1 - \cos \phi)(\mathbf{n} \cdot \mathbf{r})\mathbf{n}]\!] \tag{8}$$

(In going from 7 to 8 a couple of lines are needed in the style of eqn **9**-5.2 onwards.) On comparing (4) and (8) we obtain $\bar{\mathbf{r}}$ in exact agreement with (**9**-5.4).

It is worthwhile summarizing what we have done, which we can express as follows:

$$\mathbf{r} \mapsto \bar{\mathbf{r}} = R\mathbf{r} \qquad \Rightarrow \qquad \mathbb{r} \mapsto \bar{\mathbb{r}} = \mathbb{R}\mathbb{r}\mathbb{R}^*. \tag{9}$$

That is: when $\mathbf{r}$ is rotated into $R\mathbf{r}$, then the rotation $\mathbb{r}$ (of which $\mathbf{r}$ is the pole) gets changed into the rotation $\bar{\mathbb{r}}$ conjugate of $\mathbb{r}$ under $\mathbb{R}$. In other words, the transformation of $\mathbf{r}$ into $\bar{\mathbf{r}}$ in configuration space induces a transformation of the corresponding rotations $\mathbb{r}$ into $\bar{\mathbb{r}}$ in operator space. The pure quaternions $\mathbb{r}$ and $\bar{\mathbb{r}}$ are not vectors, clearly, but operators (binary rotations) which are transformed in order to extract from them some information about configuration space via their poles. It should be noticed that, if $\mathbb{r}$ in (9) is a pure quaternion, as we have assumed, then $\bar{\mathbb{r}}$, from eqns (4.16) and (4.17), must also be a pure quaternion of the same norm (as we have indeed found) since the Rodrigues quaternion $\mathbb{R}$ is always normalized. It is clear that if $\mathbb{R}$ were not normalized, then the transformation entailed in (9) would change the pure quaternion $\mathbb{r}$ into a pure quaternion $\bar{\mathbb{r}}$ of different norm. Such a transformation could be described as a rotation accompanied by a 'stretching'.

Historically, the transformation (R9) of $\mathbb{r}$ by any normalized quaternion $\mathbb{R}$ was the way in which Cayley, Hamilton, and their followers made explicit the relation between quaternions and rotations. It was noticed, as in Problem 4.5 that, as long as $\mathbb{R}$ is normalized, the pure quaternion $\mathbb{r}$ transforms under (9) into another pure quaternion $\bar{\mathbb{r}}$ of the same norm (or as it was loosely said, the vector $\mathbf{r}$ transforms into the vector $\bar{\mathbf{r}}$, an expression which, when taken literally, can cause some confusion). Hence (9) must be a rotation. In doing this, the relation between $\mathbb{R}$ and the Rodrigues quaternion was ignored. Rather, (R9) was taken as the definition of the rotation of $\mathbb{r}$ into $\bar{\mathbb{r}}$ by $\mathbb{R}$ (although $\mathbb{r}$ and $\bar{\mathbb{r}}$ were in fact written as plain vectors $\mathbf{r}$ and $\bar{\mathbf{r}}$ respectively), and then the product of two rotations was obtained as follows. It was observed that, if $\mathbb{R}_2\mathbb{r}\mathbb{R}_2^*$ is a rotation of $\mathbb{r}$ into $\bar{\mathbb{r}}$ for $\mathbb{R}_2$ normalized, then for $\mathbb{R}_1$ normalized, $\mathbb{R}_1\bar{\mathbb{r}}\mathbb{R}_1^*$ must be $\bar{\bar{\mathbb{r}}}$ say. But $\mathbb{R}_1\bar{\mathbb{r}}\mathbb{R}_1^*$ is $\mathbb{R}_1\mathbb{R}_2\mathbb{r}\mathbb{R}_2^*\mathbb{R}_1^*$ which, from (4.2) is $\mathbb{R}_1\mathbb{R}_2\mathbb{r}(\mathbb{R}_1\mathbb{R}_2)^*$. Thus, the product of the quaternions $\mathbb{R}_1\mathbb{R}_2$ is a quaternion $\mathbb{R}_3$, say, that takes $\mathbb{r}$ into $\bar{\bar{\mathbb{r}}}$. Because $\mathbb{R}_1\mathbb{R}_2$, being a product of normalized quaternions, must be itself normalized from Problem 4.2, then $\bar{\bar{\mathbb{r}}}$ must be a pure quaternion of the same norm as $\mathbb{r}$. Thus, $\mathbb{R}_1\mathbb{R}_2$ must itself be a rotation. Naturally, the problem remained of determining the angle and axis of the rotation associated with the quaternion $\mathbb{R}$ but this, if done at all, was done by reversing the proof leading from (6) to (8), although most mathematical treatments would not bother with such practical details.

This work was paralleled in SU(2) with the transformations of the traceless SU(2) matrix $\check{\mathbb{r}}$ of (6.7). It is clear that this is the SU(2) representative of the pure quaternion $\mathbb{r}$ with vector $\mathbf{r}$ of components x, y, z, which is the reason for the symbol $\check{\mathbb{r}}$ that we have adopted. The matrix $\check{\mathbb{r}}$ can in fact be considered as a realization of the vector $|x, y, z\rangle$. In the same manner, the matrix $\check{R}^{1/2}(\lambda; \mathbf{\Lambda})$ of (6.6) is the representative of the quaternion $\mathbb{R}=[\![\lambda, \mathbf{\Lambda}]\!]$ and can be written conveniently as $\check{R}$. It can now be held that any rotation $\mathbf{r}\mapsto\bar{\mathbf{r}}$, is a transformation that changes the traceless matrix $\check{\mathbb{r}}$ into another traceless matrix $\check{\bar{\mathbb{r}}}$:

$$\check{\mathbb{r}} = \begin{bmatrix} -iz & -y-ix \\ y-ix & iz \end{bmatrix} \mapsto \check{\bar{\mathbb{r}}} = \begin{bmatrix} -i\bar{z} & -\bar{y}-i\bar{x} \\ \bar{y}-i\bar{x} & i\bar{z} \end{bmatrix}. \tag{10}$$

Now, a similarity of $\check{\mathbb{r}}$ with a SU(2) matrix must preserve trace and determinant. In fact, if we write,

$$\check{\bar{\mathbb{r}}} = \check{\mathbb{R}}\check{\mathbb{r}}\check{\mathbb{R}}^{-1}, \tag{11}$$

then

$$\mathrm{Tr}\, \check{\bar{\mathbb{r}}} = \mathrm{Tr}\, (\check{\mathbb{R}}\check{\mathbb{r}}\check{\mathbb{R}}^{-1}) = \mathrm{Tr}\, (\check{\mathbb{R}}^{-1}\check{\mathbb{R}}\check{\mathbb{r}}) = \mathrm{Tr}\, \check{\mathbb{r}} = 0 \tag{12}$$

$$\det \check{\bar{\mathbb{r}}} = \det \check{\mathbb{R}} \det \check{\mathbb{r}} \det (\check{\mathbb{R}}^{-1}) = \det \check{\mathbb{r}}, \tag{13}$$

since the determinant of the SU(2) matrix $\check{\mathbb{R}}$ must be unity. Equation (12) entails that $\check{\bar{\mathbb{r}}}$ must indeed have the form (10), and (13) that $\bar{x}^2 + \bar{y}^2 + \bar{z}^2$ must equal $x^2 + y^2 + z^2$, which means that the SU(2) matrix $\check{\mathbb{R}}$ entails a rotation. Again, historically, this was the way in which the relation between SU(2) matrices and rotations was established. Notice that, from Problem 6.2, $\check{\mathbb{R}}^{-1}$ is the matrix representative of $\mathbb{R}^*$, a relation which can be written as $\check{\mathbb{R}}^{-1} = (\mathbb{R}^*)^{\vee}$. With this notation, eqn (11) takes the form $\check{\bar{\mathbb{r}}} = \check{\mathbb{R}}\check{\mathbb{r}}(\mathbb{R}^*)^{\vee}$, which exactly parallels (5), as it should be.

Problems 12-8 (p. 295)

1. Show that if $\mathbf{n}$ and $\mathbf{r}$ are orthogonal, then the conical transformation of $\mathbf{r}$ under $R(\phi\mathbf{n})$ entails a rotation of $\mathbf{r}$ by ϕ in a plane normal to $\mathbf{n}$.
2. Prove that if $\mathbf{r}$ is transformed under a binary rotation $R(\pi\mathbf{n})$ then in eqn (9), $\mathbb{R}\mathbb{r}\mathbb{R}^* = [\![0, -\mathbf{r}]\!]$ if and only if $\mathbf{r} \cdot \mathbf{n} = 0$.

9. THE RECTANGULAR TRANSFORMATION

In our counter-historical account of quaternions we now come to the point where Hamilton made his first important contact between quaternions and rotations. Remember that the existence of Rodrigues and his parametrization of rotations was unknown to Hamilton, that the conical transformation had not yet been discovered, and that, for Hamilton, the pure quaternion $\mathbb{r}$ and the vector $\mathbf{r}$ were one and the same thing. Consider now a rotation $R(\phi\mathbf{n})$, represented by a quaternion $\mathbb{R} = [\![\lambda, \mathbf{\Lambda}]\!]$ (although we must not assume this quaternion to be a Rodrigues quaternion). We ask the question: under what conditions will $\mathbb{R}\mathbb{r}$ be a pure quaternion $\bar{\mathbb{r}}$? The answer is simple. We have

$$\mathbb{R}\mathbb{r} = [\![\lambda, \mathbf{\Lambda}]\!][\![0, \mathbf{r}]\!] = [\![-\mathbf{\Lambda} \cdot \mathbf{r}, \lambda\mathbf{r} + \mathbf{\Lambda} \times \mathbf{r}]\!], \tag{1}$$

so that (R1) will be a pure quaternion if and only if $\mathbf{\Lambda} \cdot \mathbf{r} = 0$, that is $\mathbf{n} \cdot \mathbf{r} = 0$. This means that $\mathbf{r}$ must be orthogonal to $\mathbf{n}$. If this condition is satisfied, we can write (1) as

$$\mathbb{R}\mathbb{r} = \bar{\mathbb{r}}, \qquad \bar{\mathbb{r}} = [\![0, \lambda\mathbf{r} + \mathbf{\Lambda} \times \mathbf{r}]\!], \tag{2}$$

and if we take seriously the idea that $\mathbb{r}$ and $\mathbf{r}$ are the same thing, then we interpret (2) as a rectangular rotation of $\mathbf{r}$ (that is, a rotation of $\mathbf{r}$ around an axis normal to it) that takes $\mathbf{r}$ into $\bar{\mathbf{r}} = \lambda\mathbf{r} + \boldsymbol{\Lambda}\times\mathbf{r}$. It is very easy to solve this equation for λ and $\boldsymbol{\Lambda}$, which we do, for simplicity, assuming $|\mathbf{r}| = 1$:

$$\lambda = \mathbf{r}\,.\,\bar{\mathbf{r}}, \qquad \boldsymbol{\Lambda} = \mathbf{r}\times\bar{\mathbf{r}}. \tag{3}$$

Continuing in the same mood, if the angle of this alleged rotation is α (remember we know nothing about ϕ!), the angle between r and $\bar{r}$ is α, and because $\mathbf{n}$, the unit axis of rotation, is perpendicular to $\mathbf{r}$, we have from (3)

$$\lambda = \cos\alpha, \qquad \boldsymbol{\Lambda} = \mathbf{n}\sin\alpha. \tag{4}$$

If you knew nothing about our previous work, you would be perfectly happy about (4), specially if it were you who was inventing the rules of the game: you associate with the rotation R an 'angle of rotation' α, an axis of rotation $\mathbf{n}$ and a quaternion $[\![\cos\alpha, \mathbf{n}\sin\alpha]\!]$.

Let us now interpret what goes on *à la* Rodrigues. Equation (2), rather than a putative vector transformation, is a statement of the result of Problem **9**-4.7, whereby the product of an arbitrary rotation $R(\phi\mathbf{n})$, with quaternion $[\![\cos\frac{1}{2}\phi, \mathbf{n}\sin\frac{1}{2}\phi]\!]$, times a binary rotation around an axis $\mathbf{r}$, is another binary rotation around an axis $\bar{\mathbf{r}}$, if and only if, $\mathbf{n}\,.\,\mathbf{r}$ vanishes. It also follows from that problem that the angle between $\mathbf{r}$ and $\bar{\mathbf{r}}$ is $\frac{1}{2}\phi$: since we are not rotating any vectors but rather multiplying rotations, there is no reason whatsoever why $\mathbf{r}$ should be rotated into $\bar{\mathbf{r}}$ by the rotation angle ϕ of the rotation. This is the reason why the angle α in the Hamilton parametrization is not the true rotation angle that must be associated with the quaternion. In fact, if we rotate $\mathbf{r}$ now correctly by the conical transformation, taking the special case for $\mathbf{r}\,.\,\mathbf{n} = 0$, we do find that the angle between $\mathbf{r}$ and $\bar{\mathbf{r}}$ is ϕ; that is, twice Hamilton's 'angle of rotation' α. (Problem 8.1.)

We can see in (2) how dangerous it is to identify a pure quaternion $\mathbb{r}$ with its vector $\mathbf{r}$. If this is done, one is led to an ungeometrical angle of rotation and therefore to clumsy multiplication rules for rotations. Hamilton, however, preferred to stick to the parametrization (4) whereby he had to leave unexplained why the conical rotation rotates a vector by twice the 'angle of rotation'. One can see, though, why Hamilton liked so much this approach. First, eqn (2) gave him a geometrical picture of a quaternion as a quotient of two vectors: $\mathbb{R} = \bar{\mathbf{r}}/\mathbf{r}$. Hamilton could thus say that, given two vectors $\bar{\mathbf{r}}$ and $\mathbf{r}$, the quaternion $\mathbb{R}$ is the operator that rotates $\mathbf{r}$ into $\bar{\mathbf{r}}$, and he was, of course, aware of the important algebraic significance of being able to divide vectors. The second reason he had for liking the parametrization (4) is that it identifies a quaternion unit such as $\mathbb{i}$, for which $\lambda = 0$, $\boldsymbol{\Lambda} = \mathbf{i}$, with a rotation by $\frac{1}{2}\pi$ thus 'explaining' why $\mathbb{i}^2 = -1$ since, in (2) (with $\mathbb{R}$ replaced by $\mathbb{i}$)$\mathbb{i}^2$, that is a rotation by π in his parametrization, changes $\mathbf{r}$ into $-\mathbf{r}$. Although there were good reasons for

Hamilton's parametrization, his identification of the rotation angle caused a great deal of trouble and it was probably responsible for the very little use that was made of quaternions in the rotation group. As for the identification of pure quaternions with vectors, eqn (2) and its consequences are a warning that this cannot be done without some care, in which case a roundabout technique, such as that employed in the conical transformation, must be used.

10. QUATERNION ALGEBRA AND THE CLIFFORD ALGEBRA

We shall now show that the quaternion algebra is a particular case of a more general algebra invented by Clifford (1878). The purpose of this excursion is not merely cultural: at the end of this section we shall obtain an important insight into the work of the whole book. The motivation for the introduction of the Clifford algebra is the following. Consider an n-dimensional vector space with vectors of real components $x_1, x_2, \ldots, x_n$. The square of the norm, N^2, of such a vector is $\sum x_i^2$. N is given by the square root of this summation and the goal we have in mind is to *linearize* N, for which purpose the summation $\sum x_i^2$ has to be written as a perfect square. The strategy we adopt is to introduce unit vectors $e_1, e_2, \ldots, e_n$, the properties of which will be chosen so as to force the result wanted. That is, whereas we had

$$N^2 = \sum_i x_i^2, \tag{1}$$

we attempt to write it as

$$N^2 = \left(\sum_i x_i e_i\right)^2 = \sum_i x_i^2 e_i^2 + \sum_{i<j} x_i x_j (e_i e_j + e_j e_i). \tag{2}$$

We are here, of course, inventing the rules of the game as we play it, and what we are inventing is some 'vector product' under which $(\sum x_i e_i)^2$, e_i^2 and $e_i e_j$ all acquire meaning. (Notice that we do not assume it to be commutative.) In order to get (R2) to coincide with (R1) it is sufficient to choose this vector product (which has nothing to do with $\mathbf{a} \times \mathbf{b}$!) so that

$$e_i^2 = 1, \qquad e_i e_j + e_j e_i = 0, \qquad \forall i, j. \tag{3}$$

Equation (3) is equivalent to

$$e_i e_j + e_j e_i = 2\delta_{ij}. \tag{4}$$

It is clear that with this multiplication rule we have achieved our goal since N can now be written as $\pm \sum x_i e_i$.

The next step in this invention is to form a hierarchy of objects very much like the tensorial hierarchy defined in Chapter **5**. Given the unit vectors $e_1, e_2, \ldots, e_n$, we can form products E_m of m factors of them:

$$E_m = e_1 e_2 \ldots e_n. \tag{5}$$

Let us call $|E_m|$ the number of such products, which is the number of combinations of m elements chosen from n elements, $\binom{n}{m}$. If we call $\{E_m\}$ the set of all elements E_m with $m = 0, 1, 2 \ldots n$, and $E_0 =_{\mathrm{def}} 1$, then the set formed by $\{E_0\}, \{E_1\}, \ldots, \{E_n\}$ contains a total number of elements that we shall call $|\mathscr{C}_n|$ given by

$$|\mathscr{C}_n| = \sum_{m=0}^{n} \binom{n}{m} = 2^n. \tag{6}$$

Given this set of 2^n elements, we can form linear combinations

$$a_0 1 + a_1 e_1 + \ldots + a_n e_n + a_{n+1} e_1 e_2 + a_{n+2} e_1 e_3 + \ldots + a_{|\mathscr{C}_n|-1}\, e_1 e_2 \ldots e_n. \tag{7}$$

The set of all such linear combinations over all the real numbers a_i is called the *Clifford algebra* $\mathscr{C}_n$. If we want to visualize what is going on, we can assume that a basis is being formed by the 2^n *basic elements* of the algebra $1, e_1, e_2, \ldots, e_n, e_1 e_2, e_2 e_3, \ldots, e_1 e_2 \ldots e_n$, as a direct sum of the sets $\{E_0\}, \{E_1\}, \ldots, \{E_n\}$ and that this basis spans the algebra, very much as a basis spans a representation in group theory.

Consider as an example $\mathscr{C}_3$. This Clifford algebra is spanned by the following elements:

$$\{E_0\}: \quad 1, \tag{8}$$

$$\{E_1\}: \quad e_1, e_2, e_3, \tag{9}$$

$$\{E_2\}: \quad e_1 e_2, e_2 e_3, e_3 e_1, \tag{10}$$

$$\{E_3\}: \quad e_1 e_2 e_3. \tag{11}$$

Notice that the total number of the basic elements of the algebra is 8, i.e. 2^3, as required by (6).

Given the Clifford algebra $\mathscr{C}_n$ with basis $\{E_0\} \oplus \{E_1\} \oplus \{E_2\} \oplus \ldots \oplus \{E_n\}$, the sets $\{E_0\}, \{E_2\}, \{E_4\} \ldots$ span a sub-algebra of $\mathscr{C}_n$ since the product of two 'even' subsets will always be an 'even' subset. As an example, the product of two elements of $\{E_2\}$ in (10) is

4,4
$$e_1 e_2 e_2 e_3 = e_1 e_3 = -e_3 e_1, \tag{12}$$
4,4,4
$$e_1 e_2 e_3 e_1 = -e_2 e_1 e_3 e_1 = e_2 e_3 e_1 e_1 = e_2 e_3. \tag{13}$$

(Notice that in eqn 12 we use the fact that $e_3 e_1 \in \{E_2\}$ and let the negative sign be absorbed by the coefficients that appear in the algebra, as in eqn 7.) The even sub-algebra of $\mathscr{C}_n$ thus defined is called $\mathscr{C}_{n+}$.

Let us now consider the basis for $\mathscr{C}_{3+}$ which must come from (8) and (10). Since, as discussed above, we can change signs in the basis, we shall take

$$\mathscr{C}_{3+}: 1, e_2 e_1, e_3 e_2, e_1 e_3, \tag{14}$$

since, from (4), e_2e_1 equals $-e_1e_2$ and so forth. We shall now prove that the set (14) is isomorphic to the quaternion units in Table 5.1, under the mapping

$$1 \leftrightarrow 1, \qquad \mathbb{i} \leftrightarrow e_2e_1, \qquad \mathbb{j} \leftrightarrow e_3e_2, \qquad \mathbb{k} \leftrightarrow e_1e_3. \tag{15}$$

All we need to do in order to establish the isomorphism is to use the multiplication rule (4) of the Clifford algebra units, and compare with the result of the products for the quaternion units in Table 5.1:

$$\mathbb{i}^2 = -1 \mapsto e_2e_1e_2e_1 = -e_2e_2e_1e_1 = -1, \tag{16}$$

and similarly for $\mathbb{j}^2$ and $\mathbb{k}^2$. Also,

$$\mathbb{i}\mathbb{j} = \mathbb{k} \mapsto e_2e_1e_3e_2 = -e_1e_2e_3e_2 = e_1e_3e_2e_2 = e_1e_3. \tag{17}$$

$$\mathbb{j}\mathbb{k} = \mathbb{i} \mapsto e_3e_2e_1e_3 = -e_2e_3e_1e_3 = e_2e_1e_3e_3 = e_2e_1. \tag{18}$$

$$\mathbb{k}\mathbb{i} = \mathbb{j} \mapsto e_1e_3e_2e_1 = -e_3e_1e_2e_1 = e_3e_2e_1e_1 = e_3e_2. \tag{19}$$

We see that in every case the element on the right is the image of the correct quaternion unit. On permuting (R17) to (R19) the results $\mathbb{j}\mathbb{i} = -\mathbb{k}$, etc., follow at once. It can be seen that the Clifford algebra $\mathscr{C}_{3+}$, which is the set

$$a_1 1 + a_2e_2e_1 + a_3e_3e_2 + a_4e_1e_3, \tag{20}$$

is nothing else than the set

$$a_1 1 + a_2\mathbb{i} + a_3\mathbb{j} + a_4\mathbb{k}, \tag{21}$$

which is the set of all quaternions in the additive form (5.10).

In praise of mirrors

What we have proved above is that binary rotations (for the quaternion units are nothing else) span the even Clifford algebra of dimension 3, $\mathscr{C}_{3+}$, in the form (21). The elements of the algebra, i.e. the elements of the form (21) are quaternions, i.e. rotations. In other words, what we have seen above is that $\mathscr{C}_{3+}$ is realized by SO(3). We shall now show that $\mathscr{C}_3$ itself is realized by O(3).

Before we come to this, let us observe, on comparing (9) and (10), that, because the BBs realize $\mathscr{C}_{3+}$, they cannot themselves be the simplest (and thus the most fundamental) elements of $\mathscr{C}_3$, but, rather, that they must be composites of the elements that realize $\{E_1\}$ which, as we shall now prove, are the orthogonal mirrors $\sigma_x, \sigma_y, \sigma_z$ perpendicular to **x**, **y**, **z** respectively. In order to do this we recall that, from Table **11**-9.4, if we absorb the projective factor with the corresponding product, we have

$$\sigma_y\sigma_x + \sigma_x\sigma_y = 0, \tag{22}$$

and similar relations for the other two pairs obtained by cyclic permutation. Also, from this table,

$$\sigma_x^2 = \sigma_y^2 = \sigma_z^2 = 1, \tag{23}$$

where, in the manner of (15) we write the identity as unity to facilitate comparison with (3). We assert that the mapping

$$\sigma_x \leftrightarrow e_3, \qquad \sigma_y \leftrightarrow e_1, \qquad \sigma_z \leftrightarrow e_2, \tag{24}$$

satisfies the isomorphism. This is obvious, since $e_3^2 = 1$, thus satisfying the first of (23) and since $e_1e_3 + e_3e_1$ equals zero, from (4), thus satisfying (22). Notice two things: first, e_2e_1 is mapped by $\sigma_z\sigma_y$ which from Table **11**-9.4 equals C_{2x}. This operation, in its turn, can be identified with $\mathbb{i}$ which, correctly, is the image of e_2e_1 in (15). (This is the reason why the mapping was chosen in the form given by eqn 24.) Secondly, on accepting the mapping of E_0, i.e. of 1, onto the identity, the mapping (24) of the Clifford units onto the orthogonal mirrors entails the mapping of the whole of $\mathscr{C}_3$ onto O(3). $\mathscr{C}_3$, in fact, is given by (8) to (11) in terms of $\{E_0\}$, $\{E_1\}$, $\{E_2\}$ and $\{E_3\}$. $\{E_0\}$ maps onto the identity and $\{E_1\}$ maps by (24) onto the orthogonal mirrors. $\{E_2\}$ maps onto the BBs, since we have just seen that the mapping (24) entails exactly the mapping (15). All that remains is to find the mapping of $\{E_3\}$ in (11), entailed by (24):

$$e_1e_2e_3 \leftrightarrow \sigma_y\sigma_z\sigma_x = -C_{2x}\sigma_x = -C_{2x}\boldsymbol{i}C_{2x} = \boldsymbol{i}. \tag{25}$$

Here, $\sigma_y\sigma_z$ is obtained from Table **11**-9.4, on absorbing the projective factor into the product in the manner used in this section. Likewise, when we write σ_x as $\boldsymbol{i}C_{2x}$, the product $C_{2x}C_{2x}$ is written as -1. With (25), we have verified that $\mathscr{C}_3$ maps onto the whole of O(3); and all this thanks to the identification of the Clifford units with the orthogonal mirrors.

Up to this point in the book, the reader had every right to believe that BBs are the salt of the earth. He must now realize that, if this be so, then orthogonal mirrors are manna from heaven. Heaven, alas, is not what we have been primarily concerned with in this book: we have rather kept our feet firmly grounded in $\mathbb{R}^3$. If this is what one wants, there are enormous advantages in basing one's work on proper rotations via BBs, and thus constructing SO(3) in the knowledge that the whole of O(3) can be generated as the direct product of SO(3) with $\mathbf{C}_i$. It is clear, however, from the work of this section, that the soundest theoretical procedure is to take orthogonal mirrors as the basic operations, since they generate at a stroke all proper and improper rotations (see Problem **2**-1.5 and eqn 25). This approach, as we have said, is not the most practical when dealing with $\mathbb{R}^3$, but it has nevertheless fundamental advantages when spaces of other dimensions are studied. In $\mathbb{R}^2$ (a plane) the inversion does not exist, being indistinguishable from a binary rotation around a point of the plane and, in general, the inversion is not a symmetry operation in $\mathbb{R}^{2n}$, for n integer. Reflections, on the other hand, are meaningful in $\mathbb{R}^2$, as it is obvious, and in all $\mathbb{R}^n$, for any integral n. Thus, any work on symmetries in general spaces $\mathbb{R}^n$ should be based on reflections, as done by Cartan (1938), rather than rotations and the inversion. In $\mathbb{R}^3$, however, rotations are so convenient to use, via the quaternion algebra, that it pays to introduce these lower-grade operations rather than have to deal with a very general, but less practical, algebra. (It should be noticed, however, that Biedenharn's turns, which afford an interesting graphical depiction of rotations, are obtained by expressing $R(\phi\mathbf{n})$ as the product of two mirrors at an angle $\frac{1}{2}\phi$, as given in Fig. **2**-1.2. A full description of these operations is given in Biedenharn and Louck 1981, Chapter 4.) The superiority of mirrors as the fundamental symmetries, is indubitable: not only do they generate all rotations proper and improper, as well as the inversion, as mentioned above, but they also generate all translations, as shown in Problem **2**-1.12.

11. APPLICATIONS: ANGLE AND AXIS OF ROTATION AND SU(2) MATRICES IN TERMS OF EULER ANGLES

As an illustration of the power of the quaternion algebra in dealing with rotations, we shall determine the Euler–Rodrigues or quaternion parameters

$$\lambda = \cos\tfrac{1}{2}\phi, \qquad \boldsymbol{\Lambda} = \sin\tfrac{1}{2}\phi\ \mathbf{n}, \tag{1}$$

in terms of the Euler angles. From the definition of the latter (see §**3**-1), $R(\lambda;\boldsymbol{\Lambda})$ is given by the product of the three rotations $R(\alpha\mathbf{k})R(\beta\mathbf{j})R(\gamma\mathbf{k})$ in that order. The quaternion parameters for these rotations can be obtained from (1) at once whence, for the corresponding quaternions,

$$[\![\lambda, \boldsymbol{\Lambda}]\!] = [\![\cos\tfrac{1}{2}\alpha, \sin\tfrac{1}{2}\alpha\ \mathbf{k}]\!][\![\cos\tfrac{1}{2}\beta, \sin\tfrac{1}{2}\beta\ \mathbf{j}]\!][\![\cos\tfrac{1}{2}\gamma, \sin\tfrac{1}{2}\gamma\ \mathbf{k}]\!] \tag{2}$$

$$= [\![\cos\tfrac{1}{2}\alpha, \sin\tfrac{1}{2}\alpha\ \mathbf{k}]\!][\![\cos\tfrac{1}{2}\beta\cos\tfrac{1}{2}\gamma, \cos\tfrac{1}{2}\beta\sin\tfrac{1}{2}\gamma\,\mathbf{k} + \sin\tfrac{1}{2}\beta\cos\tfrac{1}{2}\gamma\ \mathbf{j} + \sin\tfrac{1}{2}\beta\sin\tfrac{1}{2}\gamma\ \mathbf{i}]\!], \tag{3}$$

where the quaternion multiplication rule (2.2) is used. On using this rule a second time and then comparing the scalar and vector parts of the quaternions on both sides of the resulting equation, one obtains

$$\lambda = \cos\tfrac{1}{2}\beta\ \cos\tfrac{1}{2}(\alpha+\gamma), \tag{4}$$

$$\Lambda_x = -\sin\tfrac{1}{2}\beta\ \sin\tfrac{1}{2}(\alpha-\gamma), \tag{5}$$

$$\Lambda_y = \sin\tfrac{1}{2}\beta\ \cos\tfrac{1}{2}(\alpha-\gamma), \tag{6}$$

$$\Lambda_z = \cos\tfrac{1}{2}\beta\ \sin\tfrac{1}{2}(\alpha+\gamma). \tag{7}$$

It is now trivial, from (1), to obtain ϕ and $\mathbf{n}$ in terms of the Euler angles:

$$\cos\tfrac{1}{2}\phi = \cos\tfrac{1}{2}\beta\,\cos\tfrac{1}{2}(\alpha+\gamma), \tag{8}$$

$$n_x = -(\sin\tfrac{1}{2}\phi)^{-1}\sin\tfrac{1}{2}\beta\,\sin\tfrac{1}{2}(\alpha-\gamma), \tag{9}$$

$$n_y = (\sin\tfrac{1}{2}\phi)^{-1}\sin\tfrac{1}{2}\beta\,\cos\tfrac{1}{2}(\alpha-\gamma), \tag{10}$$

$$n_z = (\sin\tfrac{1}{2}\phi)^{-1}\cos\tfrac{1}{2}\beta\,\sin\tfrac{1}{2}(\alpha+\gamma). \tag{11}$$

Because $-\pi<\phi\leqslant\pi$, it follows that $\cos\tfrac{1}{2}\phi$ must always be positive or zero but that $\sin\tfrac{1}{2}\phi$ is undetermined as to sign, whence $\mathbf{n}$, from (9) to (11), is also undetermined as to sign. Thus, the Euler angles cannot distinguish between $R(\phi\mathbf{n})$ and $R(-\phi, -\mathbf{n})$, as already shown in §**3**-4. Although these rotations are the same, we have seen that a distinction between them (i.e. between their poles) is necessary in order to satisfy the continuity conditions in SO(3). It is thus most important to realize that such continuity conditions in SO(3) are not necessarily satisfied when Euler angles are used.

On substituting (4) to (7) into the SU(2) matrix $\check{R}^{1/2}(\lambda; \boldsymbol{\Lambda})$ of (6.6), the corresponding matrix in terms of the Euler angles, $\check{R}^{1/2}(\alpha\beta\gamma)$ is obtained at once:

$$\check{R}^{1/2}(\alpha\beta\gamma) = \begin{bmatrix} \cos\tfrac{1}{2}\beta\,\exp\{-\tfrac{1}{2}\mathrm{i}(\alpha+\gamma)\} & -\sin\tfrac{1}{2}\beta\,\exp\{-\tfrac{1}{2}\mathrm{i}(\alpha-\gamma)\} \\ \sin\tfrac{1}{2}\beta\,\exp\{\tfrac{1}{2}\mathrm{i}(\alpha-\gamma)\} & \cos\tfrac{1}{2}\beta\,\exp\{\tfrac{1}{2}\mathrm{i}(\alpha+\gamma)\} \end{bmatrix}. \tag{12}$$

This equation coincides exactly with that given, e.g., in Biedenharn and Louck (1981), p. 24, or Messiah (1964), vol. 2, p. 1073, although other forms in other conventions also appear in the literature.

13

DOUBLE GROUPS

Groups were created by God; the double groups were invented by Hans Bethe.

Paraphrase of a famous saying by Kronecker ('The integers were created by God; all else is man-made'. See Weyl (1949), p. 33).

Within the approach of this book double groups need never be used. Everything that must be done with the double group $\tilde{G}$ can be done by using G and the projective representations corresponding to the spinor factor system, greatly gaining in precision as well as saving in time. It is natural, however, for those unfamiliar with the projective-representation method, to imagine that the savings in the work involved will be more than offset by the labour in learning the method. We shall see in the examples in Chapter **15** that this is not quite the case, for, given good spinor-representation tables obtained by the methods of this book, their use requires no special knowledge of projective representations.

Having said all this, why then do we need a chapter on double groups? There are various reasons for this. First, it would be unrealistic to assume that, however advisable, double groups might be dropped overnight. Secondly, given this realistic assumption, it is important to ensure that those who wish to use double groups become aware of the pitfalls in their use and of how to avoid these pitfalls. Thirdly, and most importantly, whereas the classical approach to double groups requires them to be treated as matrix groups, the quaternion parametrization permits their treatment as groups of operators, with considerable gain in precision. Fourthly, the Opechowski (1940) Theorem, which gives the vital class structure of double groups, can be proved in a very precise and general way by using the projective-representation method. This was first done by Backhouse (1973), but he used a factor system with a normalization condition that is unacceptable for spinor representations, which is that $[g, g^{-1}]$ must equal unity for all $g \in G$. This, of course, entails taking $[C_2, C_2]$ as unity for all binaries, contrary to our rules. On using the correct factor system from the point of view of continuity conditions, a much simpler proof of the Opechowski Theorem can be provided, as we shall see below (Altmann 1979). We have already given an introductory example of a double group in §**8**-6, and we shall start the work of this chapter with an illustration of the pitfalls in the usual treatment of double groups.

1. INTRODUCTION AND EXAMPLE

Given a point group G of operations g_i, each g_i is the image of two matrices of SU(2). In the projective-representation method, a one-to-one mapping between G and a set (not a group) of matrices of SU(2) is established, the doubling of the matrices being taken care of by the projective factors. In the double-group approach, the group G is replaced by a group $\tilde{G}$ of double the order, whence a one-to-one mapping can now be established between $\tilde{G}$ and a group of SU(2) matrices. When the matrices of a *vector* representation of $\tilde{G}$ are chosen that are the images of the operations $g \in G$, then the *projective* representations of G are obtained for the spinor factor system, so that $\tilde{G}$ is a covering group of G as defined in §**8**-6, but the double-group user does not need be aware of that. What we have stated so far is merely a sketch of a programme, which we shall now develop gradually to see as far as possible why as well as how things are done.

The doubling of a group G of elements $e, g_1, \ldots, g_i, \ldots$, is done as follows. An additional operation $\tilde{e}$ is introduced which can be regarded as a turn by 2π (and yet distinct from the identity e – see §**10**-6). It is clear that pre- or post-multiplication of any operation g with $\tilde{e}$ will not affect the result, since the only effect will be that of changing the class of homotopy of the path associated with g. (See §**10**-6.) It is convenient to assume that all elements g are defined by standard parameters (§**11**-5) and are thus associated with paths of class 0. Therefore, either $\tilde{e}g$ or $g\tilde{e}$ will indicate the operation g obtained as through a class 1 path, which will be designated with the symbol $\tilde{g}$ of a new group element. Clearly, also, the product $\tilde{e}\tilde{e}$, which entails a path with two jumps and therefore of class 0 (from the theorem in §**10**-4) must be the ordinary identity. The result of all this is that $\tilde{e}$, as the identity itself, commutes with all g:

$$\tilde{e}g = g\tilde{e} =_{\text{def}} \tilde{g}, \qquad \forall g, \tag{1}$$

and that

$$\tilde{e}\tilde{e} = e. \tag{2}$$

Hence, a new set $\tilde{G}$ is formed of order double that of G: $e, g_1, g_2, \ldots, g_i, \ldots, \tilde{e}, \tilde{g}_1, \tilde{g}_2, \ldots, \tilde{g}_i \ldots$. The multiplication rules of this set must be chosen so that it forms a group, and we shall now discuss them as far as we can. There are two basic principles that must obtain for these multiplication rules: (i) The multiplication rules for all $g \in G$ must *not* be preserved when they are treated as elements of $\tilde{G}$. (ii) The multiplication rules of $\tilde{G}$ must reduce to those of G when $\tilde{e}$ is identified with e. The second condition is pretty obvious, so we only need consider the first one. Suppose that, if the product $g_i g_j$ is g_k in G, it is also g_k in $\tilde{G}$ for all g_i, $g_j \in G$. Then G would be a subgroup of $\tilde{G}$ of index 2 which commutes with the group C defined by

$$C = e \oplus \tilde{e}. \tag{3}$$

It follows immediately that

$$\tilde{G} = G \otimes C, \qquad \text{(wrong)}, \tag{4}$$

which is impossible. The reason for this is as follows. C has only two irreducible representations, even and odd respectively with respect to $\tilde{e}$. All the irreducible representations of $\tilde{G}$ would then merely be all the irreducible (vector) representations of G doubled up as even and odd with respect to $\tilde{e}$. Thus, when going down to G (i.e. when taking $\tilde{e}$ as the identity) no new representation would arise, which is inadmissible. On combining the principles (i) and (ii) stated above, it follows that

$$g_i g_j = g_k \text{ in } G \qquad \Rightarrow \qquad \text{either } g_i g_j = g_k \text{ in } \tilde{G}, \tag{5}$$

$$\text{or} \quad g_i g_j = \tilde{g}_k \text{ in } \tilde{G}, \tag{6}$$

for some g_i, g_j belonging to G, in each case. Equations (5) and (6), though, must cover between them *all* the operations of G. This being so, it follows that the set $\tilde{G}$ closes: consider, e.g., the product $g_i g_j$ of any two operations of G. Then if (5) obtains, on multiplying both sides by $\tilde{e}$,

$$\tilde{g}_i g_j = \tilde{g}_k, \tag{7}$$

whereas, if (6) obtains, in the same way,

,2

$$\tilde{g}_i g_j = \tilde{e}\tilde{g}_k = \tilde{e}\tilde{e}g_k = g_k. \tag{8}$$

(R7) and (R8) belong to $\tilde{G}$. Similarly, it is found that the product $\tilde{g}_i \tilde{g}_j$ also belongs to $\tilde{G}$.

This is as far as anyone could go before the 1970s; the problem being that, whereas the product $g_i g_j$ was uniquely defined geometrically in G, there was no clear-cut rule to discriminate between the alternatives (5) and (6) in $\tilde{G}$. Opechowski (1940) had introduced a very simple scheme, which is at the basis of all the sound work on double groups done until that time. He established an isomorphism between $\tilde{G}$ and SU(2) matrices by associating an SU(2) matrix, $\check{R}^{1/2}(g_i)$ say, with each $g_i \in G$ and $-\check{R}^{1/2}(g_i)$ with each $\tilde{g}_i \in \tilde{G}$. Once this is done, the multiplication rules of this matrix group, which are, of course, well defined, are taken to be as the multiplication rules of the double group.

The problem with this approach, in its practical application, is how the isomorphism between $\tilde{G}$ and the SU(2) matrices is to be established. In principle, this isomorphism can be established in an entirely arbitrary way, as done for $\tilde{D}_2$ by Bradley and Cracknell (1972, p. 429):

$$\begin{matrix} E & C_{2x} & C_{2y} & C_{2z} \\ \begin{bmatrix} 1 & \\ & 1 \end{bmatrix} & \begin{bmatrix} & 1 \\ -1 & \end{bmatrix} & \begin{bmatrix} & i \\ i & \end{bmatrix} & \begin{bmatrix} i & \\ & -i \end{bmatrix}, \end{matrix} \tag{9}$$

the negative of these matrices being used for the elements with a tilde. The trouble with this approach is that the 'operations' C_{2x} etc., lose entirely all geometrical meaning, as will soon be verified. A more geometrical approach is based in assigning to each $g \in G$, with Euler angles α, β, γ, the positive matrix $\check{R}^{1/2}(\alpha\beta\gamma)$ given in (**12**-11.12). This, alas, cannot be done uniquely, as we shall now show. We list below the operations of $\mathbf{D}_2$, their Euler angles and then the positive matrix $\check{R}^{1/2}(\alpha\beta\gamma)$ (their negatives being assigned to the remaining operations of $\tilde{\mathbf{D}}_2$):

$$\mathbf{D}_2: \quad E \quad C_{2x} \quad C_{2y} \quad C_{2z} \tag{10}$$

$$\alpha\beta\gamma: \quad 000 \quad \begin{matrix}\pi\pi 0 \\ 0\pi\pi\end{matrix} \quad \begin{matrix}0\pi 0 \\ \pi\pi\pi\end{matrix} \quad 00\pi \tag{11}$$

$$\check{R}^{1/2}(\alpha\beta\gamma): \quad \begin{bmatrix} 1 & \\ & 1 \end{bmatrix} \quad \begin{bmatrix} & i \\ i & \end{bmatrix} \quad \begin{bmatrix} & -1 \\ 1 & \end{bmatrix} \quad \begin{bmatrix} -i & \\ & i \end{bmatrix} \tag{12}$$

Notice in (11) that, as it follows from §**3**-1, the Euler angles are not uniquely defined when $\beta = \pi$, the two possible alternatives for C_{2x} being shown and the matrix in (12) corresponding to the first of them, whereas the second set of Euler angles would lead to its negative. (The Euler angles for C_{2y} are also not unique, but both sets lead to the same matrix in eqn 12.) The very severe problem here is that, whereas both alternative matrices for C_{2x} in (12) lead to perfectly good multiplication rules for $\mathbf{D}_2$ – which can lure people into a false sense of security – the alternative displayed in (12) violates the Condon and Shortley convention, as I shall now show. The operations in (10) are, as we know, the infinitesimal generators $\boldsymbol{I}$ in §**6**-4, related to the Pauli matrices by the condition

6-4.7 $$\boldsymbol{\sigma} = i\boldsymbol{I}. \tag{13}$$

With this equation, we obtain the Pauli matrices from (12):

$$\begin{matrix} \boldsymbol{\sigma}_x & \boldsymbol{\sigma}_y & \boldsymbol{\sigma}_z \\ \begin{bmatrix} & -1 \\ -1 & \end{bmatrix} & \begin{bmatrix} & -i \\ i & \end{bmatrix} & \begin{bmatrix} 1 & \\ & -1 \end{bmatrix}, \end{matrix} \quad (\sigma_x \text{ wrong}). \tag{14}$$

On comparing with (**4**-9.9) it is clear that the first matrix here does not obey the Condon and Shortley convention, the correct one being obtained only when the sign of the C_{2x} matrix in (12) is changed, i.e. when the second set of Euler angles is used. It should be understood that the relation (13) is not one that is universally appreciated; hence the test that we have just described is not always made. Naturally, for some applications, departure from Condon and Shortley is unimportant and faulty choices such as (12) can be made without awesome consequences. If serious work on angular momentum coupling is attempted, in particular in relativistic calculations, it is a matter of life and death to ensure complete Condon-and-Shortley consistency and it is evident from the above that even for the simplest of groups, $\tilde{\mathbf{D}}_2$, this cannot automatically be satisfied when using the Euler-angle parametrization. Any reader who has had the patience to check the Euler angles in (11) will realize, moreover, that there are far pleasanter things in life to do. We shall now see that the quaternion parameters are not only instantly derived, but that they also give a perfect and unambiguous answer.

In order to use the quaternion parameters unambiguously, we must choose the positive hemisphere $\mathfrak{H}$, for which purpose we take the one depicted in Fig. **9**-1.1. Once this is done, the quaternion parameters $\lambda = \cos \frac{1}{2}\phi$ and $\mathbf{\Lambda} = \sin \frac{1}{2}\phi\ \mathbf{n}$ follow at once, as shown below:

$$\mathbf{D}_2: \qquad E \qquad C_{2x} \qquad C_{2y} \qquad C_{2z} \tag{15}$$

$$\lambda, \mathbf{\Lambda}: \qquad 1,(000) \qquad 0,(100) \qquad 0,(010) \qquad 0,(001) \tag{16}$$

$$\check{R}^{1/2}(\lambda;\mathbf{\Lambda}): \quad \begin{bmatrix} 1 & \\ & 1 \end{bmatrix} \quad \begin{bmatrix} & -\mathrm{i} \\ -\mathrm{i} & \end{bmatrix} \quad \begin{bmatrix} & -1 \\ 1 & \end{bmatrix} \quad \begin{bmatrix} -\mathrm{i} & \\ & \mathrm{i} \end{bmatrix} \tag{17}$$

(The reader should compare these matrices with eqn **6**-4.1.) The matrices (17) follow from (**12**-6.6) and on multiplication of (17) with i it can be seen at once that the correct Pauli matrices appear. There are two points that must be noticed. First, on comparing (9) with (17), it is clear that in the presentation of $\tilde{\mathbf{D}}_2$ given by (9) the elements C_{2x}, C_{2y}, C_{2z} cannot be given a geometrical meaning. This is possible, since after all one can always deal with the abstract group corresponding to $\tilde{\mathbf{D}}_2$, but highly impractical. Secondly, the matrices (17), augmented with the corresponding matrices for the elements with a tilde, lead to a multiplication table of $\tilde{\mathbf{D}}_2$. For instance:

$$C_{2x}C_{2y} = C_{2z}, \qquad\qquad C_{2y}C_{2x} = \tilde{C}_{2z}, \tag{18}$$

Since, in $\mathbf{D}_2$, both products equal C_{2z}, the first product here corresponds to case (5) and the second to case (6). When the multiplication table of $\tilde{\mathbf{D}}_2$ defined this way is obtained, it can be seen that it agrees entirely with Table **8**-6.2, deduced from the factor system. It can, in fact, be seen that (17) agrees with the representation in Table **8**-6.4.

2. THE DOUBLE GROUP IN THE QUATERNION PARAMETRIZATION

We have seen in §1 that it is only in the quaternion parametrization that the double group, defined *à la* Opechowski as a matrix group, can consistently be obtained. We shall now show that, even more importantly, the quaternion algebra allows the definition of the double group straightaway as an operator group, bypassing the construction of the Opechowski matrix group. This is a remarkable result: whereas the job of §1 was done, albeit badly in the Euler parametrization, the direct obtention of the double-group multiplication rules by means of the Euler parameters is impossible. This result, though, was to be expected.

We have seen (§**12**-6) that the group of quaternions Q is the covering group (or double group) of SO(3). Thus, the double groups of point groups, which are subgroups of SO(3), must be obtained as subgroups of Q, a result which we shall now derive in a constructive form. Consider a group G with a view to obtaining its double group $\tilde{G}$ as a group of quaternions. As usual, we must first define a suitable $\mathfrak{H}$ from which standard quaternions can immediately be written for all the proper rotations $g \in G$. As regards the improper rotations $\boldsymbol{i}g$ of G, they are also immediately parametrized, since the inversion is treated as the identity in forming poles, so that the quaternion for $\boldsymbol{i}g$ is chosen as the standard quaternion for g. (Notice that, therefore, within our operative rules, $\mathbf{C}_{3v}$ and $\mathbf{D}_3$, e.g., have identical parametrizations.) All that we now have to do in order to complete the parametrization of $\tilde{G}$ is to obtain the quaternion corresponding to $\tilde{e}$. Since we take this to be the turn by 2π, λ is $\cos \pi$, that is -1 and $\boldsymbol{\Lambda}$ is the null vector $\mathbf{0}$:

$$\tilde{e} \mapsto [\![-1, \mathbf{0}]\!]. \tag{1}$$

For an operation $\tilde{g} = g\tilde{e}$, therefore, if $[\![\lambda, \boldsymbol{\Lambda}]\!]$ is the (standard) quaternion for g, then the quaternion for $\tilde{g}$ is

12-2.1 $$\tilde{g} \mapsto [\![\lambda, \boldsymbol{\Lambda}]\!][\![-1, \mathbf{0}]\!] = [\![-\lambda, -\boldsymbol{\Lambda}]\!]. \tag{2}$$

We now have a one-to-one mapping of the n operations of $\tilde{G}$ onto a set of n quaternions. That this set is a group follows at once. Consider for this purpose the operations g_1 and g_2, the product of which, in G, is g_3. If $[\![\lambda, \boldsymbol{\Lambda}]\!]$ is the standard quaternion for g_3 we know (see eqn **11**-5.2 and the discussion below it) that the product of the quaternions of g_1 and g_2 can either give this standard quaternion or its negative $[\![-\lambda, -\boldsymbol{\Lambda}]\!]$. Whereas in the projective-representation method both these quaternions would be treated as the quaternions for g_3, the negative sign being used as a projective factor, we now recognize $[\![-\lambda, -\boldsymbol{\Lambda}]\!]$, from (2) as the quaternion for $\tilde{g}_3$. Thus, the quaternion set for $\tilde{G}$ forms a group isomorphic to $\tilde{G}$.

Strictly speaking, since the multiplication rules for $\tilde{G}$ cannot be independently established, what this statement means is that the multiplication rules of the group of quaternions that maps $\tilde{G}$ are chosen to define the multiplication rules of $\tilde{G}$. The major advantage of this approach over Opechowski's matrix realization of $\tilde{G}$ is that the quaternion rules have a clear geometrical significance. As an example, we give, from (1.16), the quaternion parameters for $\tilde{\mathbf{D}}_2$:

$$\begin{matrix} E & C_{2x} & C_{2y} & C_{2z} & \tilde{E} & \tilde{C}_{2x} & \tilde{C}_{2y} & \tilde{C}_{2z} \\ 1,(000) & 0,(100) & 0,(010) & 0,(001) & \bar{1},(000) & 0,(\bar{1}00) & 0,(0\bar{1}0) & 0,(00\bar{1}) \end{matrix} \qquad (3)$$

All one now needs in order to obtain the multiplication table for $\tilde{\mathbf{D}}_2$ is the quaternion multiplication rule (**12**-2.1). For the products in (1.18), e.g.,

$$C_{2x}C_{2y} \mapsto [\![0, (100)]\!][\![0, (010)]\!] = [\![0, (001)]\!] \leftrightarrow C_{2z}, \qquad (4)$$

$$C_{2y}C_{2x} \mapsto [\![0, (010)]\!][\![0, (100)]\!] = [\![0, (00\bar{1})]\!] \leftrightarrow \tilde{C}_{2z}, \qquad (5)$$

quite correctly. In the same way, the whole of Table **8**-6.2 can be constructed.

Having established the double group on a firm operational basis, we shall now study the general properties of products and conjugates in the double group, in order to determine its class structure (as a key to the number and dimensions of the irreducible representations).

3. NOTATION AND OPERATIONAL RULES

The first thing we must do is to establish an unambiguous notation, since it is clear from the work so far in this chapter, as well as from the remarks at the end of §**8**-6, that statements (such as multiplication rules, inverses and conjugates) valid for G are not valid for $\tilde{G}$. Thus, from Table **8**-1.3,

$$C_{2y}C_{2x} = C_{2z}, \qquad C_{2x}^{-1} = C_{2x}, \qquad (\text{in } \mathbf{D}_2), \qquad (1)$$

whereas from Table **8**-6.2,

$$C_{2y}C_{2x} = \tilde{C}_{2z}, \qquad C_{2x}^{-1} = \tilde{C}_{2x}, \qquad (\text{in } \tilde{\mathbf{D}}_2). \qquad (2)$$

Obviously, it would be serious if the product on the left of (1) were read as a product in $\tilde{\mathbf{D}}_2$ or if C_{2x}^{-1} in (2) were given its value in $\mathbf{D}_2$. Because products and inverses change, conjugates and classes also change in going from G to $\tilde{G}$. We introduce, for this reason, the following notation.

$g_i = g_j$	equality in G,	(3)
$g_i \simeq g_j$	equality in $\tilde{G}$,	(4)
g^{-1}	inverse in G,	(5)
$g^{\sim 1}$	inverse in $\tilde{G}$,	(6)

$$g_i \mathbf{c}\, g_j \qquad \text{conjugates in } G, \tag{7}$$

$$g_i \tilde{\mathbf{c}}\, g_j \qquad \text{conjugates in } \tilde{G}, \tag{8}$$

$$C(g_i) \qquad \text{class of } g_i \text{ in } G, \tag{9}$$

$$\tilde{C}(g_i) \qquad \text{class of } g_i \text{ in } \tilde{G}. \tag{10}$$

Thus, $C_{2y}C_{2x} = C_{2z}$ means that this result is valid in $\mathbf{D}_2$, but not necessarily in $\tilde{\mathbf{D}}_2$ (see eqns 1 and 2), whereas $C_{2x}C_{2y} \simeq C_{2z}$ (see Table **8**-6.2) means that this result is valid in $\tilde{\mathbf{D}}_2$. Likewise, C_{2x}^{-1} means C_{2x} (in $\mathbf{D}_2$) and $C_{2x}^{\sim 1}$ means $\tilde{C}_{2x}$ (in $\tilde{\mathbf{D}}_2$), as can be seen from (1) and (2). Also,

1.5 $$g_i g_j = g_k \quad \Rightarrow \quad \text{either } g_i g_j \simeq g_k, \tag{11}$$

1.6 $$\text{or} \quad g_i g_j \simeq \tilde{g}_k. \tag{12}$$

On multiplying (12) by $\tilde{e}$, as explained in §1, the tilde can be placed on top of any of the three elements therein. Rule (ii) of §1 (above eqn 1.3) entails the following results.

$$g_i g_j \simeq g_k \quad \Rightarrow \quad g_i g_j = g_k, \tag{13}$$

$$g_i g_j \simeq \tilde{g}_k \quad \Rightarrow \quad g_i g_j = g_k. \tag{14}$$

Inverses in $\tilde{G}$ are given by the following rules:

$$g^{\sim 1} \simeq g^{-1}, \qquad \text{(g not binary)}, \tag{15}$$

$$g^{\sim 1} \simeq \tilde{g}, \qquad \text{(g binary)}. \tag{16}$$

Result (15) follows from the product rule

$$gg^{-1} \simeq e, \qquad \text{(g not binary)}, \tag{17}$$

which can be verified as follows. If g is given by the quaternion $[\![\lambda, \mathbf{\Lambda}]\!]$, then g^{-1} is given by the quaternion $[\![\lambda, -\mathbf{\Lambda}]\!]$, if g is not binary, because in this case g and g^{-1} have the same angle but opposite axes of rotation (§**9**-1). Thus

$$gg^{-1} \simeq [\![\lambda, \mathbf{\Lambda}]\!][\![\lambda, -\mathbf{\Lambda}]\!] \simeq [\![\lambda^2 + \mathbf{\Lambda}^2, \mathbf{0}]\!] \simeq [\![1, \mathbf{0}]\!] \simeq e. \tag{18}$$

(The quaternion multiplication rules, as shown in §2, are valid in $\tilde{G}$ as long as eqns 2.1 and 2.2 are used.) When g is binary with axis of rotation $\mathbf{n}$ its quaternion is $[\![0, \mathbf{n}]\!]$, and, from (2.2) $\tilde{g}$ is $[\![0, -\mathbf{n}]\!]$, whence

$$g\tilde{g} \simeq [\![0, \mathbf{n}]\!][\![0, -\mathbf{n}]\!] \simeq [\![1, \mathbf{0}]\!] \simeq e, \tag{19}$$

thus verifying (16).

As regards the inverses of elements $\tilde{g}$, the following rules obtain

$$\tilde{g}^{\sim 1} \simeq \tilde{g}^{-1} \simeq_{\text{def}} \tilde{e}g^{-1}, \qquad \text{(g not binary)}, \tag{20}$$

$$\tilde{g}^{\sim 1} \simeq g, \qquad \text{(g binary)}. \tag{21}$$

(Notice the meaning of $\tilde{g}^{-1}$. This symbol cannot be interpreted as the inverse of $\tilde{g}$ in G since $\tilde{g}$ does not belong to G.) Equations (20) and (21) follow at once from (18) and (19), respectively, on multiplying both sides of these equations by $\tilde{e}\tilde{e}$. On comparing (20) and (21) with (15) and (16) it follows that, for all $\tilde{g}$ in $\tilde{G}$,

$$\tilde{g}^{-1} \simeq \tilde{e}g^{-1}. \tag{22}$$

We shall now consider conjugates. We say that $g_i \,\tilde{\mathbf{c}}\, g_j$ if:

$$\text{either} \quad g g_i g^{-1} \simeq g_j, \quad g \in G, \quad \text{or } \tilde{g} g_i \tilde{g}^{-1} \simeq g_j, \quad \tilde{g} \in \tilde{G}. \tag{23}$$

This is, of course, the usual group-theoretical definition of conjugation but some simplification can now be effected since, on applying (22) to the product $\tilde{g}g_i\tilde{g}^{-1}$ in (23), this takes the same form as (L23). Thus,

$$g_i \,\tilde{\mathbf{c}}\, g_j \quad \Leftrightarrow \quad g g_i g^{-1} \simeq g_j, \quad g \in G. \tag{24}$$

This relation will also be valid, clearly, if g_j in it is replaced by $\tilde{g}_j$. Likewise, because both sides of the equality in $\tilde{G}$ given in (24) can be multiplied by $\tilde{e}$, the following implication obtains:

$$g_i \,\tilde{\mathbf{c}}\, g_j \quad \Leftrightarrow \quad \tilde{g}_i \,\tilde{\mathbf{c}}\, \tilde{g}_j. \tag{25}$$

Conversely, the relations (11) and (12) provide the following implication:

$$g_i \,\mathbf{c}\, g_j \quad \Rightarrow \quad \text{either } g_i \,\tilde{\mathbf{c}}\, g_j, \tag{26}$$

$$\text{or} \quad g_i \,\tilde{\mathbf{c}}\, \tilde{g}_j. \tag{27}$$

Of course, (24) can be used once more on (R26) and (R27).

It readily follows from (24) that

$$g_i \,\tilde{\mathbf{c}}\, g_j \Rightarrow g_i \,\mathbf{c}\, g_j. \tag{28}$$

The reader should notice that the definitions given here of inverses and conjugates agree with the conventions adopted in §**11**-6. (See the discussion after eqn **11**-6.3.)

Intertwining

As defined in (**11**-2.1), we say that g_i and g_j are intertwined in G if $\exists g \in G$ such that $g\,g_i = g_j g$. It follows at once from (11) and (12) that the intertwining relation in G may or may not be preserved in $\tilde{G}$:

$$g\,g_i = g_j g \quad \Rightarrow \quad \text{either} \quad g\,g_i \simeq g_j g, \tag{29}$$

$$\text{or} \quad g\,g_i \simeq \tilde{g}_j g. \tag{30}$$

The following theorem establishes the conditions under which the alternatives (29) or (30) obtain.

Theorem 1. (*Intertwining theorem for double groups*). If g_i and g_j are operations of G intertwined by $g \in G$, then this intertwining is preserved in $\tilde{G}$, that is eqn (29) obtains, except in the case when g and g_i are either BB or orthogonal mirrors (irregular operations). In this latter case eqn (30) obtains and, because g and g_i commute in G, g_i, and g_j are identical.

Proof. We shall cannibalize for our proof the result of §§**11**-7 and **11**-10:

$$[g, g_i] = \pm[g_i^g, g], \tag{31}$$

where the negative sign obtains for g and g_i irregular and the positive sign obtains for g and g_i regular. First, we must notice that this result is relevant since from (L29) g_j is g_i^g. Thus, (31) is

$$[g, g_i] = \pm[g_j, g]. \tag{32}$$

Secondly, in (32), the positive and negative signs entail that the paths g, g_i and g_j, g are, or are not, in the same homotopy class. All that we need is to translate the notion of homotopy class into the double-group language. This is easy: a change in the class of homotopy entails a jump in parameter space, i.e. passing from a rotation by $-\pi$ to a rotation by π. That is, it entails a turn by 2π or, what is the same in double-group language, a factor of $\tilde{e}$. Thus, in (32) g g_i and g_jg will be equal in the double group when the positive sign obtains, i.e. for regular operations, whereas when the negative sign obtains in (32) (irregular operations) g g_i and g_jg will differ by a factor $\tilde{e}$ in the double group (eqn 30). This completes the proof. Naturally, a full proof could be provided from scratch for the double groups but that would merely entail a repetition of the quaternion work in §§**11**-7 and **11**-10.

It is important to summarize the results of Theorem 1 from the point of view of conjugation:

$$g_i \,\mathbf{c}\, g_j \quad \Rightarrow \quad g_i \,\tilde{\mathbf{c}}\, g_j, \qquad \text{(all regular and irregular } g_i, g_j), \tag{33}$$

$$g_i \,\mathbf{c}\, g_i \quad \Rightarrow \quad g_i \,\tilde{\mathbf{c}}\, \tilde{g}_i, \qquad \text{(all irregular } g_i). \tag{34}$$

Equation (33) follows from the fact that Theorem 1 states that (29) obtains as long as g and g_i are not irregular, but there is nothing here to preclude g_i and g_j being irregular. As an example: g_i and g_j can be BB and g a C_4 axis perpendicular to their plane. It is clear that this operation exchanges the poles of g_i and g_j and thus conjugates these operations, as required by the condition on (L29) that g_j be g_i^g.

As regards (34). Theorem 1 states that if g_i and g are irregular, in which case $gg_i = g_ig$, then (30) obtains in the form $gg_i \simeq \tilde{g}_ig$, which implies (34).

4. CLASS STRUCTURE: OPECHOWSKI THEOREM

In order to establish the class structure of $\tilde{\mathbf{G}}$ by means of the Opechowski Theorem we must first prove two theorems about conjugation.

Theorem 1. If $g_i \,\tilde{\mathbf{c}}\, \tilde{g}_j$, then $\exists g \in G$ that intertwines g_i and g_j in G but not in $\tilde{G}$:

$$g\, g_i = g_j g \qquad \text{and} \qquad g\, g_i \not\simeq g_j g. \tag{1}$$

(Notice that this result is valid when g_i and g_j are identical, in which case we are dealing with commutation.)

Proof. From the conjugation condition given and (3.24) it follows that $\exists g \in G$ such that

$$g\, g_i g^{\sim 1} \simeq \tilde{g}_j. \tag{2}$$

From (2),

$$g\, g_i \simeq \tilde{g}_j g, \tag{3}$$

which entails the first relation in (1). (See 3.14.) In order to prove the second relation in (1) assume its opposite:

$$g\, g_i \simeq g_j g. \tag{4}$$

On comparing (R4) and (R3), $\tilde{g}_j$ must equal g_j, which is absurd. This establishes the theorem.

Theorem 2. For any pair of operations g_i and g_j in G (regular or irregular),

$$g_i \,\mathbf{c}\, g_j \qquad \Rightarrow \qquad g_i \,\tilde{\mathbf{c}}\, g_j, \tag{5}$$

but *if and only if* g_i is irregular, then the following additional relation also holds:

$$g_i \,\mathbf{c}\, g_j \qquad \Rightarrow \qquad g_i \,\tilde{\mathbf{c}}\, \tilde{g}_j. \tag{6}$$

Proof. Equation (5) is merely a repetition of (3.33) from the intertwining theorem. If g_i is irregular, from (3.34) it is conjugate in $\tilde{G}$ to $\tilde{g}_i$ and, from (5), on using (3.25), $\tilde{g}_i$ is conjugate to $\tilde{g}_j$ whence, from transitivity of conjugation, (R6) follows. In order to prove that this happens only if g_i is irregular, assume (R6). This implies from (R5), in the form $\tilde{g}_i \,\tilde{\mathbf{c}}\, \tilde{g}_j$, that $g_i \,\tilde{\mathbf{c}}\, \tilde{g}_i$. It follows from Theorem 1 that there must be an operation of G that commutes with g_i in G but not in $\tilde{G}$. Thus, g_i must be irregular, since the only operations that commute in G and $\tilde{G}$ are coaxial rotations, whereas BBs and OMs commute only in G.

Theorem 3. (*Opechowski*). The class $C(g_i)$ of a regular operation $g_i \in G$ gives two classes in $\tilde{G}$ which are $\tilde{C}(g_i)$ and $\tilde{C}(\tilde{g}_i)$ respectively. If g_i is irregular, then $C(g_i)$ gives only one class in $\tilde{G}$, $\tilde{C}(g_i)$ which is identical with $\tilde{C}(\tilde{g}_i)$. (That is, this class contains all the g_j in $C(g_i)$ as well as their corresponding $\tilde{g}_j$ s).

Proof. If g_i is regular, then (5) obtains, whence all the elements in $C(g_i)$ are still conjugate to g_i in $\tilde{G}$ and are thus in $\tilde{C}(g_i)$. If an element g_j, say, belongs to $\tilde{C}(g_i)$, it must also belong to $C(g_i)$ from (3.28). Thus, $\tilde{C}(g_i)$ and $C(g_i)$ are identical. Also, (5) entails (R3.25), which means that all the elements $\tilde{g}_i$, for all $g_i \in C(g_i)$, form a second class. These two classes must be disjoint, of course, since the alternative (6) holds only when g_i is irregular. In this latter case, both (5) and (6) obtain, so that, from (5), $\tilde{C}(g_i)$ will contain the whole of $C(g_i)$ plus, from (6), all the elements of the form $\tilde{g}_i$, for all $g_i \in C(g_i)$.

As a corollary of the Opechowski Theorem, we can determine the number of spinor representations arising from the double group. Call $|C_G|$ and $|C_{\tilde{G}}|$ the number of classes in G and $\tilde{G}$ respectively and call $|r|$ and $|i|$ the number of regular and irregular classes, respectively, in G. That is,

$$|C_G| = |r| + |i|. \tag{7}$$

It follows from the Opechowski Theorem that

$$|C_{\tilde{G}}| = 2\,|r| + |i|. \tag{8}$$

Equation (8) gives the total number of irreducible representations of $\tilde{G}$, but not all these are spinor representations, since some of them must be the irreducible representations of G (see, e.g., Table **8**-6.3). The number of these irreducible representations of G is given by (7). Thus, in order to obtain the number of spinor representations we must subtract (7) from (8), whence *the total number of spinor representations equals the number of regular classes of G*. This result agrees with that obtained in the projective representation method in Theorem 1 of §**11**-4.

Examples of the use of double groups will be found in Chapter **15**. Before we tackle them, we need the representations of SO(3) of dimension $2j+1$, for j half-integral or integral, in all its values. These representations will be obtained in Chapter **14**.

Problem 13-4 (p. 295)

1. Use the Opechowski Theorem to enumerate the classes for the quaternion group $\tilde{\mathbf{D}}_2$ defined in Table **8**-6.2, and to count its spinor representations. Compare your results with Table **8**-6.3.

14

THE IRREDUCIBLE REPRESENTATIONS OF SO(3)

La poire est mûre, vous devez la cuillir.

Saint-Simon's last words (May 19, 1825), to Olinde Rodrigues. Quoted in Weill (1894).

The pear is ripe: you must pick it up.

We saw in §**4**-7 that the bases of the irreducible representations of SO(3) are of the form $\langle u^j_j, \ldots, u^j_{-j}|$. For each j, half-integral or integral, one has an irreducible representation of dimension $2j+1$, the dimension of the basis stated. The functions u^j_m ($m = j, j-1, \ldots, -j$) are the eigenfunctions of the angular momentum operator I_z and, if they are defined within the Condon and Shortley phase convention, then the representations are uniquely defined, that is no similarity transformation of the representation is allowed, and the representation and its basis are said to be *canonical.* So far, however, we have constructed matrices for such representations only for $j = \frac{1}{2}$ and $j = 1$. For the applications in Chapter **15** we need them for all j, and the object of this chapter will be to obtain them.

Readers who are not interested in the technicalities, and who are prepared to lift the formulae from this chapter for consumption in the next, can happily proceed directly to Chapter **15**.

1. MORE ABOUT SPINOR BASES

The fundamental concept for this chapter is that of the tensor bases (§**5**-2), in particular those constructed by the tensor product of spinors, as introduced in §**5**-3 and further used, e.g., in (**7**-3.5). Indeed, the present chapter could have been inserted immediately after Chapter **7**, but it has been placed here in order not to disrupt the main flow of ideas in the book, and also because we shall now be able to develop the work in a more profitable way, by exploiting the quaternion parametrization of rotations. In order to get the work done in as simple a way as possible, we shall overlook at first certain technicalities which will be fully discussed in §3.

One technicality that we must not overlook, however, is the following. In Chapters **5** to **7** we worked in what might be called the spinor configuration space. That is, spinors were written as 2-tuples in $\mathbb{C}^2$, $|\mu_1\mu_2\rangle$, where μ_1 and μ_2 are therefore complex numbers. We now want to work in the corresponding function space. That is, we shall write our spinors as the duals of the preceding ones, $\langle\mu_1\mu_2|$, where the μ's are now complex functions over $\mathbb{C}^2$. Strictly speaking as remarked after (**5**-3.8), we should use a sanserif μ, as we have done in going from $|xyz\rangle$ to $\langle \mathsf{xyz}|$, but it will be sufficient to remember that all our bases are now in function space, so that all the matrices must post-multiply them. (Naturally, once the matrices are obtained, one can move back to configuration space and use column bases.) With this proviso, and on using a more explicit notation, we can summarize the major result of our study of the stereographic projection in §**7**-3 (see, in particular, eqns **7**-3.1, **7**-3.2). Parametrize a rotation R of SO(3) by its Cayley–Klein parameters a, b of (**6**-4.9) as $R(ab)$ and denote the transform of the spinor basis elements, μ_1, μ_2, by $R(ab)\mu_1$ and $R(ab)\mu_2$ respectively, or, for simplicity, by $R(ab)\langle\mu_1\mu_2|$ for the whole transformed basis. Then, the following rotation of the spinor basis

$$\langle\bar{\mu}_1, \bar{\mu}_2| \equiv R(ab)\langle\mu_1\mu_2| = \langle\mu_1\mu_2|\,\check{R}^{1/2}(ab) \tag{1}$$

entails, and is entailed by, all rotations of SO(3). Here, $\check{R}^{1/2}(ab)$ is of course our usual SU(2) matrix of the standard representation,

6-7.3, **11**-8.6
$$\check{R}^{1/2}(ab) = \begin{pmatrix} a & b \\ -b^* & a^* \end{pmatrix} = \begin{pmatrix} \lambda - \mathrm{i}\Lambda_z & -\Lambda_y - \mathrm{i}\Lambda_x \\ \Lambda_y - \mathrm{i}\Lambda_x & \lambda + \mathrm{i}\Lambda_z \end{pmatrix} = \check{R}^{1/2}(\lambda;\boldsymbol{\Lambda}). \tag{2}$$

Strictly speaking, if $\check{R}^{1/2}(\lambda;\boldsymbol{\Lambda})$ is used on (R1), then the rotation should also be parametrized in the quaternion parameters, as $R(\lambda;\boldsymbol{\Lambda})$. On the other hand, it could also be parametrized by Euler angles, as $R(\alpha\beta\gamma)$ in which case $\check{R}^{1/2}(\alpha\beta\gamma)$ of (**12**-11.12) must be used on (R2). The basis $\langle\mu_1\mu_2|$ in (1) is, in the angular momentum notation of §**4**-7, the canonical basis $\langle u^{1/2}_{1/2} u^{1/2}_{-1/2}|$.

In order to form bases of higher dimension, we know that we must form spinors of higher rank by the tensor-product method, as used in (**5**-3.4). We recall, (see §**5**-2), that the tensor of components μ_{ij} is defined as follows. From (1), for two spinors of components μ_i and ν_i, and on calling A the matrix $\check{R}^{1/2}(a, b)$, for simplicity, we have:

$$\bar{\mu}_i = \sum_{I=1}^{2} \mu_I A_{Ii}, \qquad \bar{\nu}_j = \sum_{J=1}^{2} \nu_J A_{Jj}; \qquad i, j = 1, 2, \tag{3}$$

3
$$\bar{\mu}_i\bar{\nu}_j = \sum \mu_I \nu_J A_{Ii} A_{Jj} \qquad \Rightarrow \qquad \bar{\mu}_{ij} = \sum \mu_{IJ} A_{Ii} A_{Jj}, \tag{4}$$

where μ_{ij} is now a rank-two spinor. By repeated application of this technique we define a spinor of rank r, with r suffices, as the object with the following transformation rule,

$$\bar{\mu}_{ij\ldots l} = \sum \mu_{IJ\ldots L} A_{Ii} A_{Jj} \ldots A_{Ll}. \tag{5}$$

Notice that the multiple index of r terms $ij\ldots l$, defines a one-dimensional array for the components on (L5) when the r suffices $ij\ldots l$ are taken in dictionary order (see below the eqn **2**-5.21). We saw in §**5**-2 that tensors can be decomposed into symmetrical and antisymmetrical parts, a process that permits the reduction of the bases and which we shall now exploit in the present more general case.

We shall say that a spinor is totally symmetric (or, for short, symmetric) when it is invariant with respect to the permutation of any two of its suffices. We shall prove that the property of a spinor being symmetric is invariant with respect to all rotations. (See also §**5**-2.) Thus, assume that $\mu_{IJ\ldots L}$ is symmetric, so that we can rewrite (5):

$$\bar{\mu}_{ij\ldots l} = \sum \mu_{JI\ldots L} A_{Ii} A_{Jj} \ldots A_{Ll} \equiv \sum \mu_{JI\ldots L} A_{Jj} A_{Ii} \ldots A_{Ll} \tag{6}$$

5|R6
$$= \bar{\mu}_{ji\ldots l}. \tag{7}$$

Equation (7) means that if μ is symmetric, all its transformed spinors must also be symmetric. Thus, *the space of all symmetrical spinors of a given rank is invariant under* SO(3) and can therefore form a basis for the representation of SO(3).

We shall now prove that the basis formed by the symmetric spinors of rank r is of dimension $n = r+1$, for which purpose we must count the number of distinct components of the form $\mu_{ij\ldots l}$ ($i, j, \ldots, l$ being r suffices). In order to do this, it must be understood that each of the suffices $i, j, \ldots, l$ can take the two values 1 and 2 only, as explicit in (3). Because the tensor is symmetric, there is one and only one component $\mu^{(s)}$ with s suffices equal to 1 and $r-s$ suffices equal to 2, and all the possible componens are thus $\mu^{(0)}, \mu^{(1)}, \ldots, \mu^{(r)}$, $r+1$ in number.

It is useful to express the rank of a spinor basis in terms of the angular momentum quantum number j corresponding to it. The spinor $\langle \mu_1 \mu_2 |$ of rank $r=1$ and dimension 2 belongs to $j=\frac{1}{2}$, whence we shall write $r=2j$. Thus, combining the two results so far obtained, the symmetric spinors of rank $2j$ form a basis of SO(3) of dimension $2j+1$. Since this is precisely the dimension of the canonical basis $\langle u^j_j, \ldots, u^j_{-j}|$, we expect that our symmetric spinor basis coincides with it, as we shall presently prove. In doing this work we shall assume, for simplicity, that in forming the tensor product of two first-rank spinors μ_i and ν_j, they are both identical, since our previous conclusions are still valid in this case. In §3, however, this convenient simplification will be somewhat refined.

We recall that the functions u^j_m are fully defined by the fact that they are eigenfunctions of I_z with eigenvalue m:

4-5.6 $$I_z u^j_m = m u^j_m, \qquad j \geqslant m \geqslant -j. \tag{8}$$

Thus, for $j=\frac{1}{2}$,

$$I_z u^{1/2}_{1/2} = \tfrac{1}{2} u^{1/2}_{1/2}, \qquad I_z u^{1/2}_{-1/2} = -\tfrac{1}{2} u^{1/2}_{1/2}. \tag{9}$$

Let us now call v^j_m the $2j+1$ functions of the symmetric spinor basis of rank $2j$, where the label m ranges over $2j+1$ values. As explained above, the $2j$ suffices of these components take the values 1 and 2 in different numbers, as it will now be shown explicitly:

$$v^j_m = \mu_{\underbrace{11\ldots1}_{j+m}\,\underbrace{22\ldots2}_{j-m}}, \qquad j \geqslant m \geqslant -j. \tag{10}$$

Remember now that μ_{ij} in (4) is $\mu_i \nu_j$. Hence, we can likewise write, on taking μ and ν as identical,

$$v^j_m = \mu_{\underbrace{11\ldots1}_{j+m}\,\underbrace{22\ldots2}_{j-m}} = \underbrace{\mu_1\mu_1\ldots\mu_1}_{j+m}\,\underbrace{\mu_2\mu_2\ldots\mu_2}_{j-m} = \mu_1^{j+m}\mu_2^{j-m}. \tag{11}$$

In order to operate on this basis function with I_z we must remember that this is a differential operator. As an example,

$$I_z\mu_1\mu_1\mu_2 = (I_z\mu_1)\mu_1\mu_2 + \mu_1(I_z\mu_1)\mu_2 + \mu_1\mu_1(I_z\mu_2) \tag{12}$$

9|12 $$= (2\times\tfrac{1}{2} - 1\times\tfrac{1}{2})\mu_1\mu_1\mu_2. \tag{13}$$

Thus,

11 $$I_z v^j_m = I_z\mu_1\mu_1\ldots\mu_1\mu_2\mu_2\ldots\mu_2 \tag{14}$$

$$= \{(j+m)\tfrac{1}{2} - (j-m)\tfrac{1}{2}\}\mu_1\mu_1\ldots\mu_1\mu_2\mu_2\ldots\mu_2 \tag{15}$$

$$= m v^j_m. \tag{16}$$

It is clear from (16) and (8) that the functions v^j_m coincide with the u^j_m, except perhaps for a constant factor, and thus that they span an irreducible representation of SO(3). We shall choose the constant factor still available so as to ensure that the representation is unitary, whereupon the new functions must be identical with the u^j_m. We shall see that it is sufficient for this purpose to modify (11) as follows:

$$u^j_m = \{(j+m)!(j-m)!\}^{-1/2}\mu_1^{j+m}\mu_2^{j-m}, \tag{17}$$

as we shall now prove. For the representation to be unitary the Hermitian inner product of the basis with itself must be invariant with respect to the transformations:

$$\sum_m (u^j_m)^* u^j_m = \sum_m (\bar{u}^j_m)^* \bar{u}^j_m. \tag{18}$$

Form (L18):

$$17\quad \sum_m (u^j_m)^* u^j_m = \frac{1}{(2j)!}\sum_{m=j}^{-j} \frac{(2j)!}{(j+m)!(j-m)!}(\mu_1^*\mu_1)^{j+m}(\mu_2^*\mu_2)^{j-m} \tag{19}$$

$$= \{(2j)!\}^{-1}(\mu_1^*\mu_1+\mu_2^*\mu_2)^{2j}, \tag{20}$$

as follows from the binomial theorem written in a convenient way. (R18) will give exactly the same result, but barred. From (**6**-7.8) the bracket in (R20) is a spinor invariant, i.e. it will not change value on barring the μs. Thus, both sides of (18) are equal and the unitarity of the irreducible representation with the basis (17), with $j \geqslant m \geqslant -j$, is proved. One worry might remain, namely whether the basis just defined might not differ by a similarity from the canonical u^j_m. In other words, whether we might not be violating the Condon and Shortley convention. This, however, cannot be the case because the matrix $\check{R}^{1/2}(\lambda;\Lambda)$ in (2), from which the transformation properties of the functions $u^{1/2}_{1/2}$, $u^{1/2}_{-1/2}$ follow, has been carefully chosen to satisfy Condon and Shortley. (Some problems do arise, however, in the Euler angle parametrization as discussed in §**13**-1 and illustrated again §**15**-3.) All that now remains, thus, is to obtain the irreducible matrix representation, which will be done in §2.

2. THE IRREDUCIBLE REPRESENTATION

> *Warning*: In the whole of this section all superscripts are exponents, except those used for the basis functions u^j_m or the matrices $\check{R}^j$, $\hat{R}^j$.

For a rotation $R(a\ b)$ (see eqn 1.1), for which we shall leave implicit the parametrization, the irreducible rotation matrix $\check{R}^j$ is defined by

$$R\,\langle u^j_j \ldots u^j_{-j}| = \langle u^j_j \ldots u^j_{-j}|\,\check{R}^j \quad \Rightarrow \quad Ru^j_m = \sum_{m'=j}^{-j} u^j_{m'}\check{R}^j_{m'm}. \tag{1}$$

Although we use the inverted caret, the representation is a vector representation for integral j and the matrix has to be written as $\hat{R}^j$, but we shall leave this as assumed. Ru^j_m in (1) must be obtained by acting on (1.17) with R, remembering that a rotation must affect all functions of the function space:

$$Ru^j_m = \{(j+m)!(j-m)!\}^{-1/2}(R\mu_1)^{j+m}(R\mu_2)^{j-m}. \tag{2}$$

The transforms $R\mu_1$, $R\mu_2$ follow from (1.1) and (1.2):

$$Ru^j_m = \{(j+m)!(j-m)!\}^{-1/2}(a\mu_1-b^*\mu_2)^{j+m}(b\mu_1+a^*\mu_2)^{j-m}. \tag{3}$$

The last two brackets are expanded by the binomial theorem:

$$Ru^j_m = \{(j+m)!(j-m)!\}^{-1/2} \sum_{k=0}^{j+m} \frac{(j+m)!}{k!(j+m-k)!} (a\mu_1)^{j+m-k}(-b^*\mu_2)^k \times$$

$$\times \sum_{\kappa=0}^{j-m} \frac{(j-m)!}{\kappa!(j-m-\kappa)!} (b\mu_1)^{j-m-\kappa}(a^*\mu_2)^\kappa \tag{4}$$

$$= \{(j+m)!(j-m)!\}^{1/2} \sum_{\kappa=0}^{j-m} \sum_{k=0}^{j+m} \frac{a^{j+m-k}(-b^*)^k b^{j-m-\kappa}(a^*)^\kappa}{(j+m-k)!k!(j-m-\kappa)!\kappa!} \mu_1^{2j-k-\kappa} \mu_2^{k+\kappa}. \tag{5}$$

In order to bring the exponents of μ_1 and μ_2 in (R5) to the form $j+m'$, $j-m'$, respectively, that should appear in a function $u^j_{m'}$ of the basis (cf. eqn 1.17), we replace the summation index κ by another, m', defined as follows:

$$2j-k-\kappa = j+m' \quad \Rightarrow \quad m' = j-k-\kappa \quad \Rightarrow \quad k+\kappa = j-m', \tag{6}$$

which does precisely the job required. Thus, in (5), multiplying and dividing by the factors required by (1.17), we get

$$Ru^j_m = \{(j+m')!(j-m')!(j+m)!(j-m)!\}^{1/2} \times$$

$$\times \sum_{m'=j-k}^{m-k} \sum_{k=0}^{j+m} \frac{a^{j+m-k}(-b^*)^k b^{m'-m+k}(a^*)^{j-m'-k}}{(j+m-k)!k!(m'-m+k)!(j-m'-k)!} \times$$

$$\times \frac{\mu_1^{j+m'}\mu_2^{j-m'}}{\{(j+m')!(j-m')!\}^{1/2}}. \tag{7}$$

The last function on (R7) is, of course, $u^j_{m'}$ from (1.17), since we have been massaging the equations to this purpose. We also notice that the double summation starts at $m' = j-k$, $k=0$, that is at $m'=j$ and, in the same manner, that it ends at $m'=-j$. We shall therefore write the summation over m' with these limits, and leave the limits of the summation over k floating, to be chosen so as to be compatible with each value of m' used. Thus, we write (7) as follows:

$$Ru^j_m = \{(j+m')!(j-m')!(j+m)!(j-m)!\}^{1/2} \times$$

$$\times \sum_{m'=j}^{-j} u^j_{m'} \sum_k \frac{a^{j+m-k}(-b^*)^k b^{m'-m+k}(a^*)^{j-m'-k}}{(j+m-k)!k!(m'-m+k)!(j-m'-k)!}. \tag{8}$$

On comparing this equation with (1) we obtain at once the matrix element $\check{R}^j_{m'm}$:

$$\check{R}^j(a,b)_{m'm} = \{(j+m')!(j-m')!(j+m)!(j-m)!\}^{1/2} \times$$

$$\times \sum_k \frac{a^{j+m-k}(a^*)^{j-m'-k}\, b^{m'-m+k}(-b^*)^k}{(j+m-k)!(j-m'-k)!(m'-m+k)!k!}. \tag{9}$$

Here a and b are the Cayley–Klein parameters. If (9) is used in this form, in fact, the standard matrix representation should first be written (as in L1.2) and then a, a^*, b and $-b^*$ read off its corresponding elements. In practice, of course, the Cayley–Klein parameters will be derived either from Euler angles (a choice which, needless to say, is not favoured here for applications to point groups) or from the quaternion parameters. In the latter case, which through (9) will lead to $\check{R}^j(\lambda;\boldsymbol{\Lambda})$, the most practical procedure is to work out the value of the four matrix elements of the standard matrix, which appear in (9):

1.2 $$a=\lambda-\mathrm{i}\Lambda_z, \quad b=-\Lambda_y-\mathrm{i}\Lambda_x, \quad (-b^*)=\Lambda_y-\mathrm{i}\Lambda_x, \quad a^*=\lambda+\mathrm{i}\Lambda_z, \tag{10}$$

and to substitute these back into that equation. Improper rotations are always written in the form $\boldsymbol{i}g$, where g is a proper rotation, for which the parameters (10) are determined. Because the basis is ungerade for odd j, however, the matrix (9) must be in this case multiplied by a factor $(-1)^j$. This will be the case only for j integral, since for half-integral j in the Pauli gauge the inversion is treated as the identity and no change in (9) is required.

We shall return a little later to the parametrization of (9). As regards the values of k in the summation, we already know from (7) that $k \geqslant 0$ whence $k!$ in (9) will not be infinite (as factorials of negative numbers are). The range of k will be given by all the values of k for which the three brackets in the denominator of (9) are positive or zero thus leading to finite factorials. Examples of this choice appear below.

We have already seen that the representations given by (9) are irreducible and unitary. It follows from some properties of continuous groups that the set of these representations of SO(3) is complete and the proof can be found in Tinkham (1964), p. 109.

We shall now consider special cases of the matrix $\check{R}^j$ that are of interest in point groups. In all cyclic point groups, on taking the rotation axis along $\mathbf{z}$, as is usual, the quaternion parameters Λ_y and Λ_x are zero so that b in (10) vanishes. In dihedral groups, besides such operations round the $\mathbf{z}$ axis, all other operations are binary rotations normal to it, for which $\lambda=\cos\frac{1}{2}\pi$, and Λ_z vanishes. Hence, a vanishes for these operations. Thus, in dihedral groups, from (10), all operations are such that either a or b vanish. (It is only in cubic groups that there are operations, the threefold rotations, for which this property is not valid.) It is worthwhile, therefore, to study the special cases mentioned, since they will include all cyclic and dihedral groups.

When b vanishes (rotations around the $\mathbf{z}$ axis), the only non-vanishing matrix elements appear when the exponents of b and $-b^*$ in (9) vanish:

$$\{k=0, \qquad m'-m+k=0\} \qquad \Rightarrow \qquad m'=m. \tag{11}$$

11|9 $$\check{R}^j(a0)_{m'm}=a^{j+m}(a^*)^{j-m}\delta_{m'm}. \tag{12}$$

Equation (12) means that the matrix is diagonal, as we must expect (see eqn **4**-7.8).

When a vanishes (binary rotations perpendicular to **z**) the exponents of a and a^* in (9) must vanish:

$$\{j+m-k=0,\ j-m'-k=0\} \Rightarrow \{k=j+m=j-m'\} \Rightarrow m'=-m. \quad (13)$$

13|9 $$\check{R}^j(0b)_{m'm} = b^{j-m}(-b^*)^{j+m}\delta_{m',-m}. \quad (14)$$

It follows from (12) and (14) that, in a dihedral group, for $m=j$, $m'=\pm j$, that is, that u^j_j will transform into either u^j_j or u^j_{-j}. It is thus useful to consider the form of the matrices corresponding to bases of the form $\langle u^j_j u^j_{-j}|$.

We shall consider this matrix for the general case when a, b do not vanish; hence we must first of all ensure that the factorials in the denominator of (9) do not vanish:

$$m'=j: \quad j-m'-k=-k \quad \Rightarrow \quad k=0, \quad (15)$$

9 $$m'=-j: \quad j+m-k\geqslant 0, \quad (16)$$

9,16 $$m'-m+k=-j-m+k\geqslant 0 \quad \Rightarrow \quad k=j+m. \quad (17)$$

We can now obtain the four required matrix elements:

15,9 $$m'=j, \quad m=j, \quad k=0, \qquad \check{R}^j_{jj}=a^{2j}. \quad (18)$$

15,9 $$m'=j, \quad m=-j, \quad k=0, \qquad \check{R}^j_{j,-j}=b^{2j}. \quad (19)$$

17,9 $$m'=-j, \quad m=j, \quad k=2j, \qquad \check{R}^j_{-j,j}=(-b^*)^{2j}. \quad (20)$$

17,9 $$m'=-j, \quad m=-j, \quad k=0, \qquad \check{R}^j_{-j,-j}=(a^*)^{2j}. \quad (21)$$

It will be noticed that, so far, we have kept in (9) and all subsequent expressions the precise form of the four matrix elements of the standard matrix. Equations (18) to (21) show the advantage of this choice (by no means usual in the literature), for we can summarize the results of these equations as follows:

$$\text{Basis } \langle u^j_j u^j_{-j}|, \qquad \check{R}^{1/2}=\begin{bmatrix} a & b \\ -b^* & a^* \end{bmatrix}, \qquad \check{R}^j=\begin{bmatrix} a^{2j} & b^{2j} \\ (-b^*)^{2j} & (a^*)^{2j} \end{bmatrix}. \quad (22)$$

This shows how exceedingly easy it is to form the matrix representations for cyclic and dihedral groups, where only matrices of the form (22) can appear. First, for all operations, the standard representation is formed with the quaternion parameters in (10) in order to derive the Cayley–Klein a, b parameters. Then, for $\check{R}^j$, each matrix element is raised to the power $2j$.

Two remarks are useful here. First, the representations obtained above must always be checked for irreducibility since they may reduce, an easy operation to perform on 2×2 matrices. (Remember that the canonical bases, although irreducible in SO(3), reduce in point groups, and that the basis in eqn 22 is not even a complete canonical basis.) Secondly, $\check{R}^{1/2}$ is a SU(2) matrix and, although our rules fix uniquely their sign in the standard representation, there is an inherent sign indeterminacy of the matrix elements that leads to either the projective factors or to the double-group structure. Notice, however, that when j is integral this uncertainty disappears in $\check{R}^j$ which then becomes a vector representation $\hat{R}^j$. In fact, in the dihedral groups, the standard representation generates all the vector and projective (or double) representations of the group. Even in cubic groups it also plays a fundamental rôle, since the coefficients that appear in (9) are most conveniently extracted from the standard representation.

One final practical point. In eqn (9), the Cayley–Klein parameters must be constructed from the quaternion parameters in (10). The form of the latter, however, is not very convenient, because of the signs and because of the positioning of y along the real axis, which does not allow the quick reading of the **x**, **y** plane as an Argand plane. It is often therefore convenient to express the quaternion parameters in terms of the following parameters (*complex quaternion parameters*):

$$\rho = \lambda + \mathrm{i}\Lambda_z, \qquad \tau = \Lambda_x + \mathrm{i}\Lambda_y. \tag{23}$$

With these parameters, the standard matrix takes the form

23|1.2
$$\check{R}^{1/2} = \begin{bmatrix} a & b \\ -b^* & a^* \end{bmatrix} = \begin{bmatrix} \rho^* & -\mathrm{i}\tau^* \\ -\mathrm{i}\tau & \rho \end{bmatrix}. \tag{24}$$

If a form of the matrix elements of $\check{R}^j$ is desired that explicitly contains the quaternion parameters, the following form in ρ and τ may be used, although it is often more convenient to feed back into (9) the explicit values of a, b, a^*, $-b^*$ read out from (24) after the latter matrix has been calculated from (23).

$$\check{R}^j(\rho\tau)_{m'm} = \{(j+m')!(j-m')!(j+m)!(j-m)!\}^{1/2}$$
$$\times \sum_k (-\mathrm{i})^{m'-m+2k} \frac{\rho^{j-m'-k}(\rho^*)^{j+m-k}\tau^k(\tau^*)^{m'-m+k}}{(j-m'-k)!(j+m-k)!k!(m'-m+k)!}. \tag{25}$$

Problems 14-2 **(p. 296)**

1. Show that for cyclic and dihedral groups the function u_0^j will form always a one-dimensional basis and find $\check{R}^j_{00}$. Hence, show that

$$\hat{R}^1_{00} = aa^* - bb^*. \tag{26}$$

(Notice that we change the caret from inverted to normal because j is integral.)

2. Find the matrix $\hat{R}^1(ab)$ for SO(3) and compare with a matrix previously found in this book.
3. Obtain the standard matrix representation for the cubic-group rotations defined in Problem **2**-4.4 and compare your results with the tables of Onodera and Okazaki (1966).

3. THE BASES OF THE REPRESENTATIONS

Let us write, from (1.17) the basis of the irreducible representations of SO(3) for $j=1$:

$$\langle u_1^1 u_0^1 u_{-1}^1| = \langle 2^{-1/2}\mu_1^2,\ \mu_1\mu_2,\ 2^{-1/2}\mu_2^2|. \tag{1}$$

These, of course, should be the spherical harmonics $\langle Y_1^1 Y_1^0 Y_1^{-1}|$ and it will be useful to compare with a previous manifestation of this basis in §**7**-3, which will give us an opportunity to refine a little the way in which we generated the basis in §1. We did this in three steps. First, we formed the direct product of the spinor bases for $j=\frac{1}{2}$,

$$\langle \mu_1\mu_2| \otimes \langle \mu_1\mu_2| = \langle \mu_1\mu_1,\ \mu_1\mu_2,\ \mu_2\mu_1,\ \mu_2\mu_2|. \tag{2}$$

Secondly, we symmetrized the direct product (see eqn **5**-2.5):

$$\langle \mu_1\mu_2| \bar{\otimes} \langle \mu_1\mu_2| = \langle \mu_1\mu_1,\ \tfrac{1}{2}(\mu_1\mu_2+\mu_2\mu_1),\ \mu_2\mu_2|. \tag{3}$$

Thirdly, we got the factors $\{(j+m)!(j-m)!\}^{-1/2}$ for each u_m^j to ensure unitarity of the representation. This work, however, is not very satisfactory because, in (2), $\mu_1\mu_2$ and $\mu_2\mu_1$ are one and the same thing and the subsequent symmetrization, really, amounts to nothing more than dropping an irrelevant (because not independent) member of the basis. It is sounder to form (2) for two different spinors on the left, that nevertheless transform identically. From (**6**-7.6), the conjugate spinor $\langle \mu_2^*, -\mu_1^*|$ can be substituted for $\langle \mu_1\mu_2|$ without any change in transformation properties. So, we do this for the second factor on (L2), symmetrize, and obtain the new version of (3):

$$\langle \mu_1\mu_2| \bar{\otimes} \langle \mu_2^*, -\mu_1^*| = \langle \mu_1\mu_2^*,\ \tfrac{1}{2}(-\mu_1\mu_1^*+\mu_2\mu_2^*),\ -\mu_2\mu_1^*|. \tag{4}$$

We still have to multiply each function of the basis on (R4) by the factors required for the unitarity of the representation. These are those appearing on (R1). Thus, all that we have done, really is to rewrite (1) in a new form:

$$\langle u_1^1 u_0^1 u_{-1}^1| = \langle 2^{-1/2}\mu_1\mu_2^*,\ \tfrac{1}{2}(\mu_2\mu_2^*-\mu_1\mu_1^*),\ -2^{-1/2}\mu_1^*\mu_2|. \tag{5}$$

If we multiply the whole of this basis by an irrelevant factor of $2^{1/2}$ we recover precisely the basis (**7**-3.11), or rather its dual. We saw, in fact, in that section (see the Comments at the end of it) that that basis (**7**-3.11) was the dual of the basis $\langle Y_1^1 Y_1^0 Y_1^{-1}|$.

Two comments are opportune. The first is that identifying μ_2^* with μ_1 and $-\mu_1^*$ with μ_2 in the correct basis (5) changes nothing from the point of view of the matrix that is generated by the basis, which explains why we used this simpler approach in constructing the basis (1.17). The second is that the identification of the basis (1.17) for general j and m can easily be effected. Consider the basis u_m^j of (**7**-3.38) in which we disregard the coefficients that depend solely on j and m:

$$u_m^j = \sum_{k,r} C(j, m, k, r)\mu_1^{m+k}(\mu_2^*)^{j-k}\mu_2^{j-m-k}(-\mu_1^*)^k. \tag{6}$$

Here, $C(j, m, k, r)$ stands for the coefficients depending on these indices that appear in (**7**-3.38). On equating in (6) μ_2^* with μ_1 and $(-\mu_1^*)$ with μ_2, the spinor product in this equation becomes $\mu_1^{j+m}\mu_2^{j-m}$, exactly in agreement with (1.17). The coefficients in (6), like those in (4) merely reflect the correct symmetrization of the basis (1.17), but (6), after normalization, has still to be multiplied by a factor in order to ensure the unitarity of the representation, as done in (1.17). It follows from these comments that, as we would expect, the bases (1.17) of the representations, more properly in their form given by (6), are the spherical harmonic functions.

15

EXAMPLES AND APPLICATIONS

Il se moquait de moi en me voyant étudier le grec à vingt-cinq ans. 'Vous êtes sur le champ de bataille, disait-il; ce n'est plus le moment de polir votre fusil; il faut tirer.'

Prosper Mérimée (1850). *HB*. See Stendhal *Romans*. Préface et présentation de S. S. de Sacy. Éditions du Seuil, Paris, 1969, vol. I, p. xxvii.

He [Stendhal] laughed at me seeing me studying Greek at 25. 'You are on the battlefield, he used to say; it is no longer the time to polish your gun; you must shoot.'

We shall discuss three groups in this chapter, $\mathbf{D}_6$, $\mathbf{D}_3$, and $\mathbf{C}_{3v}$. We shall treat $\mathbf{D}_6$ and $\mathbf{D}_3$ by the projective-representation method and then $\mathbf{D}_3$ will also be treated as a double group, for comparison. $\mathbf{C}_{3v}$ will give us an example of an improper group. Finally, some applications will be discussed.

1. THE CHOICE OF THE POSITIVE HEMISPHERE

The essence of the method developed so far is that a positive hemisphere $\mathfrak{H}$ is chosen on the unit sphere. Then a *standard set of poles* is chosen with all positive and all binary rotations on $\mathfrak{H}$ and all negative rotations on $\bar{\mathfrak{H}}$, the negative hemisphere. This fixes uniquely a set of *standard quaternion parameters* for all the operations g of the group *G*, and thus a *standard representation* of SU(2) matrices, in which each g with standard parameters $[\![\lambda, \mathbf{\Lambda}]\!]$ is assigned a unique matrix $\check{R}^{1/2}(\lambda\,;\mathbf{\Lambda})$. The way in which the method works is such that *G* is embedded in SO(3) so that continuity in SO(3) and consistency with SO(3) is guaranteed.

The problem is that whereas it is possible correctly to embed any individual point group *G*, or some set of such groups, it is not possible to embed simultaneously all point groups. This is not a weakness of the method: it is a peculiarity of nature. It is indeed a very useful feature of the method that this peculiarity can instantly be recognized and attended to. Consideration of $\mathbf{D}_6$ and $\mathbf{D}_3$ will provide an immediate example of the problem and of the way to handle it. $\mathbf{D}_6$ is the group of proper rotations of an hexagonal prism and contains the rotations E, $C_3^{\pm}$, $C_6^{\pm}$, C_2 around the axis of the prism (**z** axis), as well as six binary rotations respectively

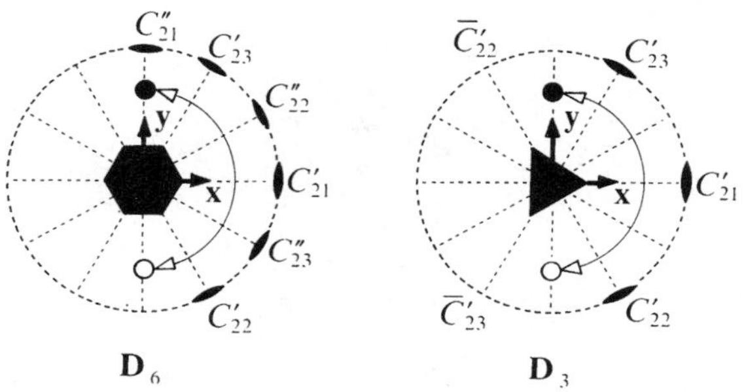

Fig. 15-1.1. The groups $\mathbf{D}_6$ and $\mathbf{D}_3$ described with the choice of the positive hemisphere depicted in Fig. **9**-1.1. The range of ϕ for $\mathfrak{H}$, for $\theta=\frac{1}{2}\pi$, is shown with the curved arrow, a full and an empty circle denoting values in or out of the range, respectively.

perpendicular to the edges and faces. $\mathbf{D}_3$ is a subgroup of $\mathbf{D}_6$, the group of proper rotations of a triangular prism. $\mathbf{D}_3$ itself will have other subgroups, such as $\mathbf{C}_3$. We thus have a *group chain* $SO(3)\supset\mathbf{D}_6\supset\mathbf{D}_3\supset\mathbf{C}_3$ and in many problems it is important to treat all groups of such a chain consistently, which requires some care.

We show in Fig. 1 the group $\mathbf{D}_6$ parametrized by the same positive hemisphere as defined in Fig. **9**-1.1. In polar coordinates a point belongs to $\mathfrak{H}$ if *either* $0\leqslant\theta<\frac{1}{2}\pi$ *or* $\theta=\frac{1}{2}\pi$ and $-\frac{1}{2}\pi<\phi\leqslant\frac{1}{2}\pi$, the latter range being displayed in the figure. The poles of the six binary rotations shown are the standard poles, since they are chosen on $\mathfrak{H}$. Consider the classes of the binary rotations shown. They are two: C'_{21}, C'_{22}, C'_{23} and C''_{21}, C''_{22}, C''_{23}. We know this because the poles of the three C'_{2i} are exchanged one into another by the $C_6^{\pm}$ operations, and the same for C''_{2i}, but there is no operation that will take a C'_{2i} pole into a C''_{2i} pole. This choice of $\mathfrak{H}$, although it is perfectly good if we are only interested in $\mathbf{D}_6$, is no good at all if we want to treat $\mathbf{D}_3$ consistently at the same time, as it can be seen at once from the picture of $\mathbf{D}_3$ in Fig. 1. In fact, there is no symmetry operation now that exchanges the three C'_{2i} poles, since the rotations $C_6^{\pm}$ that did this in $\mathbf{D}_6$ are not in $\mathbf{D}_3$. We know, however, that this would not be the case if we were to take the standard poles in the positions labelled C'_{21}, $\bar{C}'_{22}$, $\bar{C}'_{23}$, since $C_3^{\mp}$ does in fact exchange these poles. If we were to use the parametrization of the Fig. 1, $\mathbf{D}_6$ would work out all right but, because the standard poles of C'_{2i} are not conjugate in $\mathbf{D}_3$ (see §**11**-6), the character theorem fails and one would find that the matrices of this class do not all have the same character. (There are tables in the literature in which this happens for a number of subgroups.) It is easy to see how to choose $\mathfrak{H}$ so as to ensure consistency for the subgroups one wants, as we

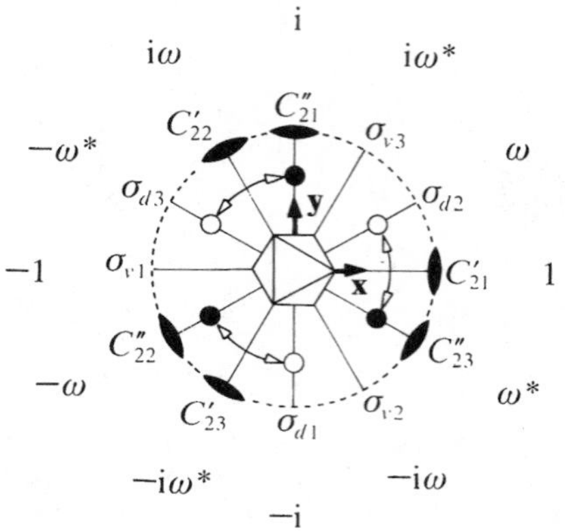

Fig. 15-1.2. The groups $\mathbf{D}_6$, $\mathbf{D}_3$, and $\mathbf{C}_{3v}$, with the correct choice of $\mathfrak{H}$. The range of ϕ for $\mathfrak{H}$, for $\theta = \frac{1}{2}\pi$, is shown with the curved arrows, a full and an empty circle denoting values in or out of the range, respectively. $\sigma_{vi} = \boldsymbol{i}C''_{2i}$, $\sigma_{di} = \boldsymbol{i}C'_{2i}$, for $i = 1, 2, 3$. The outer circle, which must be read as an Argand plane, gives the twelve points to which the real unit is taken by the twelve operations of $\mathbf{C}_{12}$, this being the cyclic group of double the order of $\mathbf{C}_6$. $\omega = \exp(\frac{1}{6}\pi\mathrm{i})$.

show in Fig. 2, which will deal consistently with $\mathbf{D}_6$, $\mathbf{D}_3$, $\mathbf{C}_6$, $\mathbf{C}_3$ and $\mathbf{C}_{3v}$, as well as various $\mathbf{C}_2$ and $\mathbf{C}_s$ subgroups. Here we take a point as belonging to $\mathfrak{H}$ *either* if $0 \leqslant \theta < \pi$ *or* $\theta = \pi$ and one of the following alternatives obtains for ϕ:

$$-\tfrac{1}{6}\pi \leqslant \phi < \tfrac{1}{6}\pi, \qquad \tfrac{1}{2}\pi \leqslant \phi < \tfrac{5}{6}\pi, \qquad \tfrac{7}{6}\pi \leqslant \phi < \tfrac{3}{2}\pi. \quad (1)$$

There are two $\mathbf{D}_3$ subgroups which can be considered, one including the three C'_{2i} and the other including the three C''_{2i}, and in both cases the standard poles are correctly conjugated. As shown by the triangle given in the picture, we shall take for our example the first of these.

As regards $\mathbf{C}_{3v}$, it comprises E, $C_3^{\pm}$ and three planes passing through the C_3 axis. We shall take them, for our example, to be the three σ_{vi} planes, which are given by the relation $\sigma_{vi} = \boldsymbol{i}C''_{2i}$ and thus the poles assigned to the three σ_{vi} are those of the three C''_{2i} (§**9**-1), which, as we have said, are correctly conjugated. It must be stressed that in the classical double-group method, a choice of the group multiplication table as entailed by a choice of SU(2) matrices is tantamount to a choice, in our method, of a standard set of poles, and that there is no way in which one can control the conjugation relations except by trial and error choices combined with laborious checks through matrix multiplications.

2. PARAMETRIZATION OF THE GROUP ELEMENTS FOR $\mathbf{D}_6$, $\mathbf{D}_3$, $\mathbf{C}_{3v}$. MULTIPLICATION TABLES AND FACTOR SYSTEMS

It is very simple to derive the group parameters from Fig. 1.2. For the improper rotations, σ_{v1}, σ_{v2}, σ_{v3}, as explained in §1 and §**9**-1, their poles are those of the proper rotation C such that $\sigma = iC$. Since $\sigma_{vi} = \boldsymbol{i}\ C''_{2i}$, from Fig. 1.2, their poles and parameters are those of the operations C''_{2i}. The group parameters are displayed in Table 1. The Euler angles shown are quite ambiguous, because β is either 0 or π, but alternatives are not shown in the table. Even for this simple group the derivation of the Euler angles is not entirely trivial, as the reader may verify. The values of ϕ and $\mathbf{n}$ are, instead, pretty instant. Notice that, in our conventions, ϕ is always positive, the sign of the rotation being carried by $\mathbf{n}$ belonging to $\mathfrak{H}$ or $\bar{\mathfrak{H}}$. The quaternion $[\![\lambda, \boldsymbol{\Lambda}]\!]$ is immediate. The complex quaternion parameters ρ and τ introduced in (**14**-2.23), in order to facilitate the writing of the representation matrices, are best derived from an Argand diagram. Since for the operation C_n of $\mathbf{C}_6$ the parameters λ and Λ_z equal respectively the cos and the sin of $\frac{1}{2}\phi$, ($\phi = 2\pi/n$), ρ is obtained by plotting on a circle in the Argand plane the twelve points to which the real unit is taken by the twelve operations of $\mathbf{C}_{12}$. Then, for C_n, ρ is read as the point on this circle determined by C_{2n}. (See Fig. 1.2; the prescription given, of course, associates with a rotation by ϕ the point on the circle given by the rotation by $\frac{1}{2}\phi$.) As regards τ, this parameter is immediately determined

Table **15**-2.1. Parametrization of $\mathbf{D}_6$, $\mathbf{D}_3$ and $\mathbf{C}_{3v}$.
α, β, γ: Euler angles. ϕ, $\mathbf{n}$: angle and axis of rotation.
$\lambda = \cos\frac{1}{2}\phi$, $\boldsymbol{\Lambda} = \sin\frac{1}{2}\phi\ \mathbf{n}$, $\rho = \lambda + \mathrm{i}\Lambda_z$, $\tau = \Lambda_x + \mathrm{i}\Lambda_y$, $\omega = \exp(\frac{1}{6}\pi\mathrm{i})$.

$\mathbf{D}_6$	$\mathbf{D}_3$	$\mathbf{C}_{3v}$	α	β	γ	ϕ	n	λ	$\boldsymbol{\Lambda}$	ρ	τ
E	E	E	0	0	0	0	(0 0 1)	$[\![1,$	$\mathbf{0}\ ]\!]$	1	0
C_6^+			0	0	$\frac{1}{3}\pi$	$\frac{1}{3}\pi$	(0 0 1)	$[\![\frac{\sqrt{3}}{2},$	$(0\ 0\ \frac{1}{2})]\!]$	ω	0
C_6^-			0	0	$-\frac{1}{3}\pi$	$\frac{1}{3}\pi$	(0 0 −1)	$[\![\frac{\sqrt{3}}{2},$	$(0\ 0\ -\frac{1}{2})]\!]$	ω^*	0
C_3^+	C_3^+	C_3^+	0	0	$\frac{2}{3}\pi$	$\frac{2}{3}\pi$	(0 0 1)	$[\![\frac{1}{2},$	$(0\ 0\ \frac{\sqrt{3}}{2})]\!]$	$\mathrm{i}\omega^*$	0
C_3^-	C_3^-	C_3^-	0	0	$-\frac{2}{3}\pi$	$\frac{2}{3}\pi$	(0 0 −1)	$[\![\frac{1}{2},$	$(0\ 0\ -\frac{\sqrt{3}}{2})]\!]$	$-\mathrm{i}\omega$	0
C_2			0	0	π	π	(0 0 1)	$[\![0,$	$(0\ 0\ 1)]\!]$	i	0
C'_{21}	C'_{21}		π	π	0	π	(1 0 0)	$[\![0,$	$(1\ 0\ 0)]\!]$	0	1
C'_{22}	C'_{22}		$\frac{1}{3}\pi$	π	0	π	$(-\frac{1}{2}\ \frac{\sqrt{3}}{2}\ 0)$	$[\![0,$	$(-\frac{1}{2}\ \frac{\sqrt{3}}{2}\ 0)]\!]$	0	$\mathrm{i}\omega$
C'_{23}	C'_{23}		$-\frac{1}{3}\pi$	π	0	π	$(-\frac{1}{2}\ -\frac{\sqrt{3}}{2}\ 0)$	$[\![0,$	$(-\frac{1}{2}\ -\frac{\sqrt{3}}{2}\ 0)]\!]$	0	$-\mathrm{i}\omega^*$
C''_{21}		σ_{v1}	0	π	0	π	(0 1 0)	$[\![0,$	$(0\ 1\ 0)]\!]$	0	i
C''_{22}		σ_{v2}	$-\frac{2}{3}\pi$	π	0	π	$(-\frac{\sqrt{3}}{2}\ -\frac{1}{2}\ 0)$	$[\![0,$	$(-\frac{\sqrt{3}}{2}\ -\frac{1}{2}\ 0)]\!]$	0	$-\omega$
C''_{23}		σ_{v3}	$\frac{2}{3}\pi$	π	0	π	$(\frac{\sqrt{3}}{2}\ -\frac{1}{2}\ 0)$	$[\![0,$	$(\frac{\sqrt{3}}{2}\ -\frac{1}{2}\ 0)]\!]$	0	ω^*

by the positions of the binary poles on the circle in the Argand plane. As shown in the figure, the **x** and **y** axes must be taken as the real and imaginary axes, respectively, of the associated Argand plane. The quaternions $[\![\lambda, \mathbf{\Lambda}]\!]$ are, from our rules, such that for all of them *either* $\lambda > 0$ *or* $\lambda = 0$ and $\mathbf{\Lambda} \in \mathfrak{H}$. They form what we call the *standard set.*

The next step in the work is to form the quaternion multiplication table, which consists in turning the handle of the quaternion multiplication rule for an hour or two. (Or seconds in a computer, for which it is ideally suited.) The work, fortunately, is largely self-checking since no two entries can be the same in each row and column. From this master table, which will not be printed here because of its bulk, the group multiplication table (Table 2) and the factor system table (Table 3) follow at once. These two tables must be constructed simultaneously in the following way. For this purpose, the rule to obtain $[g_i, g_j]$ (which gives it as $+1$ when the product quaternion of g_i, g_j is such that *either* $\lambda > 0$ *or* $\lambda = 0$ and $\mathbf{\Lambda} \in \mathfrak{H}$) can be simplified as follows: multiply the quaternions of g_i with g_j. If the resulting quaternion appears in the standard set for an operation g_k, then $g_i g_j$ equals g_k and $[g_i, g_j]$ equals $+1$. If the resulting quaternion is the negative of the quaternion of g_k in the standard set, then $g_i g_j$ equals g_k and $[g_i, g_j]$ equals -1. The standard set can be identified from Table 1. As an example, for the product of the quaternions of C_6^- and C_3^- (in that order) the result is $[\![0, (00\bar{1})]\!]$ which, from Table 1, is the negative of the quaternion of C_2. We enter $C_6^- C_3^-$ equal to C_2 in Table 2 and $[C_6^-, C_3^-]$ equal to -1 in Table 3. It should be noticed that these two tables are valid for $\mathbf{D}_6$ and all its subgroups. As regards

Table **15**-2.2. Multiplication table for $\mathbf{D}_6$ and related groups. For $\mathbf{C}_{3v}$, replace C''_{2i} by σ_{vi}. The product $g_i g_j$ appears in the intersection of the g_i row with the g_j column.

	E	C_6^+	C_6^-	C_3^+	C_3^-	C_2	C'_{21}	C'_{22}	C'_{23}	C''_{21}	C''_{22}	C''_{23}
E	E	C_6^+	C_6^-	C_3^+	C_3^-	C_2	C'_{21}	C'_{22}	C'_{23}	C''_{21}	C''_{22}	C''_{23}
C_6^+	C_6^+	C_3^+	E	C_2	C_6^-	C_3^-	C''_{22}	C''_{23}	C''_{21}	C'_{22}	C'_{23}	C'_{21}
C_6^-	C_6^-	E	C_3^-	C_6^+	C_2	C_3^+	C''_{23}	C''_{21}	C''_{22}	C'_{23}	C'_{21}	C'_{22}
C_3^+	C_3^+	C_2	C_6^+	C_3^-	E	C_6^-	C'_{23}	C'_{21}	C'_{22}	C''_{23}	C''_{21}	C''_{22}
C_3^-	C_3^-	C_6^-	C_2	E	C_3^+	C_6^+	C'_{22}	C'_{23}	C'_{21}	C''_{22}	C''_{23}	C''_{21}
C_2	C_2	C_3^-	C_3^+	C_6^-	C_6^+	E	C''_{21}	C''_{22}	C''_{23}	C'_{21}	C'_{22}	C'_{23}
C'_{21}	C'_{21}	C''_{23}	C''_{22}	C'_{22}	C'_{23}	C''_{21}	E	C_3^+	C_3^-	C_2	C_6^-	C_6^+
C'_{22}	C'_{22}	C''_{21}	C''_{23}	C'_{23}	C'_{21}	C''_{22}	C_3^-	E	C_3^+	C_6^+	C_2	C_6^-
C'_{23}	C'_{23}	C''_{22}	C''_{21}	C'_{21}	C'_{22}	C''_{23}	C_3^+	C_3^-	E	C_6^-	C_6^+	C_2
C''_{21}	C''_{21}	C'_{23}	C'_{22}	C''_{22}	C''_{23}	C'_{21}	C_2	C_6^-	C_6^+	E	C_3^+	C_3^-
C''_{22}	C''_{22}	C'_{21}	C'_{23}	C''_{23}	C''_{21}	C'_{22}	C_6^+	C_2	C_6^-	C_3^-	E	C_3^+
C''_{23}	C''_{23}	C'_{22}	C'_{21}	C''_{21}	C''_{22}	C'_{23}	C_6^-	C_6^+	C_2	C_3^+	C_3^-	E

Table **15**-2.3. Factor system for $\mathbf{D}_6$ and related groups.
For C_{3v}, replace C''_{2i} by σ_{vi}.
The factor $[g_i, g_j]$ appears in the intersection of the g_i row with the g_j column.

	E	C_6^+	C_6^-	C_3^+	C_3^-	C_2	C'_{21}	C'_{22}	C'_{23}	C''_{21}	C''_{22}	C''_{23}
E	1	1	1	1	1	1	1	1	1	1	1	1
C_6^+	1	1	1	1	1	−1	−1	−1	−1	1	1	1
C_6^-	1	1	1	1	−1	1	1	1	1	−1	−1	−1
C_3^+	1	1	1	−1	1	−1	−1	−1	−1	−1	−1	−1
C_3^-	1	1	−1	1	−1	1	−1	−1	−1	−1	−1	−1
C_2	1	−1	1	−1	1	−1	1	1	1	−1	−1	−1
C'_{21}	1	1	−1	−1	−1	−1	−1	1	1	1	1	−1
C'_{22}	1	1	−1	−1	−1	−1	1	−1	1	−1	1	1
C'_{23}	1	1	−1	−1	−1	−1	1	1	−1	1	−1	1
C''_{21}	1	−1	1	−1	−1	1	−1	−1	1	−1	1	1
C''_{22}	1	−1	1	−1	−1	1	1	−1	−1	1	−1	1
C''_{23}	1	−1	1	−1	−1	1	−1	1	−1	1	−1	−1

$\mathbf{C}_{3v}$, if the σ_{vi} are written as iC''_{2i}, it is easy to verify that $\mathbf{C}_{3v}$ is isomorphic to the $\mathbf{D}_3$ subgroup of $\mathbf{D}_6$ that is given by E, $C_3^{\pm}$, C''_{2i}, so that it is sufficient in Tables 2 and 3 to replace C''_{2i} by σ_{vi} when dealing with $\mathbf{C}_{3v}$. (If one wanted, of course, precisely the same rules would arise from the master quaternion product table, formed with the parameters for $\mathbf{C}_{3v}$ given in Table 1.) As regards the projective factors of $\mathbf{C}_{3v}$, in the Pauli gauge, from (**11**-9.18), the projective factors obtained from the quaternion multiplication table (reading C''_{2i} as σ_{vi}) are correct for this group and thus can be read off Table 3 in the same way. (Again, if desired, they could be worked out in the normal manner directly from the quaternion multiplication table.)

3. THE STANDARD REPRESENTATION

We shall now start the work necessary in order to obtain all the irreducible representations (in full, rather than given by their characters only), for the point groups. In working with point groups, for which the characters of the irreducible representations are easily available from tables, the natural thing is to borrow information from them as convenient. For the dihedral groups, however, it is so easy to derive from scratch all the necessary information, that we shall proceed in this, rather more instructive, way. In this case, the standard representation is the father and mother of all the representations and we shall first of all work it out, not only for this reason but also in order to compare the quaternion and Euler parametrizations. In the quaternion parametrization, the standard representation is most simply given in terms of the complex quaternion parameters ρ, τ, listed in Table 2.1:

14-2.24
$$\check{R}^{1/2}(\rho\tau)=\begin{bmatrix}\rho^* & -\mathrm{i}\tau^* \\ -\mathrm{i}\tau & \rho\end{bmatrix}. \qquad (1)$$

In Euler angles, for the two relevant cases in Table 2.1, $\beta=0$ and $\beta=\pi$, respectively, (**12**-11.12) gives

$$\check{R}^{1/2}(00\gamma) \qquad\qquad \check{R}^{1/2}(\alpha\pi 0)$$
$$\begin{bmatrix}\exp(-\mathrm{i}\tfrac{1}{2}\gamma) & \\ & \exp(\mathrm{i}\tfrac{1}{2}\gamma)\end{bmatrix}, \qquad \begin{bmatrix} & -\exp(-\tfrac{1}{2}\mathrm{i}\alpha) \\ \exp(\tfrac{1}{2}\mathrm{i}\alpha) & \end{bmatrix}. \qquad (2)$$

With these formulae, the matrices are very rapidly built from the parameters in Table 2.1 and they are displayed in Table 1.

It is clear from Fig. 1.2 that C_2 (normal to the **xy** plane), C'_{21} and C''_{21} are a BB set and indeed they coincide in the $\check{R}^{1/2}(\rho\tau)$ representation with $\boldsymbol{I}_z$, $\boldsymbol{I}_x$, and $\boldsymbol{I}_y$ respectively in (**12**-6.4) whence, on multiplication with i as in (**13**-1.13) they correctly give the Pauli matrices in (**4**-9.9). This is, of course, essential in order to satisfy the Condon and Shortley convention. It is clear that this is not the case for $\check{R}^{1/2}(\alpha\beta\gamma)$, where C'_{21} ($\boldsymbol{I}_x$) has the wrong sign.

In the usual double-group *cum* Euler-parametrization method, what one does is to construct $\check{R}^{1/2}(\alpha\beta\gamma)$ and then massage it a little as convenient, i.e. change the sign of the matrices judiciously. A reasonably careful author would then change the sign of C'_{21} and probably leave the

Table **15**-3.1. The standard representation.
Warning: The representation $\check{R}^{1/2}(\alpha\beta\gamma)$ should not be used.
The column χ gives the characters of the matrices. $\omega=\exp(\tfrac{1}{6}\pi\mathrm{i})$.

	$\check{R}^{1/2}(\rho\tau)$	$\check{R}^{1/2}(\alpha\beta\gamma)$	χ		$\check{R}^{1/2}(\rho\tau)$	$\check{R}^{1/2}(\alpha\beta\gamma)$	χ
E	$\begin{bmatrix}1 & \\ & 1\end{bmatrix}$	$\begin{bmatrix}1 & \\ & 1\end{bmatrix}$	2	C'_{21}	$\begin{bmatrix} & -\mathrm{i} \\ -\mathrm{i} & \end{bmatrix}$	$\begin{bmatrix} & \mathrm{i} \\ \mathrm{i} & \end{bmatrix}$	0
C_6^+	$\begin{bmatrix}\omega^* & \\ & \omega\end{bmatrix}$	$\begin{bmatrix}\omega^* & \\ & \omega\end{bmatrix}$	$\sqrt{3}$	C'_{22}	$\begin{bmatrix} & -\omega^* \\ \omega & \end{bmatrix}$	$\begin{bmatrix} & -\omega^* \\ \omega & \end{bmatrix}$	0
C_6^-	$\begin{bmatrix}\omega & \\ & \omega^*\end{bmatrix}$	$\begin{bmatrix}\omega & \\ & \omega^*\end{bmatrix}$	$\sqrt{3}$	C'_{23}	$\begin{bmatrix} & \omega \\ -\omega^* & \end{bmatrix}$	$\begin{bmatrix} & -\omega \\ \omega^* & \end{bmatrix}$	0
C_3^+	$\begin{bmatrix}-\mathrm{i}\omega & \\ & \mathrm{i}\omega^*\end{bmatrix}$	$\begin{bmatrix}-\mathrm{i}\omega & \\ & \mathrm{i}\omega^*\end{bmatrix}$	1	C''_{21}	$\begin{bmatrix} & -1 \\ 1 & \end{bmatrix}$	$\begin{bmatrix} & -1 \\ 1 & \end{bmatrix}$	0
C_3^-	$\begin{bmatrix}\mathrm{i}\omega^* & \\ & -\mathrm{i}\omega\end{bmatrix}$	$\begin{bmatrix}\mathrm{i}\omega^* & \\ & -\mathrm{i}\omega\end{bmatrix}$	1	C''_{22}	$\begin{bmatrix} & \mathrm{i}\omega^* \\ \mathrm{i}\omega & \end{bmatrix}$	$\begin{bmatrix} & -\mathrm{i}\omega^* \\ -\mathrm{i}\omega & \end{bmatrix}$	0
C_2	$\begin{bmatrix}-\mathrm{i} & \\ & \mathrm{i}\end{bmatrix}$	$\begin{bmatrix}-\mathrm{i} & \\ & \mathrm{i}\end{bmatrix}$	0	C''_{23}	$\begin{bmatrix} & -\mathrm{i}\omega \\ -\mathrm{i}\omega^* & \end{bmatrix}$	$\begin{bmatrix} & \mathrm{i}\omega \\ \mathrm{i}\omega^* & \end{bmatrix}$	0

other matrices as they are. What this means, in the Opechowski approach, is that the (modified) $\check{R}^{1/2}(\alpha\beta\gamma)$ matrices are now taken as valid for all the g operations, and their negatives for all the $\tilde{g}$ operations, and that the double-group multiplication rules are conventionally taken to be those of this matrix group. Whenever a new matrix representation is formed, the sign of its matrices has again to be suitably adjusted to satisfy the agreed multiplication rules. If the (modified) $\check{R}^{1/2}(\alpha\beta\gamma)$ representation is used in this way, although $\mathbf{D}_6$ will come up all right, $\mathbf{D}_3$ will be nonsense. In the Euler-angle approach there is no way of seeing straightaway that this will happen, but we are in a position to know that this is so. If we call C'_{2i}, C''_{2i} the poles marked in Fig. 1.2 and $\bar{C}'_{2i}$, $\bar{C}''_{2i}$ their corresponding antipoles, it is clear that when the sign of $\check{R}^{1/2}(\alpha\beta\gamma)$ is the opposite of that of $\check{R}^{1/2}(\rho\tau)$, then C, say, is replaced by $\bar{C}$. On taking, to consider the most favourable case, the modified $\check{R}^{1/2}(\alpha\beta\gamma)$, it follows at once from Table 1 that the (implicit) poles of these matrices are C'_{21}, C'_{22}, $\bar{C}'_{23}$, C''_{21}, $\bar{C}''_{22}$, $\bar{C}''_{23}$, all of them in the upper semicircle in Fig. 1.2. When subducing from $\mathbf{D}_6$ to $\mathbf{D}_3$, i.e. when dropping the $C_6^{\pm}$ operations, neither of the two sets of poles C'_{2i} or C''_{2i} remains conjugate. This means that the character theorem will not be valid for either of the two $\mathbf{D}_3$ subgroups of $\mathbf{D}_6$. (§**11**-7.) We repeat that the problem here is that Euler angles are unable to tell poles from antipoles.

It is hardly necessary to say that the representation $\check{R}^{1/2}(\rho\tau)$ satisfies the multiplication table and factor system of Tables 2.2 and 2.3 and that this is not so for $\check{R}^{1/2}(\alpha\beta\gamma)$, as the reader can quickly verify, e.g., for the product $C'_{21}C'_{22}$. This, however, is no criticism of the Euler representation, since it merely entails a change of conventions. A different choice of $\mathfrak{H}$, e.g., would also change the factor system.

4. THE IRREDUCIBLE PROJECTIVE AND VECTOR REPRESENTATIONS

Before we get the representations, it is worthwhile finding out their number and dimensions, for which we list the classes in the groups of our interest:

$\mathbf{D}_6$ (12): $E, C_6^{\pm}, C_3^{\pm}, C_2, C'_{2i}, C''_{2i}$; 6 classes (3 regular). (1)

$\mathbf{D}_3$ (6): $E, C_3^{\pm}, C'_{2i}$; 3 classes (3 regular). (2)

$\mathbf{C}_{3v}$ (6): $E, C_3^{\pm}, \sigma_{vi}$; 3 classes (3 regular). (3)

We give here in brackets the order of the groups. Notice that the last three classes of $\mathbf{D}_6$ are irregular, because C_2, C'_{21} and C''_{21} are BB, but that all classes of $\mathbf{D}_3$ are regular, which means that, whereas the characters of C'_{2i} and C''_{2i} must be all zero in $\mathbf{D}_6$ (as can be seen in Table 3.1) this need not be the case in $\mathbf{D}_3$ for the C'_{2i} class. (This is what makes the subduction from $\mathbf{D}_6$ to $\mathbf{D}_3$ chancy in other methods.)

From (1), (2), (3), the well-known theorems for the vector representations, and the results of §**11**-4 (number of projective representations equal to the number of regular classes; sum of the square of their dimensions equal to the group order), we have the following results:

$\mathbf{D}_6$:	6 vector reps.,	dimensions 1, 1, 1, 1, 2, 2,	(4)
	3 projectives,	dimensions 2, 2, 2.	(5)
$\mathbf{D}_3, \mathbf{C}_{3v}$:	3 vector reps.,	dimensions 1, 1, 2.	(6)
	3 projectives,	dimensions 1, 1, 2.	(7)

Table **15**-4.1. The representations $\check{R}^j$ for $\mathbf{D}_6$ and related groups. The superscript j has been omitted from the bases elements. The numbers of the right of the matrices are their characters. $\omega = \exp(\frac{1}{6}\pi \mathrm{i})$. $u_{3\pm} = 2^{-1/2}(u_3^3 \pm u_{-3}^3)$.

j Basis	$\frac{1}{2}$ $\langle u_{1/2} u_{-1/2}\vert$		1 $\langle u_1 u_{-1}\vert$		$\frac{3}{2}$ $\langle u_{3/2} u_{-3/2}\vert$		2 $\langle u_2 u_{-2}\vert$	
E	$\begin{bmatrix}1 & \\ & 1\end{bmatrix}$	2	$\begin{bmatrix}1 & \\ & 1\end{bmatrix}$	2	$\begin{bmatrix}1 & \\ & 1\end{bmatrix}$	2	$\begin{bmatrix}1 & \\ & 1\end{bmatrix}$	2
C_6^+	$\begin{bmatrix}\omega^* & \\ & \omega\end{bmatrix}$	$\sqrt{3}$	$\begin{bmatrix}-\mathrm{i}\omega & \\ & \mathrm{i}\omega^*\end{bmatrix}$	1	$\begin{bmatrix}-\mathrm{i} & \\ & \mathrm{i}\end{bmatrix}$	0	$\begin{bmatrix}-\mathrm{i}\omega^* & \\ & \mathrm{i}\omega\end{bmatrix}$	-1
C_6^-	$\begin{bmatrix}\omega & \\ & \omega^*\end{bmatrix}$	$\sqrt{3}$	$\begin{bmatrix}\mathrm{i}\omega^* & \\ & -\mathrm{i}\omega\end{bmatrix}$	1	$\begin{bmatrix}\mathrm{i} & \\ & -\mathrm{i}\end{bmatrix}$	0	$\begin{bmatrix}\mathrm{i}\omega & \\ & -\mathrm{i}\omega^*\end{bmatrix}$	-1
C_3^+	$\begin{bmatrix}-\mathrm{i}\omega & \\ & \mathrm{i}\omega^*\end{bmatrix}$	1	$\begin{bmatrix}-\mathrm{i}\omega^* & \\ & \mathrm{i}\omega\end{bmatrix}$	-1	$\begin{bmatrix}-1 & \\ & -1\end{bmatrix}$	-2	$\begin{bmatrix}\mathrm{i}\omega & \\ & -\mathrm{i}\omega^*\end{bmatrix}$	-1
C_3^-	$\begin{bmatrix}\mathrm{i}\omega^* & \\ & -\mathrm{i}\omega\end{bmatrix}$	1	$\begin{bmatrix}\mathrm{i}\omega & \\ & -\mathrm{i}\omega^*\end{bmatrix}$	-1	$\begin{bmatrix}-1 & \\ & -1\end{bmatrix}$	-2	$\begin{bmatrix}-\mathrm{i}\omega^* & \\ & \mathrm{i}\omega\end{bmatrix}$	-1
C_2	$\begin{bmatrix}-\mathrm{i} & \\ & \mathrm{i}\end{bmatrix}$	0	$\begin{bmatrix}-1 & \\ & -1\end{bmatrix}$	-2	$\begin{bmatrix}\mathrm{i} & \\ & -\mathrm{i}\end{bmatrix}$	0	$\begin{bmatrix}1 & \\ & 1\end{bmatrix}$	2
C'_{21}	$\begin{bmatrix} & -\mathrm{i} \\ -\mathrm{i} & \end{bmatrix}$	0	$\begin{bmatrix} & -1 \\ -1 & \end{bmatrix}$	0	$\begin{bmatrix} & \mathrm{i} \\ \mathrm{i} & \end{bmatrix}$	0	$\begin{bmatrix} & 1 \\ 1 & \end{bmatrix}$	0
C'_{22}	$\begin{bmatrix} & -\omega^* \\ \omega & \end{bmatrix}$	0	$\begin{bmatrix} & -\mathrm{i}\omega \\ \mathrm{i}\omega^* & \end{bmatrix}$	0	$\begin{bmatrix} & \mathrm{i} \\ \mathrm{i} & \end{bmatrix}$	0	$\begin{bmatrix} & -\mathrm{i}\omega^* \\ \mathrm{i}\omega & \end{bmatrix}$	0
C'_{23}	$\begin{bmatrix} & \omega \\ -\omega^* & \end{bmatrix}$	0	$\begin{bmatrix} & \mathrm{i}\omega^* \\ -\mathrm{i}\omega & \end{bmatrix}$	0	$\begin{bmatrix} & \mathrm{i} \\ \mathrm{i} & \end{bmatrix}$	0	$\begin{bmatrix} & \mathrm{i}\omega \\ -\mathrm{i}\omega^* & \end{bmatrix}$	0
C''_{21}	$\begin{bmatrix} & -1 \\ 1 & \end{bmatrix}$	0	$\begin{bmatrix} & 1 \\ 1 & \end{bmatrix}$	0	$\begin{bmatrix} & -1 \\ 1 & \end{bmatrix}$	0	$\begin{bmatrix} & 1 \\ 1 & \end{bmatrix}$	0
C''_{22}	$\begin{bmatrix} & \mathrm{i}\omega^* \\ \mathrm{i}\omega & \end{bmatrix}$	0	$\begin{bmatrix} & \mathrm{i}\omega \\ -\mathrm{i}\omega^* & \end{bmatrix}$	0	$\begin{bmatrix} & -1 \\ 1 & \end{bmatrix}$	0	$\begin{bmatrix} & -\mathrm{i}\omega^* \\ \mathrm{i}\omega & \end{bmatrix}$	0
C''_{23}	$\begin{bmatrix} & -\mathrm{i}\omega \\ -\mathrm{i}\omega^* & \end{bmatrix}$	0	$\begin{bmatrix} & -\mathrm{i}\omega^* \\ \mathrm{i}\omega & \end{bmatrix}$	0	$\begin{bmatrix} & -1 \\ 1 & \end{bmatrix}$	0	$\begin{bmatrix} & \mathrm{i}\omega \\ -\mathrm{i}\omega^* & \end{bmatrix}$	0

It would be sensible at this stage to work out characters of the representations and thus determine possible bases. (See Problem 1.) The work for dihedral groups is, however, so easy from the results of Chapter **14** that it is quicker to shoot off straightaway and construct full representations. We know from §**14**-2 that, in principle, the only bases we need consider in dihedral groups are of the form $\langle u^j_j, u^j_{-j}|$ in which case (see eqn **14**-2.22) the corresponding matrix $\check{R}^j(\rho\tau)$ is *obtained by raising each element of the standard representation* $\check{R}^{1/2}(\rho\tau)$ (as given in Table 3.1) *to the power* $2j$. (For j integral, as fairly obvious and in any case discussed in §**14**-2, the representations become vector representations.) Of course, for $j=0$, we shall have a one-dimensional basis u^0_0, which will give the trivial representation. The basis u^1_0, on the other hand (which we consider specially, since lower orders of j are always important), gives a one-

Table **15**-4.1. Continued.

$\frac{5}{2}$ $\langle u_{5/2}u_{-5/2}|$		3 $\langle u_3u_{-3}|$		0 u_0	1 u_0	3 u_{3+}	3 u_{3-}	j Basis
$\begin{bmatrix}1 & \\ & 1\end{bmatrix}$	2	$\begin{bmatrix}1 & \\ & 1\end{bmatrix}$	2	1	1	1	1	E
$\begin{bmatrix}-\omega & \\ & -\omega^*\end{bmatrix}$	$-\sqrt{3}$	$\begin{bmatrix}-1 & \\ & -1\end{bmatrix}$	-2	1	1	-1	-1	C_6^+
$\begin{bmatrix}-\omega^* & \\ & -\omega\end{bmatrix}$	$-\sqrt{3}$	$\begin{bmatrix}-1 & \\ & -1\end{bmatrix}$	-2	1	1	-1	-1	C_6^-
$\begin{bmatrix}i\omega^* & \\ & -i\omega\end{bmatrix}$	1	$\begin{bmatrix}1 & \\ & 1\end{bmatrix}$	2	1	1	1	1	C_3^+
$\begin{bmatrix}-i\omega & \\ & i\omega^*\end{bmatrix}$	1	$\begin{bmatrix}1 & \\ & 1\end{bmatrix}$	2	1	1	1	1	C_3^-
$\begin{bmatrix}-i & \\ & i\end{bmatrix}$	0	$\begin{bmatrix}-1 & \\ & -1\end{bmatrix}$	-2	1	1	-1	-1	C_2
$\begin{bmatrix} & -i\\ -i & \end{bmatrix}$	0	$\begin{bmatrix} & -1\\ -1 & \end{bmatrix}$	0	1	-1	-1	1	C'_{21}
$\begin{bmatrix} & \omega\\ -\omega^* & \end{bmatrix}$	0	$\begin{bmatrix} & -1\\ -1 & \end{bmatrix}$	0	1	-1	-1	1	C'_{22}
$\begin{bmatrix} & -\omega^*\\ \omega & \end{bmatrix}$	0	$\begin{bmatrix} & -1\\ -1 & \end{bmatrix}$	0	1	-1	-1	1	C'_{23}
$\begin{bmatrix} & -1\\ 1 & \end{bmatrix}$	0	$\begin{bmatrix} & 1\\ 1 & \end{bmatrix}$	0	1	-1	1	-1	C''_{21}
$\begin{bmatrix} & -i\omega\\ -i\omega^* & \end{bmatrix}$	0	$\begin{bmatrix} & 1\\ 1 & \end{bmatrix}$	0	1	-1	1	-1	C''_{22}
$\begin{bmatrix} & i\omega^*\\ i\omega & \end{bmatrix}$	0	$\begin{bmatrix} & 1\\ 1 & \end{bmatrix}$	0	1	-1	1	-1	C''_{23}

dimensional representation, the matrix of which, $\hat{R}^1$, is given from (**14**.2.26) by the *sum of the products of the two diagonals of the standard matrix.* We shall apply the first italicized rule above for $j=\frac{1}{2}, \frac{3}{2}, \frac{5}{2}$ (which should give us the three projectives in eqn 5) and to $j=1, 2, 3$ which should give the two two-dimensional representations in (4) plus, on possible reduction, two of the one-dimensional representations. With any luck, on adding the trivial representation and the one corresponding to u_0^1 as given by the second italicized rule above, we shall have completed all the representations. Nothing can go wrong: the worst that can happen is that we might have to add one or two extra values of j (although this will not be needed in the present case).

We do this work in Table 1, where the first column is a straight copy of the standard representation in Table 3.1. In order to construct this table all you need is a circle in the Argand plane divided in sectors of $\pi i/6$ where ω and its successive powers, $i\omega^*$, i, . . . are marked. (See Fig. 1.2.) You then start with the standard matrix and raise its elements to the power of 2, 3, 4, 5, and 6 entering the elements row by row in the six matrices corresponding to each element just as fast as you can write. The longest piece of work in forming this table is to draw a blank grid of the 72 matrices required! The column for $j=0$ is trivial and the following one for $j=1$ is obtained at once from the second italicized rule above. One then goes back to the Argand circle in order to add up the diagonal elements to get the characters and the whole thing is done.

The next thing to do is to check for irreducibility on adding up the square of the characters. This checks all representations except $j=3$, but take care: this is valid only for $\mathbf{D}_6$ and a little more work might be needed for the subgroups.

We must, before we do this, go back to $\mathbf{D}_6$ and reduce the representation $\hat{R}^3$. It is easy to guess that this is done by introducing a new basis

$$\langle u_{3+}, u_{3-}| =_{\text{def}} \langle u_3^3 u_{-3}^3| \begin{bmatrix} 2^{-1/2} & 2^{-1/2} \\ 2^{-1/2} & -2^{-1/2} \end{bmatrix} =_{\text{def}} \langle u_3^3 u_{-3}^3|\, S. \tag{8}$$

In fact, the new matrices in the new basis are $S^{-1}\hat{R}S$, which in this case is the same as $S\hat{R}S$, and it is easy to see that under this transformation the two types of matrices that appear in $\hat{R}^3$ change as follows

$$\begin{bmatrix} a & \\ & a \end{bmatrix} \mapsto \begin{bmatrix} a & \\ & a \end{bmatrix}, \qquad \begin{bmatrix} & a \\ a & \end{bmatrix} \mapsto \begin{bmatrix} a & \\ & -a \end{bmatrix}. \tag{9}$$

Equation (9) verifies, in fact, that $\hat{R}^3$ is reduced by the transformation (8). It is very easy from (9) to write the two remaining vector representations of $\mathbf{D}_6$, which is done in the last two columns of the table. (Check that they are orthogonal.)

We now have all the representations of $\mathbf{D}_6$. The reader should check a few matrix products to verify that the multiplication and factor tables are satisfied.

The representations of $\mathbf{D}_3$

In subducing to $\mathbf{D}_3$ from Table 1 we must watch for the fact that some irreducible representations of $\mathbf{D}_6$ will become reducible in $\mathbf{D}_3$, for which purpose it is best to collect first the characters of the subduced representations, which we do in Table 2. We shall not bother to subduce all representations in Table 1, since from (6) and (7) we expect three vector and three projective representations only. As we copy the characters in Table 2, at the same time checking for irreducibility, we see that all the irreducible vector representations appear, as well as one two-dimensional irreducible projective. The last representation in the table is reducible and, when looking at the matrices in Table 1, one can immediately see that all that is required in order to reduce it is precisely the same trick used in (8) and (9). It is thus extremely easy to collect the irreducible matrix representations in Table 3. It should be noticed that on reducing in

Table **15**-4.2. Subduction of the representations of $\mathbf{D}_6$ (from Table 1) down to $\mathbf{D}_3$.
Only characters are given.

Basis	E	C_3^+	C_3^-	C'_{21}	C'_{22}	C'_{23}	
u_0^0	1	1	1	1	1	1	irred.
u_0^1	1	1	1	−1	−1	−1	irred.
$\langle u_1^1 u_{-1}^1 \vert$	2	−1	−1	0	0	0	irred.
$\langle u_{1/2}^{1/2} u_{-1/2}^{1/2} \vert$	2	1	1	0	0	0	irred.
$\langle u_{3/2}^{3/2} u_{-3/2}^{3/2} \vert$	2	−2	−2	0	0	0	red.

Table **15**-4.3. The irreducible representations of $\mathbf{D}_3$.
$u_{3/2\pm} = 2^{-1/2}(u_{3/2}^{3/2} \pm u_{-3/2}^{3/2})$, $\omega = \exp(\frac{1}{6}\pi i)$.

Basis	E	C_3^+	C_3^-	C'_{21}	C'_{22}	C'_{23}
u_0^0	1	1	1	1	1	1
u_0^1	1	1	1	−1	−1	−1
$\langle u_1^1 u_{-1}^1 \vert$	$\begin{bmatrix} 1 & \\ & 1 \end{bmatrix}$	$\begin{bmatrix} -i\omega^* & \\ & i\omega \end{bmatrix}$	$\begin{bmatrix} i\omega & \\ & -i\omega^* \end{bmatrix}$	$\begin{bmatrix} & -1 \\ -1 & \end{bmatrix}$	$\begin{bmatrix} & -i\omega \\ i\omega^* & \end{bmatrix}$	$\begin{bmatrix} & i\omega^* \\ -i\omega & \end{bmatrix}$
$\langle u_{1/2}^{1/2} u_{-1/2}^{1/2} \vert$	$\begin{bmatrix} 1 & \\ & 1 \end{bmatrix}$	$\begin{bmatrix} -i\omega & \\ & i\omega^* \end{bmatrix}$	$\begin{bmatrix} i\omega^* & \\ & -i\omega \end{bmatrix}$	$\begin{bmatrix} & -i \\ -i & \end{bmatrix}$	$\begin{bmatrix} & -\omega^* \\ \omega & \end{bmatrix}$	$\begin{bmatrix} & \omega \\ -\omega^* & \end{bmatrix}$
$u_{3/2+}$	1	−1	−1	i	i	i
$u_{3/2-}$	1	−1	−1	−i	−i	−i

the way described the representation on the basis $\langle u^{3/2}_{3/2}\, u^{3/2}_{-3/2}|$ that results from the (defective) standard representation $\check{R}^{1/2}(\alpha\beta\gamma)$ of Table 3.1, the operations C'_{2i} appear with different characters, for the reasons discussed in §3. (See Problem 2.)

The representations of $\mathbf{C}_{3v}$

We have to do first the job of Table 2, replacing the C'_{2i} first by C''_{2i} (from Table 1), and then by the σ_{vi}. Moreover, in order to allow for the passage from the proper to the improper operation, the following changes have to be effected (see the discussion after eqn **14**-2.10). For integral j, the matrices for the binary rotations have to be multiplied by ± 1 for j even and odd respectively, whereas for half-integral j, in the Pauli gauge, no changes are required. With this in mind, we look at Table 2 to see what will happen in $\mathbf{C}_{3v}$. Nothing changes, except that the second representation now coincides with the first. We look at the matrices for C''_{2i} in Table 1, and we see that the basis u_{3+} will restore the negative signs for the reflections (once the factor -1 is allowed for).

For the remaining representations, the matrices for C''_{2i} from Table 1 have to be used (changed in sign for $j = 1$) and the representation for $j = \frac{3}{2}$ reduced as indicated in Problem 3. We collect the irreducible representations in Table 4. Notice that, although $\mathbf{C}_{3v}$ is isomorphic to $\mathbf{D}_3$, not all representations and bases are identical with those in Table 3.

Table **15**-4.4. The irreducible representations of $\mathbf{C}_{3v}$.
$u_{3/2\pm} = 2^{-1/2}(u^{3/2}_{3/2} \pm \mathrm{i}u^{3/2}_{-3/2})$. $\omega = \exp(\frac{1}{6}\pi\mathrm{i})$. $u_{3+} = 2^{-1/2}(u^3_3 + u^3_{-3})$.

Basis	E	C_3^+	C_3^-	σ_{v1}	σ_{v2}	σ_{v3}
u^0_0	1	1	1	1	1	1
u_{3+}	1	1	1	-1	-1	-1
$\langle u^1_1 u^1_{-1}|$	$\begin{bmatrix}1 & \\ & 1\end{bmatrix}$	$\begin{bmatrix}-\mathrm{i}\omega^* & \\ & \mathrm{i}\omega\end{bmatrix}$	$\begin{bmatrix}\mathrm{i}\omega & \\ & -\mathrm{i}\omega^*\end{bmatrix}$	$\begin{bmatrix} & -1 \\ -1 & \end{bmatrix}$	$\begin{bmatrix} & -\mathrm{i}\omega \\ \mathrm{i}\omega^* & \end{bmatrix}$	$\begin{bmatrix} & \mathrm{i}\omega^* \\ -\mathrm{i}\omega & \end{bmatrix}$
$\langle u^{1/2}_{1/2} u^{1/2}_{-1/2}|$	$\begin{bmatrix}1 & \\ & 1\end{bmatrix}$	$\begin{bmatrix}-\mathrm{i}\omega & \\ & \mathrm{i}\omega^*\end{bmatrix}$	$\begin{bmatrix}\mathrm{i}\omega^* & \\ & -\mathrm{i}\omega\end{bmatrix}$	$\begin{bmatrix} & -1 \\ 1 & \end{bmatrix}$	$\begin{bmatrix} & \mathrm{i}\omega^* \\ \mathrm{i}\omega & \end{bmatrix}$	$\begin{bmatrix} & -\mathrm{i}\omega \\ -\mathrm{i}\omega^* & \end{bmatrix}$
$u_{3/2-}$	1	1	1	i	i	i
$u_{3/2+}$	1	-1	-1	$-$i	$-$i	$-$i

Problems 15-4 (p. 296)

1. From the standard representation characters of Table 3.1 guess, by means of the orthogonality theorem, those of the two remaining projective representations of $\mathbf{D}_6$. Hence show that these two representations must be spanned by bases corresponding to $j = \frac{3}{2}$ and $j = \frac{5}{2}$ respectively.

2. From the representation $\check{R}^{1/2}(\alpha\beta\gamma)$ in Table 3.1 derive the matrix representation $\check{R}^{3/2}$ for $\mathbf{D}_3$ in the manner of Table 1. Show that it is reducible and reduce it. Is the character theorem satisfied?
3. Reduce the representation for $\mathbf{C}_{3v}$ in Table 1 (on using the matrices for E, C_3, $3C''_{2i}$) for $j=\frac{3}{2}$ and find the bases of the irreducible representations.

5. THE DOUBLE GROUP $\tilde{\mathbf{D}}_3$

We shall now treat the spinor (and incidentally the vector) representations of $\mathbf{D}_3$ by the double-group method. Of course, rather than using the Opechowski approach, in which a matrix double group is first formed in order to derive the multiplication rules, we shall use the quaternion parametrization method. The quaternion parameters for all operations g of $\tilde{\mathbf{D}}_3$ such that $g \in \mathbf{D}_3$ are obtained straightaway from Table 2.1. (We use, of course, the geometry defined by Fig. 1.2.) The quaternion parameter of $\tilde{g}$ is simply the negative of that of g. The parameters are given in Table 1 and with them we construct a quaternion multiplication table from which the group multiplication table (Table 2) is constructed. In principle, all that is required is to multiply the quaternions corresponding to the operations in Table 1 and then identify the product through the resulting quaternion. The work, however, can be simplified as follows. Table 2 contains four quadrants that correspond to products of the form gg, g$\tilde{g}$, $\tilde{g}$g, $\tilde{g}\tilde{g}$. We first form the products in the first quadrant. The product of the two quaternions for C_3^+ and C'_{21}, $\llbracket\frac{1}{2}, (0\ 0\ \frac{\sqrt{3}}{2}\rrbracket$, $\llbracket 0, (100)\rrbracket$ is $\llbracket 0, (\frac{1}{2}\ \frac{\sqrt{3}}{2}\ 0)\rrbracket$ which, from Table 1, is the quaternion for $\tilde{C}'_{23}$. In the same manner, we get the rest of this quadrant. Once this quadrant is filled in, the other quadrants of Table 2 are written at once, g$\tilde{g}$ being the 'complement' of gg, in the sense that each g and $\tilde{g}$ in it are replaced by $\tilde{g}$ and g respectively. Then, the quadrants $\tilde{g}$g and $\tilde{g}\tilde{g}$ are identical copies of the diagonally opposed quadrants g$\tilde{g}$ and gg respectively.

The representations themselves are constructed precisely as in Table 4.1, the matrices for g being identical to those for $\mathbf{D}_3$ and the matrices for $\tilde{g}$ being their negatives. It can be seen at once that the representation

Table **15**-5.1. Quaternion parameters for $\tilde{\mathbf{D}}_3$.

E	$\llbracket 1,\ \mathbf{0}\rrbracket$	$\tilde{E}$	$\llbracket -1,\ \mathbf{0}\rrbracket$
C_3^+	$\llbracket \frac{1}{2}, (0\ \ 0\ \ \frac{\sqrt{3}}{2}\rrbracket$	$\tilde{C}_3^+$	$\llbracket -\frac{1}{2}, (0\ \ 0\ -\frac{\sqrt{3}}{2})\rrbracket$
C_3^-	$\llbracket \frac{1}{2}, (0\ \ 0\ -\frac{\sqrt{3}}{2})\rrbracket$	$\tilde{C}_3^-$	$\llbracket -\frac{1}{2}, (0\ \ 0\ \ \frac{\sqrt{3}}{2})\rrbracket$
C'_{21}	$\llbracket 0, (1\ \ 0\ \ 0)\rrbracket$	$\tilde{C}'_{21}$	$\llbracket 0, (-1\ \ 0\ \ 0)\rrbracket$
C'_{22}	$\llbracket 0, (-\frac{1}{2}\ \ \frac{\sqrt{3}}{2}\ \ 0)\rrbracket$	$\tilde{C}'_{22}$	$\llbracket 0, (\frac{1}{2}\ -\frac{\sqrt{3}}{2}\ \ 0)\rrbracket$
C'_{23}	$\llbracket 0, (-\frac{1}{2}\ -\frac{\sqrt{3}}{2}\ \ 0)\rrbracket$	$\tilde{C}'_{23}$	$\llbracket 0, (\frac{1}{2}\ \ \frac{\sqrt{3}}{2}\ \ 0)\rrbracket$

Table **15**-5.2. Multiplication table for $\tilde{\mathbf{D}}_3$.
The product $g_i g_j$ appears in the intersection of the g_i row with the g_j column.

	E	C_3^+	C_3^-	C'_{21}	C'_{22}	C'_{23}	$\tilde{E}$	$\tilde{C}_3^+$	$\tilde{C}_3^-$	$\tilde{C}'_{21}$	$\tilde{C}'_{22}$	$\tilde{C}'_{23}$
E	E	C_3^+	C_3^-	C'_{21}	C'_{22}	C'_{23}	$\tilde{E}$	$\tilde{C}_3^+$	$\tilde{C}_3^-$	$\tilde{C}'_{21}$	$\tilde{C}'_{22}$	$\tilde{C}'_{23}$
C_3^+	C_3^+	$\tilde{C}_3^-$	E	$\tilde{C}'_{23}$	$\tilde{C}'_{21}$	$\tilde{C}'_{22}$	$\tilde{C}_3^+$	C_3^-	$\tilde{E}$	C'_{23}	C'_{21}	C'_{22}
C_3^-	C_3^-	E	$\tilde{C}_3^+$	$\tilde{C}'_{22}$	$\tilde{C}'_{23}$	$\tilde{C}'_{21}$	$\tilde{C}_3^-$	$\tilde{E}$	C_3^+	C'_{22}	C'_{23}	C'_{21}
C'_{21}	C'_{21}	$\tilde{C}'_{22}$	$\tilde{C}'_{23}$	$\tilde{E}$	C_3^+	C_3^-	$\tilde{C}'_{21}$	C'_{22}	C'_{23}	E	$\tilde{C}_3^+$	$\tilde{C}_3^-$
C'_{22}	C'_{22}	$\tilde{C}'_{23}$	$\tilde{C}'_{21}$	C_3^-	$\tilde{E}$	C_3^+	$\tilde{C}'_{22}$	C'_{23}	C'_{21}	$\tilde{C}_3^-$	E	$\tilde{C}_3^+$
C'_{23}	C'_{23}	$\tilde{C}'_{21}$	$\tilde{C}'_{22}$	C_3^+	C_3^-	$\tilde{E}$	$\tilde{C}'_{23}$	C'_{21}	C'_{22}	$\tilde{C}_3^+$	$\tilde{C}_3^-$	E
$\tilde{E}$	$\tilde{E}$	$\tilde{C}_3^+$	$\tilde{C}_3^-$	$\tilde{C}'_{21}$	$\tilde{C}'_{22}$	$\tilde{C}'_{23}$	E	C_3^+	C_3^-	C'_{21}	C'_{22}	C'_{23}
$\tilde{C}_3^+$	$\tilde{C}_3^+$	C_3^-	$\tilde{E}$	C'_{23}	C'_{21}	C'_{22}	C_3^+	$\tilde{C}_3^-$	E	$\tilde{C}'_{23}$	$\tilde{C}'_{21}$	$\tilde{C}'_{22}$
$\tilde{C}_3^-$	$\tilde{C}_3^-$	$\tilde{E}$	C_3^+	C'_{22}	C'_{23}	C'_{21}	C_3^-	E	$\tilde{C}_3^+$	$\tilde{C}'_{22}$	$\tilde{C}'_{23}$	$\tilde{C}'_{21}$
$\tilde{C}'_{21}$	$\tilde{C}'_{21}$	C'_{22}	C'_{23}	E	$\tilde{C}_3^+$	$\tilde{C}_3^-$	C'_{21}	$\tilde{C}'_{22}$	$\tilde{C}'_{23}$	$\tilde{E}$	C_3^+	C_3^-
$\tilde{C}'_{22}$	$\tilde{C}'_{22}$	C'_{23}	C'_{21}	$\tilde{C}_3^-$	E	$\tilde{C}_3^+$	C'_{22}	$\tilde{C}'_{23}$	$\tilde{C}'_{21}$	C_3^-	$\tilde{E}$	C_3^+
$\tilde{C}'_{23}$	$\tilde{C}'_{23}$	C'_{21}	C'_{22}	$\tilde{C}_3^+$	$\tilde{C}_3^-$	E	C'_{23}	$\tilde{C}'_{21}$	$\tilde{C}'_{22}$	C_3^+	C_3^-	$\tilde{E}$

table for $\tilde{\mathbf{D}}_3$ will be the same as that for $\mathbf{D}_3$ in Table 4.3, but repeated twice, the second half, for the $\tilde{g}$ operations, with changed signs in tilde representations. We shall for brevity collect the corresponding characters in Table 6.3, the first column of which gives the usual conventional labels for the representations. Notice that because all the classes of $\tilde{\mathbf{D}}_3$ are regular, each class of $\mathbf{D}_3$ splits in two, as shown, by the Opechowski Theorem. Thus, the double group, with six classes, has six representations, three vector and three spinor.

Problem 15-5 **(p. 297)**

1. List the classes of $\mathbf{D}_6$ and hence derive the number and nature of its representations.

6. SOME APPLICATIONS

The range of applications of spinor representations and double groups which we can discuss here is very limited. Applications to solid-state and relativistic problems require a theoretical background beyond the scope of this book. So, we shall only be able to discuss some applications to molecular and atomic problems. These applications, however, show precisely the way in which these concepts can be applied to the groups which appear in solid-state problems. It must be realized, on the other hand, that the main purpose of this section is not to provide a comprehensive review of applications but, rather, to provide a comparison of the use of the projective-representation and double-group methods when a black-box approach to representation tables is adopted. Most people interested

Table **15**-6.1. Characters of the irreducible representations of $\mathbf{D}_6$, and applications.

$$u_{3\pm} = 2^{-1/2}(u_3^3 \pm u_{-3}^3)$$

	E	$2C_6$	$2C_3$	C_2	$3C_2'$	$3C_2''$	Basis
A_1	1	1	1	1	1	1	u_0^0
A_2	1	1	1	1	−1	−1	u_0^1
B_1	1	−1	1	−1	1	−1	u_{3-}
B_2	1	−1	1	−1	−1	1	u_{3+}
E_1	2	1	−1	−2	0	0	$\langle u_1^1 u_{-1}^1 \|$
E_2	2	−1	−1	2	0	0	$\langle u_2^2 u_{-2}^2 \|$
$\bar{E}_1$	2	$\sqrt{3}$	1	0	0	0	$\langle u_{1/2}^{1/2} u_{-1/2}^{1/2} \|$
$\bar{E}_2$	2	$-\sqrt{3}$	1	0	0	0	$\langle u_{5/2}^{5/2} u_{-5/2}^{5/2} \|$
$\bar{E}_3$	2	0	−2	0	0	0	$\langle u_{3/2}^{3/2} u_{-3/2}^{3/2} \|$
$\sin \frac{1}{2}\phi$		$\frac{1}{2}$	$\frac{\sqrt{3}}{2}$	1	1	1	
$\sin 4\phi$		$-\frac{\sqrt{3}}{2}$	$\frac{\sqrt{3}}{2}$	0	0	0	
$\chi^{7/2}$	8	$-\sqrt{3}$	1	0	0	0	$\bar{E}_1 + 2\bar{E}_2 + \bar{E}_3$
$\bar{E}_1 \otimes \bar{E}_2$	4	−3	1	0	0	0	$B_1 + B_2 + E_2$

in these problems will, in fact, use published character tables and they will not be very much interested in how the tables were obtained. So we have collected in Tables 1, 2, and 3 the character tables for $\mathbf{D}_6$ (from Table 4.1), for $\mathbf{D}_3$ (from Table 4.3) and for $\tilde{\mathbf{D}}_3$ (as described in §5), respectively.

We shall consider the following two problems. (1) How an atomic state of angular momentum $j = \frac{7}{2}$ splits in a field of symmetry $\mathbf{D}_6$ or $\mathbf{D}_3$. (2) How a two-electron state with wave function $\psi_i(1)\psi_j(2)$ splits in a field of

Table **15**-6.2. Characters of the irreducible representations of $\mathbf{D}_3$, and applications.

$$u_{3/2\pm} = 2^{-1/2}(u_{3/2}^{3/2} \pm u_{-3/2}^{3/2})$$

	E	$2C_3$	$3C_2'$	Basis
A_1	1	1	1	u_0^0
A_2	1	1	−1	u_0^1
E	2	−1	0	$\langle u_1^1 u_{-1}^1 \|$
$\bar{E}_1$	2	1	0	$\langle u_{1/2}^{1/2} u_{-1/2}^{1/2} \|$
$^1\bar{E}$	1	−1	i	$u_{3/2+}$
$^2\bar{E}$	1	−1	−i	$u_{3/2-}$
$\sin \frac{1}{2}\phi$		$\frac{\sqrt{3}}{2}$	1	
$\sin 4\phi$		$\frac{\sqrt{3}}{2}$	0	
$\chi^{7/2}$	8	1	0	$3\bar{E}_1 + {}^1\bar{E} + {}^2\bar{E}$
$E \otimes \bar{E}_1$	4	−1	0	$\bar{E}_1 + {}^1\bar{E} + {}^2\bar{E}$

Table **15**-6.3. Characters of the irreducible representations of $\tilde{\mathbf{D}}_3$, and applications.
$u_{3/2\pm} = 2^{-1/2}(u^{3/2}_{3/2} \pm u^{3/2}_{-3/2})$

	E	$2C_3$	$3C_2'$	$\bar{E}$	$2\bar{C}_3$	$3\bar{C}_2'$	Basis
A_1	1	1	1	1	1	1	u^0_0
A_2	1	1	-1	1	1	-1	u^1_0
E	2	-1	0	2	-1	0	$\langle u^1_1 u^1_{-1}\|$
$\bar{E}_1$	2	1	0	-2	-1	0	$\langle u^{1/2}_{1/2} u^{1/2}_{-1/2}\|$
${}^1\bar{E}$	1	-1	i	-1	1	$-$i	$u_{3/2+}$
${}^2\bar{E}$	1	-1	$-$i	-1	1	i	$u_{3/2-}$
$\sin \frac{1}{2}\phi$		$\frac{\sqrt{3}}{2}$	1		$-\frac{\sqrt{3}}{2}$	-1	
$\sin 4\phi$		$\frac{\sqrt{3}}{2}$	0		$\frac{\sqrt{3}}{2}$	0	
$\chi^{7/2}$	8	1	0	-8	-1	0	$3\bar{E}_1 + {}^1\bar{E} + {}^2\bar{E}$
$E \otimes \bar{E}_1$	4	-1	0	-4	1	0	$\bar{E}_1 + {}^1\bar{E} + {}^2\bar{E}$

symmetry $\mathbf{D}_6$ (for $\psi_i \in \bar{E}_1$, $\psi_j \in \bar{E}_2$), and in a field of symmetry $\mathbf{D}_3$ (for $\psi_i \in E$, $\psi_j \in \bar{E}_1$). For comparison, the cases with $\mathbf{D}_3$ symmetry will be treated both by the projective-representation and by the double-group method.

In order to deal with Problem 1 we need the character χ^j corresponding to the basis $\langle u^j_j \ldots u^j_{-j}|$ which was given in (**4**-7.15) and (**4**-7.16), which we now transcribe with an obvious change for application to double groups:

$$\chi^j(\phi) = \sin(j+\tfrac{1}{2})\phi(\sin \tfrac{1}{2}\phi)^{-1}, \qquad \phi \neq 0, \tag{1}$$

$$= 2j+1, \qquad \phi = 0, \tag{2}$$

$$= -(2j+1), \qquad \phi = 2\pi. \tag{3}$$

In (1), the rotation angle ϕ for an operation $\tilde{g}$ is taken to be that of the operation g plus 2π. These formulae are applied in Tables 1 to 3, and once the characters χ^j are obtained they are split over the irreducible representations, as shown on the right-hand column.

For Problem 2, the direct product of the representations involved has to be worked out and then split over the irreducible group representations, as shown in the last line of the table.

It is obvious from this work that neither in the projective nor in the double-group method, the user of the tables needs be concerned about the nature or meaning of the representations that he uses. It is also evident that the black-box approach which is common in double-group applications can equally well be adopted in the use of the projective representation tables with a minimum of trouble and with some manifest advantages. First, it is clear from Tables 2 and 3 that, in using the double group, one is printing redundant information and duplicating unnecessary work. Secondly, the use of the double groups requires some understanding of their somewhat complicated class structure, which is entirely avoided in the projective-representation method, since the group to be studied requires no modification whatsoever in this case. (All that an unsophisticated user of the tables has to understand in this method is what the regular classes are, and that their number coincides with the number of the spinor representations.) Thirdly, there is a conceptual advantage in avoiding the use of the operation $\tilde{E}$, since, if it is naïvely used as a rotation by 2π, the understanding of a rotation as a linear coordinate transformation and nothing else might become blurred. On the other hand, the delicate distinction which we have introduced between a rotation by 2π (identical with the identity in this book) and a turn by 2π, can be entirely avoided in the black-box approach to the projective-representation tables.

16

SOLUTIONS TO PROBLEMS

I knew one, that when he wrote a Letter, he would put that which was most Materiall, in the Postscript, as if it had beene a By-matter.

Francis Bacon, Lord Verulam, Viscount St Alban (1625). *The Essayes or Counsells, Civill and Morall.* John Havilland, London. Essay XXII, p. 130.

All cross references in this chapter are given by Problem and equation number. **P2**-6.1 is Problem **2**-6.1, whereas (P**2**-6.1) is eqn (1) of the section headed Problems **2**-6. The figures are numbered serially, with no section number.

PROBLEMS 2-1

1. $gg^{-1}=E \Rightarrow \{gg^{-1}g=g$ *and* $ggg^{-1}=g\} \Rightarrow gg^{-1}g=ggg^{-1} \Rightarrow gE=g(g^{-1}g) \Rightarrow g^{-1}g=E,$ because the identity is unique.
2. $\{g_j^{-1}g_i^{-1}g_ig_j=g_j^{-1}g_j=E\} \Rightarrow g_j^{-1}g_i^{-1}=(g_ig_j)^{-1}$.
3. Figure 1, *a* and *b* shows respectively that $\boldsymbol{i}$ commutes with $R(\alpha\mathbf{z})$ and σ. Draw similar figures for other operations. For inversion and translation, draw a figure performing them along one single line. They do not commute.
4. See Fig. 1*c*. The point marked × here is in the plane of the drawing. No third dimension exists.
5. From Fig. 1*d*, $\sigma=\boldsymbol{i}C_2=\boldsymbol{i}R(\pi\mathbf{n})$, $\mathbf{n}\perp\sigma$.

   $$S_m=C_m\sigma=R(\tfrac{2\pi}{m}\mathbf{n})\boldsymbol{i}R(\pi\mathbf{n})=R(\tfrac{(2+m)\pi}{m}\mathbf{n})\boldsymbol{i}.$$

6. Draw a cube and operate as in Fig. 1.
7. From Fig. 1*e*, $\sigma_1\sigma_2=\sigma_2\sigma_1$. Do the same for σ_1 and σ_2 not normal.
8. (i) Coaxial: operate as in Fig. 1. (ii) Bilateral binary, from Fig. 1*f*.
9. Draw a figure like Fig. 1*f* with $\mathbf{n}\parallel\mathbf{x}$ and $\mathbf{p}$ normal to the drawing.
10. See Fig. **2**-1.2.
11. See Fig. 1g. (σ_x and σ_y commute).
12. Draw two parallel reflection planes separated by $\frac{1}{2}$t.
13. From Fig. 1*h*, $C_3^+\sigma_{v1}=\sigma_{v3}$. Draw similar figures for the other products. Compare with results in Table **15**-2.2.
14. As in Fig. 1*f*.

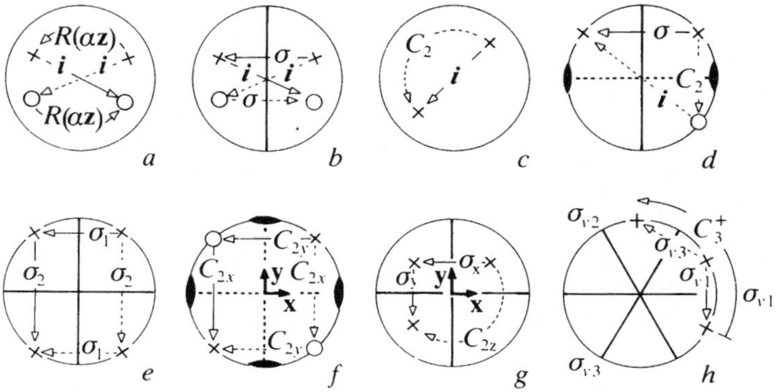

Fig. 16.1. Problems **2**-1. Enter the figures at the point marked in the first quadrant and follow first the circuit depicted with full arrows and then the circuit in broken lines. × point above, ○ point below the drawing. **z** is normal to and above the drawing.

PROBLEMS 2-2

1. Call σ_z a reflection plane (mirror) in the plane of the drawing of Fig. 1g. **y** and **z** are eigenvectors of σ_x (eigenvalue +1) and **x** is an eigenvector of σ_x (eigenvalue −1). The vectors **x**, **y**, **z** are eigenvectors common to σ_x, σ_y, σ_z. (Although their eigenvalues may differ from mirror to mirror.)
2. All vectors of the configuration space are eigenvectors of $\boldsymbol{i}$ with eigenvalue −1. Therefore $\boldsymbol{i}$ shares all eigenvectors with all other symmetry operations.
3. Coaxial: share single eigenvector (rotation axis). Bilateral binary: C_{2x}, C_{2y} in Fig. 1*f* share eigenvectors **x**, **y**, **z**. For example, eigenvectors of C_{2x} with eigenvalues in brackets: **x** (+1), **y** (−1), **z** (−1). Similarly for the others.
4. Take $R(\phi\mathbf{z})$. For $\phi = \pi$, the eigenvectors, with eigenvalues in brackets, are: **z** (+1), **x** (−1), **y** (−1). If σ is the **xy** plane its eigenvectors are **x** (+1), **y** (+1), **z** (−1) and coincide with those of $R(\pi\mathbf{z})$.
5. Only then they have common eigenvectors.

PROBLEM 2-3

1. Because θ is constant under g, take $p_x = \cos\varphi$, $p_y = \sin\varphi$.

$$g\cos\varphi = \boldsymbol{R}(\tfrac{1}{2}\pi\mathbf{z})\cos\varphi = \cos(g^{-1}\varphi) = \cos(\varphi - \tfrac{1}{2}\pi) = \sin\varphi = p_y.$$

$$g\sin\varphi = \boldsymbol{R}(\tfrac{1}{2}\pi\mathbf{z})\sin\varphi = \sin(g^{-1}\varphi) = \sin(\varphi - \tfrac{1}{2}\pi) = -\cos\varphi = -p_x.$$

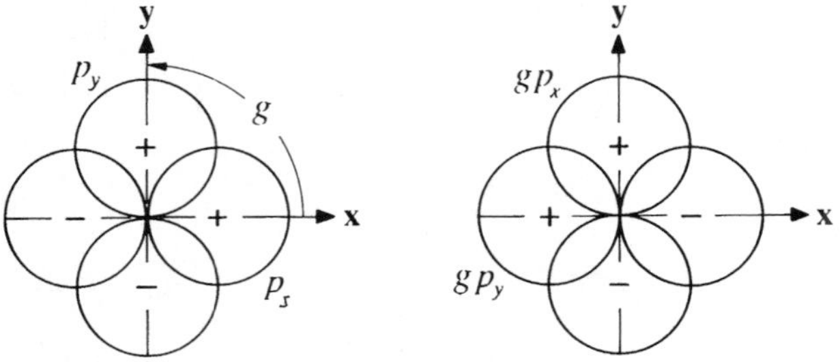

Fig. 16.2. The functions p_x, p_y, and their transforms gp_x, gp_y, for $g = R(\frac{1}{2}\pi\mathbf{z})$.

Compare with Fig. 2, where the functions are actively transformed. This agrees with above results but agreement is lost if g^{-1} is replaced by g in the brackets.

PROBLEMS **2**-4

1.
$$\sigma_x =_{\text{def}} \mathbf{xz} \text{ plane.}$$

P2-1.10 $\Rightarrow \quad \sigma_\alpha\sigma_x = R(2\alpha\mathbf{z}) \quad \Rightarrow \quad \sigma_\alpha = R(2\alpha\mathbf{z})\sigma_x. \quad (1)$

$$\sigma_x\, |xy\rangle = |x, -y\rangle = [{}^1{}_{-1}]\, |xy\rangle =_{\text{def}} \hat{\sigma}_x\, |xy\rangle. \quad (2)$$

2,**2**-4.1|1

$$\hat{\sigma}_\alpha = \begin{bmatrix} \cos 2\alpha & -\sin 2\alpha \\ \sin 2\alpha & \cos 2\alpha \end{bmatrix}\begin{bmatrix} 1 & \\ & -1 \end{bmatrix} = \begin{bmatrix} \cos 2\alpha & \sin 2\alpha \\ \sin 2\alpha & -\cos 2\alpha \end{bmatrix}. \quad (3)$$

2. (**2**-4.13) follows from Fig. 3. Because θ is constant under σ_α, take $\mathsf{x} = \cos\varphi$, $\mathsf{y} = \sin\varphi$. $\sigma_\alpha^{-1} = \sigma_\alpha$.

$$\dot{\sigma}_\alpha \mathsf{x} = \cos(\sigma_\alpha\varphi) = \cos(2\alpha - \varphi) = \cos 2\alpha \ \cos\varphi + \sin 2\alpha \sin\varphi. \quad (4)$$

$$\dot{\sigma}_\alpha \mathsf{y} = \sin(\sigma_\alpha\varphi) = \sin(2\alpha - \varphi) = \sin 2\alpha \cos\varphi - \cos 2\alpha \sin\varphi. \quad (5)$$

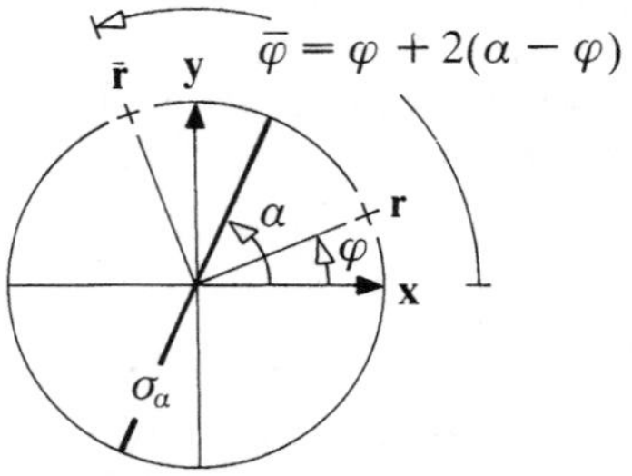

Fig. 16.3. Transform of the azimuth φ under a reflection σ_α.

Write (4) and (5) as $\acute{\sigma}_\alpha \langle \mathsf{xy}| = \langle \mathsf{xy}|\, \hat{\sigma}_\alpha$ and verify that $\hat{\sigma}_\alpha$ coincides with (3). Note change from $|xy\rangle$ in Problem 1 to $\langle \mathsf{xy}|$ in Problem 2.

3.
$$|\bar{\mathbf{r}}\rangle = \mathrm{A}\,|\mathbf{r}\rangle \quad \Rightarrow \quad \bar{r}_i = \textstyle\sum_j A_{ij} r_j. \tag{6}$$

The unit vectors $\mathbf{e}_s$ and their transforms decompose on the fixed axes as follows:

$$\mathbf{e}_s = \textstyle\sum_i (\mathbf{e}_s)_i \mathbf{E}_i = \sum_i (\mathbf{e}_s)_i \mathbf{e}_i \quad \Rightarrow \quad (\mathbf{e}_s)_i = \delta_{si}. \tag{7}$$

$$\bar{\mathbf{e}}_s = \textstyle\sum_i (\bar{\mathbf{e}}_s)_i \mathbf{E}_i = \sum_i (\bar{\mathbf{e}}_s)_i \mathbf{e}_i. \tag{8}$$

On applying the linear transformation (6) to the component $(\bar{\mathbf{e}}_s)_i$:

,7
$$(\bar{\mathbf{e}}_s)_i = \textstyle\sum_j A_{ij} (\mathbf{e}_s)_j = \sum_j A_{ij}\delta_{sj} = A_{is}. \tag{9}$$

9|8
$$\bar{\mathbf{e}}_s = \textstyle\sum_i A_{is}\mathbf{e}_i \equiv \sum_i \mathbf{e}_i A_{is} \quad \Rightarrow \quad \langle \bar{\mathbf{e}}| = \langle \mathbf{e}|\, A. \tag{10}$$

4. Take $R(-\frac{2}{3}\pi(1\bar{1}1))$ as an example. It is inconvenient to transform $|xyz\rangle$ directly but, from a drawing, the transform of $\langle \mathbf{ijk}|$ is easy to obtain:

$$\mathbf{i} \mapsto -\mathbf{j},\ -\mathbf{j} \mapsto \mathbf{k},\ \mathbf{k} \mapsto \mathbf{i} \quad \Rightarrow \quad \langle \mathbf{ijk}| \mapsto \langle \bar{\mathbf{j}}\bar{\mathbf{k}}\mathbf{i}| = \langle \mathbf{ijk}|\, A,$$

with $A = \begin{bmatrix} & & 1 \\ -1 & & \\ & -1 & \end{bmatrix}$.

Therefore:

$$R\,|xyz\rangle = \begin{bmatrix} & & 1 \\ -1 & & \\ & -1 & \end{bmatrix} |xyz\rangle = |z\bar{x}\bar{y}\rangle.$$

This agrees with Onodera and Okazaki's transform as listed in their Table II for the inverse operation, i.e. the rotation by $+\frac{2}{3}\pi$. (See the heading to their Table II.)

PROBLEMS 2-5

1. Form the multiplication table as in **P2**-1.13 and verify the group properties.
2. The class of σ_{v1}, e.g., $C(\sigma_{v1})$ is given by

$$C(\sigma_{v1}) = E\sigma_{v1}E \oplus C_3^+\sigma_{v1}C_3^- \oplus C_3^-\sigma_{v1}C_3^+ \oplus \sigma_{v1}\sigma_{v1}\sigma_{v1}$$
$$\oplus\, \sigma_{v2}\sigma_{v1}\sigma_{v2} \oplus \sigma_{v3}\sigma_{v1}\sigma_{v3}$$

T15-2.2
$$= \sigma_{v1} \oplus C_3^+\sigma_{v3} \oplus C_3^-\sigma_{v2} \oplus \sigma_{v1}E \oplus \sigma_{v2}C_3^+ \oplus \sigma_{v3}C_3^-$$

T15-2.2
$$= \sigma_{v1} \oplus \sigma_{v2} \oplus \sigma_{v3} \oplus \sigma_{v1} \oplus \sigma_{v3} \oplus \sigma_{v2}.$$

You will find three classes:

$$C(E) = E, \quad C(C_3^+) = C_3^+ \oplus C_3^-, \quad C(\sigma_{v1}) = \sigma_{v1} \oplus \sigma_{v2} \oplus \sigma_{v3}.$$

This ghastly work is superseded by **P9**-1.1.

3. The necessary matrices on the basis $|xy\rangle$ are

L**2**-4.1, **2**-4.12
$$\hat{R}(\alpha\mathbf{z}) = \begin{bmatrix} \cos\alpha & -\sin\alpha \\ \sin\alpha & \cos\alpha \end{bmatrix}, \qquad \hat{\sigma}_\alpha = \begin{bmatrix} \cos 2\alpha & \sin 2\alpha \\ \sin 2\alpha & -\cos 2\alpha \end{bmatrix}. \quad (1)$$

These matrices are written as the angles are read from Fig. **2**-1.3:

$$\begin{matrix} \hat{E} & \hat{C}_3^+ & \hat{C}_3^- & \hat{\sigma}_{v1} & \hat{\sigma}_{v2} & \hat{\sigma}_{v3} \\ \begin{bmatrix} 1 & \\ & 1 \end{bmatrix} & \begin{bmatrix} -\frac{1}{2} & -\frac{\sqrt{3}}{2} \\ \frac{\sqrt{3}}{2} & -\frac{1}{2} \end{bmatrix} & \begin{bmatrix} -\frac{1}{2} & \frac{\sqrt{3}}{2} \\ -\frac{\sqrt{3}}{2} & -\frac{1}{2} \end{bmatrix} & \begin{bmatrix} 1 & \\ & -1 \end{bmatrix} & \begin{bmatrix} -\frac{1}{2} & -\frac{\sqrt{3}}{2} \\ -\frac{\sqrt{3}}{2} & \frac{1}{2} \end{bmatrix} & \begin{bmatrix} -\frac{1}{2} & \frac{\sqrt{3}}{2} \\ \frac{\sqrt{3}}{2} & \frac{1}{2} \end{bmatrix} \end{matrix} \quad (2)$$

Verify, e.g., that $\hat{C}_3^-\hat{\sigma}_{v2} = \hat{\sigma}_{v3}$ as required by Table **15**-2.2.

4. Write $R(\alpha\mathbf{z}) \cos 2\varphi = \cos 2(\varphi - \alpha) = \cos 2\varphi \; \cos 2\alpha + \sin 2\varphi \; \sin 2\alpha$. For σ_α use (**2**-4.13): $\sigma_\alpha \cos 2\varphi = \cos\{2(2\alpha - \varphi)\}$. The same for $\sin 2\varphi$. Thus obtain the following matrices on $\langle d_{x^2-y^2}, d_{xy}|$, and then proceed as for (2):

$$\hat{R}(\alpha\mathbf{z}) = \begin{bmatrix} \cos 2\alpha & -\sin 2\alpha \\ \sin 2\alpha & \cos 2\alpha \end{bmatrix}, \qquad \hat{\sigma}_\alpha = \begin{bmatrix} \cos 4\alpha & \sin 4\alpha \\ \sin 4\alpha & -\cos 4\alpha \end{bmatrix}. \quad (3)$$

$$\begin{matrix} \hat{E} & \hat{C}_3^+ & \hat{C}_3^- & \hat{\sigma}_{v1} & \hat{\sigma}_{v2} & \hat{\sigma}_{v2} \\ \begin{bmatrix} 1 & \\ & 1 \end{bmatrix} & \begin{bmatrix} -\frac{1}{2} & \frac{\sqrt{3}}{2} \\ -\frac{\sqrt{3}}{2} & -\frac{1}{2} \end{bmatrix} & \begin{bmatrix} -\frac{1}{2} & -\frac{\sqrt{3}}{2} \\ \frac{\sqrt{3}}{2} & -\frac{1}{2} \end{bmatrix} & \begin{bmatrix} 1 & \\ & -1 \end{bmatrix} & \begin{bmatrix} -\frac{1}{2} & \frac{\sqrt{3}}{2} \\ \frac{\sqrt{3}}{2} & \frac{1}{2} \end{bmatrix} & \begin{bmatrix} -\frac{1}{2} & -\frac{\sqrt{3}}{2} \\ -\frac{\sqrt{3}}{2} & \frac{1}{2} \end{bmatrix} \end{matrix}. \quad (4)$$

The characters of (2) and (4) are the same: the representations are equivalent. The sum of the square of the characters in (4) is six, equal to the order of the group: the representation is irreducible.

5. Consider first the basis $\langle xy|$. Use the matrix (**2**-4.11) for $\hat{R}(\alpha\mathbf{z})$. For $\hat{\sigma}_\alpha$ form $\sigma_\alpha \cos\varphi = \cos(2\alpha - \varphi)$, from (**2**-4.13), expand, do the same for $\sin\varphi$ and obtain $\hat{\sigma}_\alpha$. You will find that $\hat{R}(\alpha\mathbf{z})$ and $\hat{\sigma}_\alpha$ on this basis coincide with (1). (Because this basis is the dual of $|xy\rangle$.) Under all the operations of $\mathbf{C}_{3v}$, θ is left invariant, so that $z = \cos\theta$ is also invariant. Thus

$$R(\alpha\mathbf{z})\langle zxy| = \langle zxy| \begin{bmatrix} 1 & 0 & 0 \\ 0 & \cos\alpha & -\sin\alpha \\ 0 & \sin\alpha & \cos\alpha \end{bmatrix},$$

and similarly for $\hat{\sigma}_\alpha$. The representation reduces into the trivial representation and (2).

6. Both representations are irreducible because the sum of the characters squared is six. From Problem 2, one more irreducible representation exists because there are three classes.

7. Use the sum of the square of the characters for irreducibility (not forgetting to take into account the number of elements of each class listed on the first line). Also, the result of Problem 6 for the number of irreducible representations.

8. $$\hat{G}'(g_i)\hat{G}'(g_j) = L\hat{G}(g_i)L^{-1}L\hat{G}(g_j)L^{-1} = L\hat{G}(g_i)\hat{G}(g_j)L^{-1} = L\hat{G}(g_ig_j)L^{-1} = \hat{G}'(g_ig_j).$$

9. $\mathrm{Tr}\,(AB) = \sum_i (AB)_{ii} = \sum_{ij} A_{ij}B_{ji} = \sum_{ji} B_{ji}A_{ij} = \sum_j (BA)_{jj} = \mathrm{Tr}\,(BA)$.

10. Any element of $C(g_i)$ is of the form g_i^g:

$$\chi(g_i^g) = \mathrm{Tr}\,\hat{G}(g_i^g) = \mathrm{Tr}\,\hat{G}(gg_ig^{-1}) = \mathrm{Tr}\,\{\hat{G}(g)\hat{G}(g_i)\hat{G}(g^{-1})\}. \tag{5}$$

$$gg^{-1} = E \;\Rightarrow\; \hat{G}(g)\hat{G}(g^{-1}) = \mathbf{1} \;\Rightarrow\; \hat{G}(g^{-1}) = \{\hat{G}(g)\}^{-1}. \tag{6}$$

Introduce (6) into (5) and, from Problem 9, interchange the last matrix with the first factor, $\{\hat{G}(g)\hat{G}(g_i)\}$. The result will be:

$$\chi(g_i^g) = \mathrm{Tr}\,\hat{G}(g_i) = \chi(g_i), \qquad \forall g.$$

11. $\boldsymbol{i}$ is its own inverse, $\boldsymbol{i}^2 = E$ (closure). Representation:

$$\boldsymbol{i}\,\langle f, g| = \boldsymbol{i}\,\langle f, \boldsymbol{i}f| = \langle g, f| \;\Rightarrow\; \hat{\boldsymbol{i}} = \begin{bmatrix} & 1 \\ 1 & \end{bmatrix}. \quad \hat{E} = \mathbf{1}_2. \tag{7}$$

This representation is reduced by pre- and post-multiplication with $(2)^{-1/2}\left[\begin{smallmatrix} 1 & 1 \\ 1 & -1 \end{smallmatrix}\right]$. (This matrix is its own inverse.) The irreducible bases are $2^{-1/2}(f \pm g)$. The +sign corresponds to the so-called *gerade* representation A_g (basis even with respect to $\boldsymbol{i}$) and the − sign corresponds to the *ungerade* representation A_u (basis odd with respect to $\boldsymbol{i}$).

12. Multiply the operations as in Fig. 1*f*. All the operations commute because they are bilateral binary rotations. (See Problem **2**-2.3.)

13. Verify multiplication rules. Irreducibility: sum of the square of the characters must be four.

14. $C_{2x}\,\langle \mathbf{ijk}| = \langle \mathbf{i}\bar{\mathbf{j}}\bar{\mathbf{k}}|, \qquad C_{2y}\,\langle \mathbf{ijk}| = \langle \bar{\mathbf{i}}\mathbf{j}\bar{\mathbf{k}}|, \qquad C_{2z}\,\langle \mathbf{ijk}| = \langle \bar{\mathbf{i}}\bar{\mathbf{j}}\mathbf{k}|.$

(see Fig. **2**-5.1). From here, the representation is:

$$\overset{E}{\begin{bmatrix} 1 & & \\ & 1 & \\ & & 1 \end{bmatrix}} \overset{C_{2x}}{\begin{bmatrix} 1 & & \\ & -1 & \\ & & -1 \end{bmatrix}} \overset{C_{2y}}{\begin{bmatrix} -1 & & \\ & 1 & \\ & & -1 \end{bmatrix}} \overset{C_{2z}}{\begin{bmatrix} -1 & & \\ & -1 & \\ & & 1 \end{bmatrix}}. \tag{8}$$

From the characters of these matrices and Table **2**-5.3 deduce that this representation reduces into $B_1 \oplus B_2 \oplus B_3$. In fact, it is already reduced, the first line being B_3 (basis **i**), the second B_2 (basis **j**) and the third B_1 (basis **k**).

15. ∇^2 is invariant under the transformation $x \mapsto -x$, $y \mapsto -y$, $z \mapsto -z$, i.e. it is invariant with respect to $\boldsymbol{i}$. Thus its group is $\mathbf{C}_i$ which from Problem 11 has only two irreducible representations, gerade and ungerade respectively.

16. Both matrices on (R2-5.8) must pertain to the same basis, as is the case in the proof given of that equation. On the other hand, in the expression $g_i\{g_j \langle\varphi|\}$, g_i acts on the basis expressed by the curly brackets, say $\langle\varphi^1|$, and $g_i \langle\varphi^1|$ is $\langle\varphi^1|\,{}^1\hat{G}(g_i)$. Since $\langle\varphi^1|$, that is $g_j \langle\varphi|$, can be written as $\langle\varphi|\,\hat{G}(g_j)$, we get $\langle\varphi|\,\hat{G}(g_j)^1\hat{G}(g_i)$, which is not now in contradiction with (8), since the matrices are now different. The apparent fallacy is due to leaving out the label denoting the basis.
17. The set contains identity and inverses, and it is easy to prove from Table **15**-2.2 that it closes.
18. Because of the closure of H, the set gH, with $g \in H$, contains only elements of H, $|H|$ in number, which must be all distinct (if gh_i equals gh_j, then, on multiplying by g^{-1}, the absurd equality of two elements of H follows). Thus, gH equals H.
19. Assume that the cosets g_iH, g_jH have an element in common:

$$g_ih_r = g_jh_s \qquad \Rightarrow \qquad g_j = g_ih_rh_s^{-1} =_{\text{def}} g_ih', \quad h' \in H. \tag{9}$$

Thus, the coset g_jH equals $g_ih'H$, which, from Problem 18, equals g_iH. Thus the two cosets either have an element in common, in which case the cosets are identical, or they have no common element.
20. The total number of elements in $g_1H, \ldots, g_nH$, equals $|G|$, and on choosing the g's so that no two cosets are identical, these $|G|$ elements must all be distinct, from the result of Problem 19.
21. $\mathbf{E}\mathbf{C}_3$ equals E, C_3^+, C_3^-, and from Table **15**-2.2 the set $\sigma_{v1}E$, $\sigma_{v1}C_3^+$, $\sigma_{v1}C_3^-$ is σ_{v1}, σ_{v2}, σ_{v3}. Notice that the coset representative σ_{v1} can be replaced by σ_{v2} or σ_{v3}. (Check this.)
22. Form the conjugates of E, C_3^+, C_3^- by all operations of $\mathbf{C}_{3v}$ and argue that each class must be invariant under conjugation.
23. If $H \lhd G$, then, for all $g \in H$,

$$H \lhd G \qquad \Rightarrow \qquad gHg^{-1} = H \qquad \Rightarrow \qquad gH = Hg.$$

24. Because H is of index two, it can only have two distinct cosets in G, whether left or right. Thus, for $g \notin H$,

$$G = EH \oplus gH = HE \oplus Hg.$$

Since Hg cannot equal EH, it must equal gH. From this equality, we have:

$$Hg = gH \qquad \Rightarrow \qquad H = gHg^{-1} \qquad \Rightarrow \qquad H \lhd G.$$

PROBLEMS 2-6

1. From (**2**-6.3) and (**2**-6.5),

$$\det(LAL^{-1}) = \det L \;\det A \det(L^{-1}) = \det L \;\det A\,(\det L)^{-1} = \det A.$$

2. Diagonalize A into A'. $\mathrm{Tr}\, A = \mathrm{Tr}\, A'$ and $\det A = \det A'$. The diagonal elements of A' are the eigenvalues of A. Thus $\mathrm{Tr}\, A'$ is the sum of these eigenvalues and $\det A'$ is its product.
3. $\mathbf{r} = |1\mathrm{i}\rangle$, $\mathbf{r}^{\mathsf{T}} = \langle 1\mathrm{i}|$, $\mathbf{r}^{\dagger} = \langle 1, -\mathrm{i}|$. $\mathbf{r}^{\mathsf{T}}\mathbf{r} = 1 + \mathrm{i}^2 = 0$. $\mathbf{r}^{\dagger}\mathbf{r} = 1 - \mathrm{i}^2 = 2$.

4. $$(AB)^{\mathsf{T}}_{ij} = (AB)_{ji} = \textstyle\sum_r A_{jr}B_{ri} = \sum_r B^{\mathsf{T}}_{ir}A^{\mathsf{T}}_{rj} = (B^{\mathsf{T}}A^{\mathsf{T}})_{ij}. \quad (1)$$

1 $$(AB)^{\mathsf{T}} = B^{\mathsf{T}}A^{\mathsf{T}}. \quad (2)$$

,2 $$(AB)^{\dagger} = \{(AB)^*\}^{\mathsf{T}} = (A^*B^*)^{\mathsf{T}} = (B^*)^{\mathsf{T}}(A^*)^{\mathsf{T}} = B^{\dagger}A^{\dagger}. \quad (3)$$

$$\{B^{-1}A^{-1}AB = \mathbf{1} \textit{ and } (AB)^{-1}AB = \mathbf{1}\} \Rightarrow (AB)^{-1} = B^{-1}A^{-1}. \quad (4)$$

5. $$A^{\mathsf{T}}A = \mathbf{1} \Rightarrow AA^{\mathsf{T}}A = A \Rightarrow AA^{\mathsf{T}}AA^{\mathsf{T}} = AA^{\mathsf{T}} \Rightarrow AA^{\mathsf{T}} = \mathbf{1}. \quad (5)$$

The last entailment in (5) requires multiplication by $(AA^{\mathsf{T}})^{-1}$, i.e. that AA^{T} be non-singular, i.e. $\det(AA^{\mathsf{T}}) \neq 0$, i.e. $(\det A)^2 \neq 0$, i.e. $\det A \neq 0$. Hence, A must be non-singular.

6. $$A^{\mathsf{T}}A = \mathbf{1}, \quad B^{\mathsf{T}}B = \mathbf{1}. \quad (6)$$

,6 $$(AB)^{\mathsf{T}}AB = B^{\mathsf{T}}A^{\mathsf{T}}AB = B^{\mathsf{T}}B = \mathbf{1}. \quad (7)$$

7. $$A^{\mathsf{T}}A = \mathbf{1} \quad \Rightarrow \quad \sum A^{\mathsf{T}}_{ir}A_{rj} = \delta_{ij} \quad \Rightarrow \quad \sum A_{ri}A_{rj} = \delta_{ij}. \quad (8)$$

$$AA^{\mathsf{T}} = \mathbf{1} \quad \Rightarrow \quad \sum A_{ir}A^{\mathsf{T}}_{rj} = \delta_{ij} \quad \Rightarrow \quad \sum A_{ir}A_{jr} = \delta_{ij}. \quad (9)$$

(8) $\Rightarrow$ columns orthonormal; (9) $\Rightarrow$ rows orthonormal.

8. $$A^{\dagger}A = \mathbf{1} \Rightarrow AA^{\dagger}A = A \Rightarrow AA^{\dagger}AA^{\dagger} = AA^{\dagger} \Rightarrow AA^{\dagger} = \mathbf{1}. \quad (10)$$

A must be non-singular. See Problem 5.

9. $$A^{\dagger}A = 1 \quad \Rightarrow \quad \sum A^{\dagger}_{ir}A_{rj} = \delta_{ij} \quad \Rightarrow \quad \sum A^*_{ri}A_{rj} = \delta_{ij}. \quad (11)$$

$$AA^{\dagger} = 1 \quad \Rightarrow \quad \sum A_{ir}A^{\dagger}_{rj} = \delta_{ij} \quad \Rightarrow \quad \sum A_{ir}A^*_{jr} = \delta_{ij}. \quad (12)$$

10. $$A^{\dagger}A = 1, \quad B^{\dagger}B = 1. \quad (13)$$

3,13 $$(AB)^{\dagger}AB = B^{\dagger}A^{\dagger}AB = B^{\dagger}B = \mathbf{1}. \quad (14)$$

11. $$A = \begin{bmatrix} a & b \\ c & d \end{bmatrix}. \qquad A^{-1} = (\det A)^{-1}\begin{bmatrix} d & -b \\ -c & a \end{bmatrix}, \quad (15)$$

because $AA^{-1} = 1$ (verify). Call $\det A = \Delta$. The unitary condition $AA^{\dagger} = \mathbf{1}$ entails $A^{\dagger} = A^{-1}$ and $\Delta\Delta^* = 1$, $(\Delta^{-1} = \Delta^*)$:

$$\Delta^*\begin{bmatrix} d & -b \\ -c & a \end{bmatrix} = \begin{bmatrix} a^* & c^* \\ b^* & d^* \end{bmatrix} \quad \Rightarrow \quad \begin{aligned} c^* &= -b\Delta^*, & (16) \\ d^* &= a\Delta^*. & (17) \end{aligned}$$

(Verify that eqns 16 and 17 entail the equalities for a^* and b^*.)

16,17|15 $$A = \begin{bmatrix} a & b \\ -b^*\Delta & a^*\Delta \end{bmatrix}, \quad aa^* + bb^* = 1, \quad \Delta\Delta^* = 1. \quad (18)$$

(The two conditions on R18 follow from $AA^\dagger = \mathbf{1}$; Δ in eqn 18 can be any arbitrary complex number of unit modulus.)

12. $$A = A^\dagger, \quad B = B^\dagger. \tag{19}$$

3,19 $$(AB)^\dagger = B^\dagger A^\dagger = BA = AB \text{ if and only if } AB - BA = 0. \tag{20}$$

13. $$A^\dagger = \pm A. \tag{21}$$

2-6.42 $$A\mathbf{r} = a\mathbf{r} \quad \Rightarrow \quad \mathbf{r}^\dagger A\mathbf{r} = a\mathbf{r}^\dagger\mathbf{r}. \tag{22}$$

Form the adjoint of both sides of (22) by (3):

,21 $$\mathbf{r}^\dagger A^\dagger \mathbf{r} = a^*\mathbf{r}^\dagger\mathbf{r} \quad \Rightarrow \quad \pm\mathbf{r}^\dagger A\mathbf{r} = a^*\mathbf{r}^\dagger\mathbf{r}. \tag{23}$$

On comparing the last two eqns in (22) and (23): $a^* = \pm a$. (24)

14. $$A^\dagger = \pm A; \qquad (24) \quad \Rightarrow \quad \{a^* = \pm a, \, b^* = \pm b\}. \tag{25}$$

$$A\mathbf{r} = a\mathbf{r} \quad \Rightarrow \quad \mathbf{s}^\dagger A\mathbf{r} = a\mathbf{s}^\dagger\mathbf{r}. \tag{26}$$

$$A\mathbf{s} = b\mathbf{s} \quad \Rightarrow \quad \mathbf{r}^\dagger A\mathbf{s} = b\mathbf{r}^\dagger\mathbf{s} \quad \overset{\dagger}{\Rightarrow} \quad \mathbf{s}^\dagger A^\dagger\mathbf{r} = b^*\mathbf{s}^\dagger\mathbf{r}.$$

25 $$\pm\mathbf{s}^\dagger A\mathbf{r} = \pm b\mathbf{s}^\dagger\mathbf{r}. \tag{27}$$

In the entailment marked †, take adjoints of both sides of the previous equation. The last equations in (26) and in (27) entail $(a-b)\mathbf{s}^\dagger\mathbf{r} = 0$, whence the result follows for $a \neq b$.

15. $$A \otimes B = \begin{bmatrix} 2 & 1 & 2i & i \\ -1 & 2 & -i & 2i \\ 2i & i & -2 & -1 \\ -i & 2i & 1 & -2 \end{bmatrix}.$$

16. $$\{(A \otimes B)(C \otimes D)\}_{[ij]} = \sum (A \otimes B)_{[ir]}(C \otimes D)_{[rj]}$$

2-6.54 $$= \sum A_{ir}BC_{rj}D = (AC)_{ij}BD$$

2-6.54 $$= (AC \otimes BD)_{[ij]}. \tag{28}$$

17.

,**2**-5.28

$$\text{Tr}\,(A \otimes B) = \textstyle\sum_{ij} (A \otimes B)_{ij,ij} = \sum_{ij} A_{ii}B_{jj} = \text{Tr}\,A\,\text{Tr}\,B. \tag{29}$$

$$\chi(g_ig_j \mid {}^r\hat{G} \otimes {}^s\hat{G}) = \text{Tr}\,\hat{G}(g_ig_j) \text{ for } \hat{G}(g_ig_j) = {}^r\hat{G}(g_i) \otimes {}^s\hat{G}(g_j) \tag{30}$$

29 $$= \text{Tr}\,{}^r\hat{G}(g_i)\,\text{Tr}\,{}^s\hat{G}(g_j) = \chi(g_i \mid {}^r\hat{G})\chi(g_j \mid {}^s\hat{G}). \tag{31}$$

18. From $A^\dagger_{mn} = A^*_{nm}$ and (**2**-5.28),

$$(A \otimes B)^\dagger_{ij,kl} = (A \otimes B)^*_{kl,ij} = A^*_{ki}B^*_{lj} = A^\dagger_{ik}B^\dagger_{jl} = (A^\dagger \otimes B^\dagger)_{ij,kl}. \tag{32}$$

19. A and B are square matrices of dimension m and n respectively. From (**2**-6.63)

$$(A \otimes B)(A \otimes B)^{\dagger} = (A \otimes B)(A^{\dagger} \otimes B^{\dagger})$$
$$= AA^{\dagger} \otimes BB^{\dagger} = \mathbf{1}_m \otimes \mathbf{1}_n = \mathbf{1}_{mn}. \qquad (33)$$

Equation (28) is used here. $\mathbf{1}_{mn}$ is the unit matrix of dimension mn.

20. *Sufficiency*. I assume known that the matrix L that diagonalizes A under the transformation $L^{-1}AL$ is the matrix the columns of which are the (normalized) eigenvectors of A. Thus, if A and B have all eigenvectors in common, $A' = L^{-1}AL$ and $B' = L^{-1}BL$ are simultaneously diagonal, whence $[A', B'] = 0$. Similarity preserves multiplication rules, whence $[A', B'] = 0 \quad \Rightarrow \quad [A, B] = 0$.

Necessity. Assume that A and B commute and that L is the matrix of the eigenvectors of A but not of B. The diagonal matrix $A' = L^{-1}AL$ must be a diagonal supermatrix with constant diagonal matrix elements $a\mathbf{1}_n$ where n is the degeneracy of the eigenvalue a. $B' = L^{-1}BL$ must commute with A' and from the theorem italicized before (**2**-6.56) it must be a diagonal supermatrix of the same structure as A'. Do now a similarity of A' and B' under the matrix M of the eigenvectors of B'. This will diagonalize the diagonal blocks of B' and, because those of A' are of the form $a\mathbf{1}_n$, it will leave them invariant. Thus, both A and B have now being fully diagonalized simultaneously by the same matrix, the columns of which must therefore be eigenvectors both of A and of B.

21.
$$A^{\mathsf{T}} = -A.$$
$$(\exp A)^{\mathsf{T}}(\exp A) = \exp(A^{\mathsf{T}}) \exp A = \exp(-A) \exp A = \mathbf{1}. \qquad (34)$$

22. First entailment: follows from (34) with $A = A^*$. Second entailment: write $\exp A \equiv \exp\{\mathrm{i}(-\mathrm{i}A)\}$, whence, from (**2**-6.62), $\exp A$ unitary $\Leftrightarrow$ $-\mathrm{i}A$ Hermitian. therefore:

$$(-\mathrm{i}A)^{\dagger} = -\mathrm{i}A \quad \Rightarrow \quad \mathrm{i}A^{\dagger} = -\mathrm{i}A \quad \Rightarrow \quad A^{\dagger} = -A. \qquad (35)$$

23. $$A = \mathrm{i}\mathbf{1} \quad \Rightarrow \quad A^2 = -\mathbf{1}, \quad A^3 = -A, \quad A^4 = \mathbf{1}, \quad A^5 = A, \ldots \qquad (36)$$

$$\exp(\phi A) = 1 + \phi A + \tfrac{1}{2}\phi^2 A^2 + \frac{1}{3!}\phi^3 A^3 + \frac{1}{4!}\phi^4 A^4 + \frac{1}{5!}\phi^5 A^5 + \ldots \qquad (37)$$

36|37
$$= \mathbf{1}\left(1 - \tfrac{1}{2}\phi^2 + \frac{1}{4!}\phi^4 \ldots\right) + A\left(1 - \frac{1}{3!}\phi^3 + \frac{1}{5!}\phi^5 \ldots\right) \qquad (38)$$

$$= \mathbf{1} \cos\phi + A \sin\phi. \qquad (39)$$

24. All eigenvalues of a real symmetric matrix are real (statement below eqn **2**-6.49), whereas general rotation matrices can have only one real eigenvalue corresponding to the rotation axis as eigenvector. (The exceptions, with three real eigenvalues, are the identity and binary rotations. See eqns **2**-6.33, **2**-6.34.)

PROBLEMS 2-7

1. From (**2**-7.8) and $\partial A/\partial t = 0$,

$$\langle A\rangle = \langle\psi|\, A\, |\psi\rangle \quad \Rightarrow \quad \partial\langle A\rangle/\partial t = \langle\partial\psi/\partial t|\, A\, |\psi\rangle + \langle\psi|\, A\, |\partial\psi/\partial t\rangle. \tag{1}$$

Use here the Schrödinger equation with time, $H\psi = \hbar \mathrm{i}\, \partial\psi/\partial t$, which gives $\partial\psi/\partial t = (\hbar\mathrm{i})^{-1}H\psi$. (Remember that terms from the left of the bracket must be removed from it, if constant, *after conjugation.* Also, use the Hermitian property of H.)

$$\begin{aligned}\partial\langle A\rangle/\partial t &= -(\hbar\mathrm{i})^{-1}\,\langle H\psi|\, A\, |\psi\rangle + (\hbar\mathrm{i})^{-1}\,\langle\psi|\, AH\, |\psi\rangle\\ &= -(\hbar\mathrm{i})^{-1}\,\langle\psi|\, HA\, |\psi\rangle + (\hbar\mathrm{i})^{-1}\,\langle\psi|\, AH\, |\psi\rangle\\ &= (\hbar\mathrm{i})^{-1}\,\langle\psi|\, AH - HA\, |\psi\rangle = (\hbar\mathrm{i})^{-1}\,\langle[A, H]\rangle.\end{aligned} \tag{2}$$

2. As in (**2**-7.22), first for δ and then for $t = n\delta$,

$$\tau f(x) = f(x+\delta) = f(x) + \delta\frac{\mathrm{d}}{\mathrm{d}x}f(x) = \left(1 + \delta\frac{\mathrm{d}}{\mathrm{d}x}\right)f(x). \tag{3}$$

$$\begin{aligned}f(x+t) &= \lim_{n\to\infty}\left(1+\delta\frac{\mathrm{d}}{\mathrm{d}x}\right)^n f(x) = \lim\left(1+\frac{t}{n}\frac{\mathrm{d}}{\mathrm{d}x}\right)^n f(x)\\ &= \exp\left(t\frac{\mathrm{d}}{\mathrm{d}x}\right)f(x) = f(x) + t\frac{\mathrm{d}f(x)}{\mathrm{d}x} + \frac{1}{2!}t^2\frac{\mathrm{d}^2f(x)}{\mathrm{d}x^2} + \ldots\end{aligned} \tag{4}$$

This is precisely the Taylor series re-derived. Notice the change of sign in §**2**-7 required by the correct definition of the function-space operator in (**2**-7.23).

3. $[uv, w] = uvw - wuv = uvw - uwv + uwv - wuv = u[v, w] + [u, w]v.$

$[u, vw] = uvw - vwu = uvw - vuw + vuw - vwu = [u, v]w + v[u, w].$

4. The argument should run as follows:

$$\boldsymbol{i}\psi(x) = \psi(-x) \qquad \Rightarrow \qquad \boldsymbol{i}^2\psi(x) = \psi(x) \equiv \omega\psi(x),$$

for an arbitrary phase factor ω. Hence $\boldsymbol{i}^2$ is not necessarily E. Thus the whole argument about $\mathbf{C}_i$ breaks down. (See §**11**-9.)

PROBLEMS 3-1

1. Use (**2**-1.6) in the form:

$$R(\pi\mathbf{y})R(\omega\mathbf{z}) = R(-\omega\mathbf{z})R(\pi\mathbf{y}). \qquad (1)$$

$$R(\alpha\pi\gamma) = R(\alpha\mathbf{z})R(\omega\mathbf{z})R(-\omega\mathbf{z})R(\pi\mathbf{y})R(\gamma\mathbf{z}) \qquad (2)$$

1|2

$$= R(\alpha\mathbf{z})R(\omega\mathbf{z})R(\pi\mathbf{y})R(\omega\mathbf{z})R(\gamma\mathbf{z}) \qquad (3)$$

$$= \mathrm{R}(\alpha+\omega, \mathbf{z})R(\pi\mathbf{y})R(\gamma+\omega, \mathbf{z}). \qquad (4)$$

2. Draw a figure like Figs **3**-1.1 and **3**-1.2. The effect of $R(\pi\mathbf{n})$ is: $\mathbf{k}\mapsto-\mathbf{k}$, $\mathbf{i}\mapsto\mathbf{j}$. The second set of parameters is obtained from the first with eqn (4) for $\omega=\frac{1}{4}\pi$. The inverse of the first set, from (**3**-1.3), is $R(\pi, \pi, \frac{3}{2}\pi) = R(\pi, \pi, -\frac{1}{2}\pi)$. On using $\omega=-\frac{3}{2}\pi$ in (4), this inverse becomes $R(-\frac{1}{2}\pi, \pi, 0)$; that is, the original rotation, which, being binary, is its own inverse.
3. Substitute β for α in $\hat{R}(\alpha\mathbf{z})$ of (**3**-1.14). On cycling $|xyz\rangle\mapsto|zxy\rangle$ this matrix must be $\hat{R}'(\beta\mathbf{y})$ on the basis $|zxy\rangle$. Write down the three equations of transformation of this basis with $\hat{R}'(\beta\mathbf{y})$. From them, write down the matrix A such that $R(\beta\mathbf{y})\,|xyz\rangle = A\,|xyz\rangle$. A is $\hat{R}(\beta\mathbf{y})$.
4. The required changes in (**3**-1.15) are: $\cos\gamma\mapsto-\cos\alpha$, $\sin\gamma\mapsto\sin\alpha$, $\cos\alpha\mapsto-\cos\gamma$, $\sin\alpha\mapsto\sin\gamma$. With these changes you will obtain the transpose of (**3**-1.15), i.e. its inverse.
5. Exercise care in reading Fig. **3**-1.1. In the active conventions, **i**, **j**, **k** in b should be understood as $\bar{\mathbf{i}}$, $\bar{\mathbf{j}}$, $\bar{\mathbf{k}}$. (Compare with Fig. **2**-4.1 for $\alpha=\frac{1}{2}\pi$ and eqn **2**-4.2 for this value.) Thus

$$\langle\bar{\mathbf{i}}\bar{\mathbf{j}}\bar{\mathbf{k}}| = \langle\mathbf{kij}| = \langle\mathbf{ijk}|\begin{bmatrix} & 1 & \\ & & 1 \\ 1 & & \end{bmatrix}. \qquad (5)$$

(Notice that if you read the figure as meaning 'transform of **i** is **j** – a mistake all too often made – you are putting the bars on the vectors in a, which is absurd.) $\hat{R}^1_{\mathbf{r}}(\frac{1}{2}\pi, \frac{1}{2}\pi, \pi)$ from (**3**-1.15) agrees with (5).
6. Draw figures like **3**-1.1 and **3**-1.2. You will find $R(0, \frac{1}{2}\pi, \frac{1}{2}\pi)$. The rotation is the inverse of $R(\frac{1}{2}\pi, \frac{1}{2}\pi, \pi)$ in (**3**-1.6). (**3**-1.13) on the latter gives $R(0, \frac{1}{2}\pi, \frac{1}{2}\pi)$. The transformation matrix is given by

$$\langle\bar{\mathbf{i}}\bar{\mathbf{j}}\bar{\mathbf{k}}| = \langle\mathbf{jki}| = \langle\mathbf{ijk}|\begin{bmatrix} & & 1 \\ 1 & & \\ & 1 & \end{bmatrix}. \qquad (6)$$

7. The mirrors σ must be written as $\sigma = iC_2$ (axis of $C_2\perp\sigma$, see Fig. **15**-1.2). The results are given in Table **15**-2.1.

PROBLEMS **3**-2

1. To obtain $R(\phi\mathbf{x})$ write $R(\phi\mathbf{z})$ as in (**3**-1.14) and interchange the 2×2 and 1×1 matrices along its diagonal. From (**3**-2.19),

$$n_x = (\sin\phi + \sin\phi)/2\sin\phi = 1, \qquad n_y = n_z = 0.$$

$R(\phi\mathbf{y})$, in the same manner, gives correctly $n_x = 0$, $n_y = 1$, $n_z = 0$.
2. Use (**3**.2.18), (**3**.2.19). $\cos\phi = -\frac{1}{2} \Rightarrow \phi = \frac{2}{3}\pi$; $\sin\phi = \frac{1}{2}3^{1/2}$; $n_x = \{2\sin(\frac{2}{3}\pi)\}^{-1} = 3^{-1/2}$; $\mathbf{n} = 3^{-1/2}(111)$. This agrees with (P **3**-1.6), which has the same matrix.
3. Their trace is equal from P**2**-5.9, hence also their angle of rotation from (**3**-2.3).

PROBLEMS **3**-3

1. Form Z, Z^2, Z^3. You will find $Z^3 = -Z$. Then:

$$\begin{aligned}\exp(\phi Z) &= \mathbf{1} + \phi Z + \tfrac{1}{2}\phi^2 Z^2 - \frac{1}{3!}\phi^3 Z - \frac{1}{4!}\phi^4 Z^2 + \frac{1}{5!}\phi^5 Z + \frac{1}{6!}\phi^6 Z^2 - \dots \\ &= \mathbf{1} + \left(\phi - \frac{1}{3!}\phi^3 + \frac{1}{5!}\phi^5 - \dots\right) Z + \left(\tfrac{1}{2}\phi^2 - \frac{1}{4!}\phi^4 + \frac{1}{6!}\phi^6 - \dots\right) Z^2 \\ &= \mathbf{1} + (\sin\phi) Z + (1 - \cos\phi) Z^2.\end{aligned}$$

2. Form, from (**3**-3.15), with $n^2 = 1$,

$$\begin{aligned}\bar{\mathbf{r}} &= \mathbf{r} + \sin\phi(\mathbf{n}\times\mathbf{r}) + 2\sin^2\tfrac{1}{2}\phi\{(\mathbf{n}\cdot\mathbf{r})\mathbf{n} - \mathbf{r}\} \\ &= (1 - 2\sin^2\tfrac{1}{2}\phi)\mathbf{r} + \sin\phi(\mathbf{n}\times\mathbf{r}) + 2\sin^2\tfrac{1}{2}\phi(\mathbf{n}\cdot\mathbf{r})\mathbf{n}. \qquad (1)\end{aligned}$$

From (1):

$$\begin{aligned}\bar{x} &= (1 - 2\sin^2\tfrac{1}{2}\phi)x + \sin\phi(n_y z - n_z y) + 2\sin^2\tfrac{1}{2}\phi(n_x^2 x + n_x n_y y + n_x n_z z) \\ &= \{1 + 2\sin^2\tfrac{1}{2}\phi(n_x^2 - 1)\}x + (-n_z\sin\phi + 2n_x n_y\sin^2\tfrac{1}{2}\phi)y \\ &\quad + (n_y\sin\phi + 2n_x n_z\sin^2\tfrac{1}{2}\phi)z.\end{aligned}$$

From here, on writing $n_x^2 - 1$ as $-(n_y^2 + n_z^2)$ from the normalization condition, the first row of the matrix follows and similarly the others.
3. $\mathbf{r} = (100)$, $\sin\phi = \sin\frac{1}{2}\phi = \frac{\sqrt{3}}{2}$, $(\mathbf{n}\times\mathbf{r}) = \frac{\sqrt{3}}{3}(01\bar{1})$, $\mathbf{n}\times(\mathbf{n}\times\mathbf{r}) = \frac{1}{3}(\bar{2}11)$.

3-3.15 $\qquad \bar{\mathbf{r}} = \bar{\mathbf{i}} = (100) + \frac{1}{2}(01\bar{1}) + \frac{1}{2}(\bar{2}11) = (010) = \mathbf{j},$

as shown in (P **3**-1.6).

PROBLEM **3**-4

1. $\cos\beta = 0 \Rightarrow \beta = \frac{1}{2}\pi$. $\{\tan\alpha = \infty, \sin\alpha = 1\} \Rightarrow \alpha = \frac{1}{2}\pi$.

$$\{\tan\gamma = 0, \cos\gamma = -1\} \Rightarrow \gamma = \pi.$$

PROBLEMS 3-5

1. Matrix from P 3-1.6.

3-5.3 $\tan \xi = 0/0 \quad \Rightarrow \quad \xi$ undetermined.
Take (i) $\xi = 0$; (ii) $\xi = \frac{1}{2}\pi$.

3-5.6 $\tan \eta = 1/0 \quad \Rightarrow \quad \eta = \frac{1}{2}\pi$.

3-5.7,8 (i) $\cos \zeta = 0, \sin \zeta = 1 \quad \Rightarrow \quad \zeta = \frac{1}{2}\pi$;
(ii) $\cos \zeta = 1, \sin \zeta = 0 \quad \Rightarrow \quad \zeta = 0$.

Feedback ξ, η, ζ into (3-5.1) and multiply both sets of matrices.

2. A: $\phi = \frac{1}{2}\pi$, $\mathbf{n} = -2^{-1/2}(110)$ $\xi = -\frac{1}{2}\pi$, $\eta = \frac{3}{4}\pi$, $\zeta = -\frac{1}{4}\pi$.
B: $\phi = \frac{1}{3}\pi$, $n = -3^{-1/2}(111)$. $\xi = \frac{3}{4}\pi$; $\eta = 19.47°$, $\zeta = \frac{3}{4}\pi$.

3.

3-5.11
$$\hat{R}^1_{\mathbf{r}}(\phi\mathbf{n}) = \mathbf{1} + \phi\mathbf{n}\,.\,\hat{\mathbf{J}} = \begin{bmatrix} 1 & -n_z\phi & n_y\phi \\ n_z\phi & 1 & -n_x\phi \\ -n_y\phi & n_x\phi & 1 \end{bmatrix}. \tag{1}$$

1|3-5.3 $\tan \xi = n_x\phi \quad \Rightarrow \quad \xi = n_x\phi$. (2)

1|3-5.6 $\tan \eta = n_y\phi \quad \Rightarrow \quad \eta = n_y\phi$. (3)

1|3-5.7,8 $\cos \zeta = 1, \sin \zeta = n_z\phi \quad \Rightarrow \quad \zeta = n_z\phi$. (4)

$$\hat{R}^1_{\mathbf{r}}(\phi\mathbf{n}) = \hat{X}(\phi n_x)\hat{Y}(\phi n_y)\hat{Z}(\phi n_z), \quad \phi \to 0. \tag{5}$$

Notice that the expansion (5) is now unique, and that here ϕn_x is the rotation angle around the $\mathbf{x}$ axis, that is, $X(\phi n_x) = R(\phi n_x \mathbf{x}) = R(\phi \mathbf{n}_x)$, and similarly for the others.

PROBLEMS 4-2

1. Use standard change to polars to obtain

$$\partial/\partial x = \sin\theta\ \cos\varphi\ \partial/\partial r + r^{-1}\cos\theta\ \cos\varphi\ \partial/\partial\theta - (r\sin\theta)^{-1}\sin\varphi\ \partial/\partial\varphi, \tag{1}$$

$$\partial/\partial y = \sin\theta\ \sin\varphi\ \partial/\partial r + r^{-1}\cos\theta\ \sin\varphi\ \partial/\partial\theta + (r\sin\theta)^{-1}\cos\varphi\ \partial/\partial\varphi, \tag{2}$$

$$\partial/\partial z = \cos\theta\ \partial/\partial r - r^{-1}\sin\theta\ \partial/\partial\theta. \tag{3}$$

Write now $I_x = -\mathrm{i}\nabla_x = -\mathrm{i}(y\ \partial/\partial z - z\ \partial/\partial y)$ and similarly for the other components. To obtain $\mathbf{I}^2$, I_x^2 must be obtained by double application of I_x on some function u, say, and similarly for the other components.

2.

4-2.7 $$\nabla^2 = (\partial^2/\partial\xi^2, \partial^2/\partial\eta^2, \partial^2/\partial\zeta^2). \quad (4)$$

This is a Laplacian on the variables ξ, η, ζ. Since they are angular coordinates, ∇^2 does not act on $\mathbf{r}$.

PROBLEM 4.4

1. $$[\mathbf{I}^2, I_x] = [I_x^2, I_x] + [I_y^2, I_x] + [I_z^2, I_x] = [I_y^2, I_x] + [I_z^2, I_x]. \quad (1)$$

2-7.29, **4**-4.13 $$[I_y^2, I_x] = I_y[I_y, I_x] + [I_y, I_x]I_y = -\mathrm{i}I_yI_z - \mathrm{i}I_zI_y. \quad (2)$$

2-7.29, **4**-4.13 $$[I_z^2, I_x] = I_z[I_z, I_x] + [I_z, I_x]I_z = \mathrm{i}I_zI_y + \mathrm{i}I_yI_z. \quad (3)$$

PROBLEMS 4-5

1.

,**4**-4.13 $$[I_z, I_\pm] = [I_z, I_x] \pm \mathrm{i}[I_z, I_y] = \mathrm{i}I_y \pm I_x = \pm I_\pm. \quad (1)$$

,**4**-4.13 $$[I_+, I_-] = [I_+, I_x] - \mathrm{i}[I_+, I_y] = \mathrm{i}[I_y, I_x] - \mathrm{i}[I_x, I_y] = 2I_z. \quad (2)$$

4-4.15 $$[\mathbf{I}^2, I_\pm] = [\mathbf{I}^2, I_z] = 0. \quad (3)$$

,**4**-4.13 $$I_+I_- = (I_x + iI_y)(I_x - iI_y) = I_x^2 + I_y^2 + \mathrm{i}[I_y, I_x] = I_x^2 + I_y^2 + I_z \quad (4)$$

$$= \mathbf{I}^2 - I_z^2 + I_z. \quad (5)$$

2. From the Hermitian property of I_x, I_y, and remembering that i changes sign on hopping from right to left of a bracket,

$$\langle u_1 | (I_x \pm \mathrm{i}I_y)u_2 \rangle = \langle (I_x \mp iI_y)u_1 \mid u_2 \rangle \quad \Rightarrow \quad I_\pm^\dagger = I_\mp.$$

3.

4-5.5, **4**-4.15 $$[I_+I_-, I_z] = [\mathbf{I}^2, I_z] + [I_z, I_z] - [I_z^2, I_z] = 0.$$

4. Use **4**-2.16, **4**-2.17.

PROBLEMS 4-6

1.

4-6.16 $$\begin{aligned} C_{m+1} &= C_m - 2m \\ C_m &= C_{m-1} - 2m + 2 \\ C_{m-1} &= C_{m-2} - 2m + 4 \\ &\vdots \\ C_{m-r} &= C_{m-r-1} - 2m + 2r + 2. \end{aligned}$$

Add up both sides of the above equations:

$$C_{m+1} = C_{m-r-1} - 2m(r+2) + 2\{1+2+\ldots+(r+1)\}.$$
$$= C_{m-r-1} - 2m(r+2) + (r+1)(r+2). \qquad (1)$$

4-6.14,1 $\qquad C_{j+1} = 0 \;\Rightarrow\; C_{j-r-1} = 2j(r+2) + (r+1)(r+2) = 0. \qquad (2)$

Take $j - r - 1 = m$ in (2) and the result follows after some algebra.

2.

4-5.5 $\qquad I^2 u^j_m = (I_+ I_- + I_z^2 - I_z) u^j_m. \qquad (3)$

4-6.26 $\qquad I^2 u^j_m =$

$$= [\{j(j+1) - (m-1)m\}^{1/2}\{j(j+1) - m(m-1)\}^{1/2} + m^2 - m]u^j_m \qquad (4)$$
$$= j(j+1)u^j_m. \qquad (5)$$

PROBLEMS **4**-7

1.

$$R(\phi z)Y^m_l(\theta\varphi) = \mathrm{i}^{m+|m|}P^{|m|}_l(\cos\theta)\exp\{\mathrm{i}m(\varphi-\phi)\}$$
$$= \exp(-\mathrm{i}m\phi)Y^m_l(\theta\varphi). \qquad (1)$$

2.

,**4**-7.8 $\qquad \mathrm{Tr}\,\hat{R}^j(\phi\mathbf{n}) = \mathrm{Tr}\,\hat{R}^j(\phi\mathbf{z}) = \sum_{m=-j}^{j}\exp(\mathrm{i}m\phi) \qquad (2)$

$\phi \neq 0$‡ $\qquad = [\exp\{\mathrm{i}(j+1)\phi\} - \exp(-\mathrm{i}j\phi)]\{\exp(\mathrm{i}\phi) - 1\}^{-1} \qquad (3)$

$$= \exp(-\mathrm{i}\tfrac{1}{2}\phi)[\exp\{\mathrm{i}(j+1)\phi\} - \exp(-\mathrm{i}j\phi)]$$
$$\times\{\exp(\mathrm{i}\tfrac{1}{2}\phi) - \exp(-\mathrm{i}\tfrac{1}{2}\phi)\}^{-1} \qquad (4)$$
$$= [\exp\{\mathrm{i}(j+\tfrac{1}{2})\phi\} - \exp\{-\mathrm{i}(j+\tfrac{1}{2})\phi\}](2\mathrm{i}\sin\tfrac{1}{2}\phi)^{-1}. \qquad (5)$$

‡ For $\phi \neq 0$, (2) is a geometric progression. For $\phi = 0$ its sum is $2j+1$. The result for $\phi \neq 0$ follows from (5).

3. From (**4**-7.5).
4. On using (**4**-7.15), (**4**-7.16), and Problem 3 (with $\phi = \pi$), the characters of E, $2C_3$ and $3\sigma_v$ are, respectively, 7, 1, and 1. Express these characters as linear combinations of the irreducible characters.
5. The complete list of the homogeneous polynomials of degree j is:

$$x^j;\; x^{j-1}y, x^{j-1}z;\; x^{j-2}y^2, x^{j-2}yz, x^{j-2}z^2;\; \ldots;\; y^j, y^{j-1}z, y^{j-2}z^2, \ldots, z^j. \qquad (6)$$

The number of monomials in each section of (6) is $1, 2, 3, \ldots, j+1$, respectively, whence their total number is $\tfrac{1}{2}(j+1)(j+2)$. The second derivatives that follow from Laplace's equation are homogeneous polynomials of degree $j-2$ and thus $\tfrac{1}{2}j(j-1)$ in number. Hence, the total number of independent monomials is $\frac{1}{2}(j+1)(j+2)-\frac{1}{2}j(j-1)$, which is $2j+1$.

6. Straightforward change of variables.
7. (i) Notice that for $j=1$, $m=1, 0, -1$, r is respectively 0, 0, 1.
(ii) Introduce (**4**-7.17) into (**4**-7.18):

$$2\partial^2 u^j_m/\partial u_1 \partial u_{-1} = 2\sum_r c_r r(m+r)(u_1)^{m+r-1}(u_{-1})^{r-1}(u_0)^{j-m-2r}. \tag{7}$$

$$\partial^2 u^j_m/\partial(u_0)^2 = $$

$$= \sum_r c_r(j-m-2r)(j-m-2r-1)(u_1)^{m+r}(u_{-1})^r u_0^{j-m-2r-2}. \tag{8}$$

Because the term for $r=0$ in (7) vanishes, r in this equation can be changed into $r+1$, whereupon the exponents in (7) and (8) all agree and, on equating the corresponding coefficients, (**4**-7.19) follows. In order to obtain (**4**-7.20) apply (**4**-7.19) for $c_r, c_{r-1}, \ldots c_1$ and multiply the r equations obtained.

PROBLEM 4-8

1. $I'_+\langle u_m \mid u_{m'}\rangle = \langle (I'_+)^\dagger u_m \mid I'_+ u_{m'}\rangle = e^{-i\alpha}\langle I^\dagger_+ u_m \mid e^{i\alpha} I_+ u_{m'}\rangle = \langle I^\dagger_+ u_m \mid I_+ u_{m'}\rangle$.

PROBLEMS 4-9

1. From (**4**-9.2),

$$I_x u_1 = 2^{-1/2} u_0, \qquad I_x u_0 = 2^{-1/2} u_1 + 2^{-1/2} u_{-1}, \qquad I_x u_1 = 2^{-1/2} u_0.$$

The matrix $\hat{I}_x$ now follows. The others similarly.
2. Straightforward matrix multiplication.
3.
4-9.2 For $j=\frac{1}{2}$: $I_x u_{1/2} = \frac{1}{2} u_{-1/2}$, $I_x u_{-1/2} = \frac{1}{2} u_{1/2}$.

The matrix $\hat{I}_x$ now follows. The others similarly.
4.
4-9.9 $$\mathbf{n}\cdot\boldsymbol{\sigma} = n_x\boldsymbol{\sigma}_x + n_y\boldsymbol{\sigma}_y + n_z\boldsymbol{\sigma}_z = \begin{bmatrix} n_z & n_x - in_y \\ n_x + in_y & -n_z \end{bmatrix} =_{\text{def}} A. \tag{1}$$

$$A^2 = \mathbf{1}_2. \tag{2}$$

2|**4**-9.10 $\hat{R}^{1/2}(\phi\mathbf{n}) =$

$$= \exp(-\tfrac{1}{2}i\phi A) = 1 - \tfrac{1}{2}i\phi A + \frac{1}{2!}(-\tfrac{1}{2}i\phi A)^2 + \frac{1}{3!}(-\tfrac{1}{2}i\phi A)^3 + \ldots$$

$$= \left\{1 - \frac{1}{2!}(\tfrac{1}{2}\phi)^2 + \frac{1}{4!}(\tfrac{1}{2}\phi)^4 \ldots\right\}\mathbf{1}_2 - i\left\{\tfrac{1}{2}\phi - \frac{1}{3!}(\tfrac{1}{2}\phi)^3 + \ldots\right\}A$$

$$= (\cos\tfrac{1}{2}\phi)\mathbf{1}_2 - i(\sin\tfrac{1}{2}\phi)A. \tag{3}$$

(**4**-9.12) follows from (1) and (3).

PROBLEMS 5-1

1. From (**2**-4.1),

$$R(\phi\mathbf{z})x = x\cos\phi - y\sin\phi, \quad R(\phi\mathbf{z})y = x\sin\phi + y\cos\phi. \tag{1}$$

1 $$R(\phi\mathbf{z})(x \pm \mathrm{i}y) = \exp(\pm\mathrm{i}\phi)x \pm (\mathrm{i}\cos\phi \mp \sin\phi)y$$
$$= \exp(\pm\mathrm{i}\phi)x \pm (\cos\phi \pm \mathrm{i}\sin\phi)\mathrm{i}y$$
$$= \exp(\pm\mathrm{i}\phi)(x \pm \mathrm{i}y). \tag{2}$$

2.

2 $$R(\phi\mathbf{z})\mathbb{U}_1 = R(\phi\mathbf{z})(x - \mathrm{i}y) = \exp(-\mathrm{i}\phi)(x - \mathrm{i}y)$$
$$= \exp(-\mathrm{i}\phi)\mathbb{U}_1. \tag{3}$$
$$R(\phi\mathbf{z})\mathbb{U}_0 = R(\phi\mathbf{z})(-z) = -z = \mathbb{U}_0. \tag{4}$$

2 $$R(\phi\mathbf{z})\mathbb{U}_{-1} = R(\phi z)\{-(x + \mathrm{i}y)\} = \exp(\mathrm{i}\phi)\{-(x + \mathrm{i}y)\}$$
$$= \exp(\mathrm{i}\phi)\mathbb{U}_{-1}. \tag{5}$$

3. Polars: $x = r\sin\theta\cos\varphi, \quad y = r\sin\theta\sin\varphi, \quad z = r\cos\theta.$ (6)

Call $\bar{x}$, $\bar{y}$, $\bar{z}$ the transforms of x, y, z under a rotation of *minus* β around the **y** axis. (To correspond to $R(\beta\mathbf{y})$ in function space.)

3-1.14 $$\bar{z} = x\sin\beta + z\cos\beta. \tag{7}$$

7|6 $$\bar{Y}_1^0 = \cos\bar{\theta} = r^{-1}(x\sin\beta + z\cos\beta)$$
$$= \sin\theta\cos\varphi\sin\beta + \cos\theta\cos\beta. \tag{8}$$

8 $$\bar{Y}_1^0 = Y_1^0\cos\beta + \sin\beta\sin\theta\cos\varphi. \tag{9}$$

Here, Y_1^0 equals z from (**5**-1.2), and $\sin\theta\cos\varphi$ is x.

PROBLEMS 5-2

1. From (**P2**-5.31), the character for the basis $\langle u^j| \otimes \langle u^{j'}|$ is $\chi^j(\phi)\chi^{j'}(\phi)$. Strategy: to find this character and then to find out how it splits over the irreducible characters of SO(3).

(**P4**-7.3) $$\chi^j(\phi) = [\exp\{\mathrm{i}(j+1)\phi\} - \exp(-\mathrm{i}j\phi)][\exp(\mathrm{i}\phi) - 1\}^{-1}. \tag{1}$$

(**P4**-7.2) $$\chi^{j'}(\phi) = \sum_{m=-j'}^{j'} \exp(\mathrm{i}m\phi)$$
$$= 1 + \sum_{m=1}^{j'} \{\exp(\mathrm{i}m\phi) + \exp(-\mathrm{i}m\phi)\}. \tag{2}$$

1,2 $$\chi^j(\phi)\chi^{j'}(\phi)=\chi^j(\phi)+$$

$$+\sum_{m=1}^{j'}[\exp\{\mathrm{i}(j+m+1)\phi\}-\exp\{-\mathrm{i}(j+m)\phi\}]\{\exp(\mathrm{i}\phi)-1\}^{-1}$$

$$+\sum_{m=1}^{j'}[\exp\{\mathrm{i}(j-m+1)\phi\}-\exp\{-\mathrm{i}(j-m)\phi\}]\{\exp(\mathrm{i}\phi)-1\}^{-1} \quad (3)$$

1|3 $$=\chi^j(\phi)+\sum_{m=1}^{j'}\{\chi^{j+m}(\phi)+\chi^{j-m}(\phi)\} \quad (4)$$

$$=\chi^{j+j'}(\phi)+\chi^{j+j'-1}(\phi)+\ldots+\chi^j(\phi)+\ldots+\chi^{j-j'}(\phi),\quad (j>j'). \quad (5)$$

This means that the basis $\langle u^j|\otimes\langle u^{j'}|$ reduces into bases belonging to the representations listed on (R5), as given in (**5**-2.2).

2. From (**5**-1.1),

$$R(\phi\mathbf{z})u_{\pm1}=\exp(\mp\mathrm{i}\phi)u_{\pm1},\quad R(\phi\mathbf{z})u_0=u_0. \quad (6)$$

6 $$R(\phi\mathbf{z})(u_1v_0-u_0v_1)=\exp(-\mathrm{i}\phi)u_1v_0-u_0\exp(-\mathrm{i}\phi)v_1$$

$$=\exp(-\mathrm{i}\phi)(u_1v_0-u_0v_1). \quad (7)$$

6 $$R(\phi\mathbf{z})(u_1v_{-1}-u_{-1}v_1)=\exp(-\mathrm{i}\phi)u_1\exp(\mathrm{i}\phi)v_{-1}$$

$$-\exp(\mathrm{i}\phi)u_{-1}\exp(-\mathrm{i}\phi)v_1$$

$$=u_1v_{-1}-u_{-1}v_1. \quad (8)$$

6 $$R(\phi\mathbf{z})(u_{-1}v_0-u_0v_{-1})=\exp(\mathrm{i}\phi)u_{-1}v_0-u_0\exp(\mathrm{i}\phi)v_{-1}$$

$$=\exp(\mathrm{i}\phi)(u_{-1}v_0-u_0v_{-1}). \quad (9)$$

Equations (7), (9), and (8) exhibit the same behaviour as the three equations in (6), in the same order. Thus the basis (**5**-2.11) behaves like a spherical vector under rotations round the **z** axis. In order to deal with arbitrary rotations the matrix (**7**-3.10) would have to be substituted for (**5**-1.1) in this treatment. (Heavy work!)

PROBLEM 5-3

1.

5-3.3 $$R(\phi\mathbf{z})\mu_1=\exp(-\mathrm{i}\tfrac{1}{2}\phi)\mu_1,\qquad R(\phi\mathbf{z})\mu_2=\exp(\mathrm{i}\tfrac{1}{2}\phi)\mu_2. \quad (1)$$

5-3.10 $$\boldsymbol{i}\mu_1=-\mathrm{i}\mu_1,\qquad \boldsymbol{i}\mu_2=-\mathrm{i}\mu_2. \quad (2)$$

1 $$R(\phi\mathbf{z})(\mu_1\nu_2-\mu_2\nu_1)=\exp(-\mathrm{i}\tfrac{1}{2}\phi)\mu_1\exp(\mathrm{i}\tfrac{1}{2}\phi)\nu_2$$

$$-\exp(\mathrm{i}\tfrac{1}{2}\phi)\mu_2\exp(-\mathrm{i}\tfrac{1}{2}\phi)\nu_1$$

$$=\mu_1\nu_2-\mu_2\nu_1. \quad (3)$$

2 $$\boldsymbol{i}(\mu_1\nu_2-\mu_2\nu_1)=(-\mathrm{i})\mu_1(-\mathrm{i})\nu_2-(-\mathrm{i})\mu_2(-\mathrm{i})\nu_1$$

$$=-(\mu_1\nu_2-\mu_2\nu_1). \quad (4)$$

On assuming that the correct behaviour given by (3) is valid for all rotations as discussed in P **5**-2.2, (3) and (4) entail that $\mu_1\nu_2 - \mu_2\nu_1$ is a pseudoscalar.

PROBLEMS **6**-4

1. $\boldsymbol{I}^2 = -\mathbf{1}$ entails, for $n = 0, 1, 2 \ldots,$

$$\boldsymbol{I}^{4n} = \mathbf{1}, \quad \boldsymbol{I}^{4n+1} = \boldsymbol{I}, \quad \boldsymbol{I}^{4n+2} = -\mathbf{1}, \quad \boldsymbol{I}^{4n+3} = -\boldsymbol{I}. \tag{1}$$

$$\exp\left(\tfrac{1}{2}\pi\boldsymbol{I}\right) = \textstyle\sum_n (n!)^{-1}\left(\tfrac{1}{2}\pi\boldsymbol{I}\right)^n.$$

$$= \mathbf{1} + \tfrac{1}{2}\pi\boldsymbol{I} - \frac{1}{2!}\left(\tfrac{1}{2}\pi\right)^2\mathbf{1} - \frac{1}{3!}\left(\tfrac{1}{2}\pi\right)^3\boldsymbol{I} + \frac{1}{4!}\left(\tfrac{1}{2}\pi\right)^4\mathbf{1} + \ldots$$

$$= \mathbf{1}\cos\tfrac{1}{2}\pi + \boldsymbol{I}\sin\tfrac{1}{2}\pi = \boldsymbol{I}. \tag{2}$$

2. Binary rotations $\Rightarrow$ $\phi = \pi$. (**4**-7.15) $\Rightarrow$ $\chi^{1/2}(\pi) = 0$.
3. Compare (**6**-2.5) with (**4**-9.12). The elements a and b are sufficient to determine the whole of the SU(2) matrix.

PROBLEMS **6**-6

1.

6-6.4 Case 2:

$$\{bb^* = 1, a = 0\} \quad \Rightarrow \quad \bar{w} = A \circ w = b(b^*w)^{-1} = -b^2w^*. \tag{1}$$

(Use A from eqn **6**-6.3 and eqn **6**-3.7 in the last step.) w from (**6**-3.5):

$$\sigma_x w = \sigma_x(x - \mathrm{i}y)z^{-1} = (-x - \mathrm{i}y)z^{-1} = -w^*, \tag{2}$$

$$\sigma_y w = (x + \mathrm{i}y)z^{-1} = w^*. \tag{3}$$

1,2 $\quad b^2 = 1 \quad \Rightarrow \quad \hat{\sigma}_x = \begin{bmatrix} & 1 \\ 1 & \end{bmatrix},$ (4)

1,3 $\quad b^2 = -1 \quad \Rightarrow \quad \hat{\sigma}_y = \begin{bmatrix} & \mathrm{i} \\ -\mathrm{i} & \end{bmatrix}.$ (5)

2.

6-4.1, **6**-6.9 $\quad \hat{\sigma}_x = \boldsymbol{I}_i\boldsymbol{I}_x = \begin{bmatrix} & -1 \\ -1 & \end{bmatrix}.$ (6)

$$\hat{\sigma}_y = \begin{bmatrix} & \mathrm{i} \\ -\mathrm{i} & \end{bmatrix}. \tag{7}$$

To get complete agreement the phase in (4) has to be changed.

PROBLEM 6-7

1.

6-7.6, **6**-4.1 $\quad |\mu_1^*\mu_2^*\rangle = |-\mu_2\mu_1\rangle = \begin{bmatrix} & -1 \\ 1 & \end{bmatrix} |\mu_1\mu_2\rangle = \boldsymbol{I}_y\, |\mu_1\mu_2\rangle.$

PROBLEM 7-2

1.

,**7**-2.8

$$x, y, z \in \text{sphere} \quad \Rightarrow \quad x^2+y^2+z^2=1 \quad \Rightarrow \quad uu^*+z^2=1. \tag{1}$$

7-2.11, **7**-2.10

$$uu^*+z^2=4UU^*(1+UU^*)^{-2}+(UU^*-1)^2(1+UU^*)^{-2} \equiv 1. \tag{2}$$

Hence, for any U, eqn (1) is satisfied.

PROBLEMS 7-3

1.

7-3.10 $\quad \det \hat{R}^1(A) = (aa^*)^2(aa^*-bb^*)+4aa^*(b^*)^2b^2$

$$-(bb^*)^2(aa^*-bb^*)+4(a^*)^2a^2bb^*$$

$$=(aa^*-bb^*)(aa^*-bb^*)(aa^*+bb^*)$$

$$+4aa^*bb^*(aa^*+bb^*) \tag{1}$$

7-3.2|1 $\quad =(aa^*-bb^*)^2+4aa^*bb^*=(aa^*+bb^*)^2=1.$

2. From (**7**-2.2), determine M so that $M\,|xyz\rangle = |\text{u}_1\text{u}_0\text{u}_{-1}\rangle$. Verify that it is unitary and that $\det M=-\text{i}$. From (**7**-3.17) $\hat{R}^1_{\mathbf{r}}(A)$ is the product of unitary matrices and thus it must be unitary but, since it must leave invariant the quadratic form $x^2+y^2+z^2=1$ it must be orthogonal. This is confirmed by getting its determinant which, from (**7**-3.17) is $\det M^\dagger \det \hat{R}^1(A) \det M = \text{i}1(-\text{i})=1$. Therefore, $\hat{R}^i_{\mathbf{r}}(A)$ is a proper rotation and belongs to SO(3), as expected.
3. Use (**7**-3.17) with M from Problem 2.
4. Introduce a and b from (**6**-4.9) into (**7**-3.32).
5.

6-4.9 $\quad a=\cos\tfrac{1}{2}\beta, \qquad b=-\sin\tfrac{1}{2}\beta. \qquad (2)$

Introduce (2) into (**7**-3.10).

6.

7-3.33 $\quad R(\beta\mathbf{y})\,Y_1^0 = Y_1^0\cos\beta+\sin\beta\; 2^{-1/2}(-Y_1^1+Y_1^{-1}). \qquad (3)$

Introduce (**4**-7.11) to (**4**-7.13), disregarding the constant factor $(3/4\pi)^{1/2}$. Notice that (**7**-3.34) and (**5**-1.4) coincide for $\varphi=0$, for which the latter equation is correct. Notice also that (**7**-3.34) agrees with (**P5**-1.8). See the warning at the end of §**5**-1.

7. Write $v=x+iy$ in polars ($r=1$) as $\sin\theta\exp(i\varphi)$. Write $\sin\theta$ as $(1-\cos^2\theta)^{1/2}$ and, on identifying $\cos\theta$ with z, (**7**-3.35) follows.

8.

7-3.36 $\qquad \mu_1\mu_1^* = \tfrac{1}{2}(1-z), \qquad \mu_2\mu_2^* = \tfrac{1}{2}(1+z).$ (4)

4 $\qquad \mu_1\mu_1^* - \mu_2\mu_2^* = -z, \qquad \mu_1\mu_1^* + \mu_2\mu_2^* = 1.$ (5)

9. Write in (**7**-3.36),

$$\exp(i\tfrac{1}{2}\varphi) = (\cos\varphi + i\sin\varphi)^{1/2} = \{x(\sin\theta)^{-1} + iy(\sin\theta)^{-1}\}^{1/2}$$
$$= (\sin\theta)^{-1/2}(x+iy)^{1/2}. \tag{6}$$

Write $\sin\theta$ as in Problem 7.

10. Apply the function-space operator on (**7**-3.36).

11.

4-7.20|**4**-7.18 $\qquad u_m^j = C_{jm}\sum_r \dfrac{(u_1)^{m+r}(u_{-1})^r(u_0)^{j-m-2r}}{2^r(m+r)!\,r!\,(j-m-2r)!}.$ (7)

Replace u_1, u_0, u_{-1} by their values from (**7**-3.26), expanding $(\mu_2\mu_2^* - \mu_1\mu_1^*)^{j-m-2r}$ by the binomial theorem, and collect terms together:

$$u_m^j = 2^{\frac{1}{2}m}C_{jm}$$
$$\times\sum_r\sum_{s=0}^{j-m-2r}(-1)^{r+s}\binom{j-m-2r}{s}\frac{(\mu_1)^{m+r+s}(\mu_2^*)^{j-r-s}(\mu_2)^{j-m-r-s}(\mu_1^*)^{r+s}}{(m+r)!\,r!\,(j-m-2r)!}. \tag{8}$$

Call $r+s=k$, expand the binomial coefficient, and the result follows.

PROBLEMS 8-3

1.

T**8**-1.3|**8**-3.10 $\qquad [C_{2x}, C_{2y}][C_{2z}, C_{2z}] = [C_{2x}, C_{2x}][C_{2y}, C_{2z}].$ (1)

T**8**-1.3|1 $\qquad 1 \quad (-1) \quad = \quad (-1) \quad 1$ (2)

2.

8-2.18 $\qquad [g_i, g_j]' = \omega_{g_i}\omega_{g_j}\omega_{g_ig_j}^{-1}[g_i, g_j].$ (3)

8-2.18 $\qquad [g_i, g_j]'' = \omega'_{g_i}\omega'_{g_j}(\omega'_{g_ig_j})^{-1}[g_i, g_j].$ (4)

3|4 $\qquad [g_i, g_j]'' = \omega'_{g_i}\omega_{g_i}^{-1}\omega'_{g_j}\omega_{g_j}^{-1}(\omega'_{g_ig_j})^{-1}\omega_{g_ig_j}[g_i, g_j]'.$ (5)

$\qquad = \omega''_{g_i}\omega''_{g_j}(\omega''_{g_ig_j})^{-1}[g_i, g_j]'.$ (6)

3.

8-3.1	$[g, e][ge, e] = [g, ee][e, e].$	(7)
7	$[g, e][g, e] = [g, e][e, e] \Rightarrow [g, e] = [e, e].$	(8)
8-3.1	$[g, g^{-1}][gg^{-1}, g] = [g, g^{-1}g][g^{-1}, g].$	(9)
9	$[g, g^{-1}][e, g] = [g, e][g^{-1}, g].$	(10)
8,8-3.7|10	$[g, g^{-1}] = [g^{-1}, g].$	(11)

PROBLEMS 8-5

1. We must prove, from (8-3.1) on (L8-5.3), eqn (1) below, from which (2) is obtained from $gh = hg$:

$$[g_ih_k, g_jh_l][g_ih_kg_jh_l, g_mh_n] = [g_ih_k, g_jh_lg_mh_n][g_jh_l, g_mh_n]. \quad (1)$$

$$[g_ih_k, g_jh_l][g_ig_jh_kh_l, g_mh_n] = [g_ih_k, g_jg_mh_lh_n][g_jh_l, g_mh_n]. \quad (2)$$

Apply (8-5.3) on both sides of (2):

$$[g_i, g_j][h_k, h_l][g_ig_j, h_kh_l][g_i, h_k]^{-1}[g_j, h_l]^{-1} \times [g_ig_j, g_m][h_kh_l, h_n][g_ig_jg_m, h_kh_lh_n][g_ig_j, h_kh_l]^{-1}[g_m, h_n]^{-1}$$
$$= [g_i, g_jg_m][h_k, h_lh_n][g_ig_jg_m, h_kh_lh_n][g_i, h_k]^{-1}[g_jg_m, h_lh_n]^{-1} \times [g_j, g_m][h_l, h_n][g_jg_m, h_lh_n][g_j, h_l]^{-1}[g_m, h_n]^{-1}. \quad (3)$$

3 $$[g_i, g_j][g_ig_j, g_m][h_k, h_l][h_kh_l, h_n] = [g_i, g_jg_m][g_j, g_m][h_k, h_lh_n][h_l, h_n]. \quad (4)$$

(Both sides of (4) are equal, from (8-3.1).

2.

8-5.10	$\check{G}(g_i)\check{G}(g_j) = \{{}^r\check{G}(g_i) \otimes {}^s\check{G}(g_i)\}\{{}^r\check{G}(g_j) \otimes {}^s\check{G}(g_j)\}$	(5)
2-6.55|5	$= {}^r\check{G}(g_i){}^r\check{G}(g_j) \otimes {}^s\check{G}(g_i){}^s\check{G}(g_j)$	(6)
8-5.9|6	$= {}^r[g_i, g_j]{}^s[g_i, g_j]{}^r\check{G}(g_ig_j) \otimes {}^s\check{G}(g_ig_j)$	(7)
8-5.10|7	$= {}^r[g_i, g_j]{}^s[g_i, g_j]\check{G}(g_ig_j).$	(8)

PROBLEMS 8-6

1. Straightforward matrix multiplication, to be compared with group products from Table 8-1.3.
2. Matrix multiplication as in Problem 1.

PROBLEMS 9-1

1. If **z** coincides with C_3, then the poles of the three σ_v (written as iC_2) are in the **x**, **y** plane and they are interchanged under C_3. Therefore

the three σ_v are one class. The axis C_3 is bilateral, therefore C_3^+ and C_3^- are in the same class. E is the third class.

2. Both in **O** and in **T** there are four C_3 axes. In **O** their poles are exchanged under C_4 (along the **z** axis) and they are also bilateral. Thus, in **O**, there is a class that contains $4C_3^+$ and $4C_3^-$. In **T**, C_4 has gone and the C_3 are no longer bilateral. The four C_3^+ poles are all interchanged by any of the C_3 operations plus the C_2 along **z**. Thus, the $4C_3^+$ will form one class and the $4C_3^-$ will form another class.
3. An invariant subgroup is invariant under conjugation so that it must be a sum of classes of the supergroup, i.e. all its poles must be interchanged under operations of the supergroup. All operations of **O** exchange poles of $\mathbf{D}_2$ into poles of $\mathbf{D}_2$ (i.e. they all transform the set of axes **ijk** in the usual setting into other sets in the identical setting).

PROBLEMS 9-3

1.

$$\mathbf{l}^* = K\mathbf{m}\times\mathbf{p} \quad \Rightarrow \quad \mathbf{m}\,.\,\mathbf{l}^* = 0. \tag{1}$$

$$\mathbf{l}\,.\,\mathbf{l}^* = K\mathbf{l}\,.\,\mathbf{m}\times\mathbf{p} = 1 \quad \Rightarrow \quad K = (\mathbf{l}\,.\,\mathbf{m}\times\mathbf{p})^{-1}. \tag{2}$$

Other vectors and equalities by cycling.

9-3.11 $$\mathbf{n}\,.\,\mathbf{l}^* = f\mathbf{l}\,.\,\mathbf{l}^* + g\mathbf{m}\,.\,\mathbf{l}^* + h\mathbf{p}\,.\,\mathbf{l}^* = f. \tag{3}$$

2.

$$\mathbf{l}^* = K\mathbf{m}\times\mathbf{p} \qquad \mathbf{m}^* = K\mathbf{p}\times\mathbf{l}, \qquad \mathbf{p}^* = K\mathbf{l}\times\mathbf{m}, \tag{4}$$

$$K = (\mathbf{l}\,.\,\mathbf{m}\times\mathbf{p})^{-1}, \qquad \mathbf{p} = \mathbf{l}\times\mathbf{m}. \tag{5}$$

$$K = \{\mathbf{l}\,.\,\mathbf{m}\times(\mathbf{l}\times\mathbf{m})\}^{-1}$$
$$= \{\mathbf{l}\,.\,(m^2\mathbf{l} - \mathbf{l}\,.\,\mathbf{m}\mathbf{m})\}^{-1} = \{l^2m^2 - (\mathbf{l}\,.\,\mathbf{m})^2\}^{-1} = \{1-(\mathbf{l}\,.\,\mathbf{m})^2\}^{-1}. \tag{6}$$

$$\mathbf{l}^* = K\mathbf{m}\times(\mathbf{l}\times\mathbf{m}) = K(m^2\mathbf{l} - \mathbf{l}\,.\,\mathbf{m}\mathbf{m}) = K(\mathbf{l} - \mathbf{l}\,.\,\mathbf{m}\mathbf{m}). \tag{7}$$

$$\mathbf{m}^* = K(\mathbf{l}\times\mathbf{m})\times\mathbf{l} = K(\mathbf{m} - \mathbf{l}\,.\,\mathbf{m}\mathbf{l}). \tag{8}$$

PROBLEMS 9-4

1.

$$\lambda_3^2 = \lambda_1^2\lambda_2^2 + (\mathbf{\Lambda}_1\,.\,\mathbf{\Lambda}_2)^2 - 2\lambda_1\lambda_2\mathbf{\Lambda}_1\,.\,\mathbf{\Lambda}_2. \tag{1}$$

$$\mathbf{\Lambda}_3^2 = \lambda_1^2\mathbf{\Lambda}_2^2 + \lambda_2^2\mathbf{\Lambda}_1^2 + (\mathbf{\Lambda}_1\times\mathbf{\Lambda}_2)^2 + 2\lambda_1\lambda_2\mathbf{\Lambda}_1\,.\,\mathbf{\Lambda}_2. \tag{2}$$

1,2 $$\lambda_3^2 + \mathbf{\Lambda}_3^2 = \lambda_1^2 + (\mathbf{\Lambda}_1\,.\,\mathbf{\Lambda}_2)^2 + \lambda_2^2\mathbf{\Lambda}_1^2 + \mathbf{\Lambda}_1\,.\,\mathbf{\Lambda}_2\times(\mathbf{\Lambda}_1\times\mathbf{\Lambda}_2) \tag{3}$$

$$= \lambda_1^2 + (\mathbf{\Lambda}_1\,.\,\mathbf{\Lambda}_2)^2 + \lambda_2^2\mathbf{\Lambda}_1^2 + \mathbf{\Lambda}_1^2\mathbf{\Lambda}_2^2 - (\mathbf{\Lambda}_1\,.\,\mathbf{\Lambda}_2)^2 = 1. \tag{4}$$

2.

9-4.11 $\lambda_1 = 1, \quad \mathbf{\Lambda}_1 = \frac{1}{2}\phi_1\mathbf{n}_1; \quad \lambda_2 = 1, \quad \mathbf{\Lambda}_2 = \frac{1}{2}\phi\mathbf{n}_2.$ (5)

5|9-4.7 $\lambda_3 = 1, \qquad \mathbf{\Lambda}_3 = \frac{1}{2}\phi_1\mathbf{n}_1 + \frac{1}{2}\phi\mathbf{n}_2 = \mathbf{\Lambda}_1 + \mathbf{\Lambda}_2.$ (6)

Equation (6) shows that the vector addition rule is satisfied. Also that the rotations commute.

3. From (**9**-4.7):

$$\lambda_{12} = \lambda_1\lambda_2 - \mathbf{\Lambda}_1 \cdot \mathbf{\Lambda}_2; \qquad \lambda_{21} = \lambda_2\lambda_1 - \mathbf{\Lambda}_2 \cdot \mathbf{\Lambda}_1. \tag{7}$$

$$\mathbf{\Lambda}_{12} = \lambda_1\mathbf{\Lambda}_2 + \lambda_2\mathbf{\Lambda}_1 + \mathbf{\Lambda}_1 \times \mathbf{\Lambda}_2; \qquad \mathbf{\Lambda}_{21} = \lambda_2\mathbf{\Lambda}_1 + \lambda_1\mathbf{\Lambda}_2 + \mathbf{\Lambda}_2 \times \mathbf{\Lambda}_1. \tag{8}$$

4. Follows from (7) with $\lambda = \cos\frac{1}{2}\phi$.
5. From (**9**-4.12) two general rotations do not commute. For coaxial rotations $\mathbf{\Lambda}_1 \times \mathbf{\Lambda}_2$ vanishes.
6. From (7), (8), and condition (**9**-4.8), the necessary and sufficient conditions for commutation are:

$$\text{either} \qquad R(\lambda_{12}, \mathbf{\Lambda}_{12}) = R(\lambda_{21}, \mathbf{\Lambda}_{21}), \tag{9}$$

$$\text{or} \qquad R(\lambda_{12}, \mathbf{\Lambda}_{12}) = R(-\lambda_{21}, -\mathbf{\Lambda}_{21}). \tag{10}$$

In case (9), since, from (7), λ_{12} always equals λ_{21}, the condition is $\mathbf{\Lambda}_{12} = \mathbf{\Lambda}_{21}$. This requires, from (8),

$$\mathbf{\Lambda}_1 \times \mathbf{\Lambda}_2 = \mathbf{0}. \tag{11}$$

This will be the case, and only will be the case, when $\mathbf{\Lambda}_1$ and $\mathbf{\Lambda}_2$ are parallel (coaxial rotations). When this is not so, commutation requires (10) to be valid, that is, from (7) and (8),

$$\lambda_1\lambda_2 - \mathbf{\Lambda}_1 \cdot \mathbf{\Lambda}_2 = -\lambda_2\lambda_1 + \mathbf{\Lambda}_2 \cdot \mathbf{\Lambda}_1. \tag{12}$$

$$\lambda_1\mathbf{\Lambda}_2 + \lambda_2\mathbf{\Lambda}_1 + \mathbf{\Lambda}_1 \times \mathbf{\Lambda}_2 = -\lambda_2\mathbf{\Lambda}_1 - \lambda_1\mathbf{\Lambda}_2 + \mathbf{\Lambda}_1 \times \mathbf{\Lambda}_2. \tag{13}$$

Equation (13) requires $\lambda_1 = \lambda_2 = 0$, $(\phi = \pi)$, and consequently, from (12), we must have $\mathbf{\Lambda}_1 \cdot \mathbf{\Lambda}_2 = 0$, that is $\mathbf{n}_1 \perp \mathbf{n}_2$. These rotations are bilateral binary (BB). Notice, from (10), that commutation of BB rotations, as a difference with coaxial, necessarily entails a change of sign of the quaternion parameters. This will lead to a phase change, as shown in Fig. **11**-1.1, a feature of major importance as regards the structure of SO(3).

7. First rotation factor: $R(\lambda_1; \mathbf{\Lambda}_1)$; $\mathbf{\Lambda}_1 \parallel \mathbf{n}_1$. Second rotation factor:

$$\lambda = \cos\tfrac{1}{2}\pi = 0; \qquad R(0; \mathbf{\Lambda}_2), \qquad \mathbf{\Lambda}_2 \perp \mathbf{n}_1:$$

9-4.7 $R(\lambda_1; \mathbf{\Lambda}_1)R(0; \mathbf{\Lambda}_2) = R(-\mathbf{\Lambda}_1 \cdot \mathbf{\Lambda}_2; \lambda_1\mathbf{\Lambda}_2 + \mathbf{\Lambda}_1 \times \mathbf{\Lambda}_2)$

$$= R(0; \lambda_1\mathbf{\Lambda}_2 + \mathbf{\Lambda}_1 \times \mathbf{\Lambda}_2) =_{\text{def}} R(0; \mathbf{\Lambda}). \tag{14}$$

Notice that $\mathbf{\Lambda}_1 . \mathbf{\Lambda} = 0$. From (14), $\mathbf{\Lambda} . \mathbf{\Lambda}_2 = \lambda_1 = \cos \frac{1}{2}\phi$.

8. Introduce (**9**-4.5) and (**9**-4.9) into (**3**-3.11).

PROBLEMS **9**-5

1. From (**9**-5.4),

$$Rx = \cos \phi x + \sin \phi (n_y z - n_z y) + (1 - \cos \phi)(n_x x + n_y y + n_z z) n_x$$
$$= \{\cos \phi + (1 - \cos \phi) n_x^2\} x + \{-\sin \phi n_z + (1 - \cos \phi) n_x n_y\} y$$
$$+ \{\sin \phi n_y + (1 - \cos \phi) n_x n_z\} z. \quad (1)$$

Cycle (1) to obtain Ry, Rz; then write:

$$R\,|xyz\rangle = \left\{ \cos \phi \begin{bmatrix} 1 & & \\ & 1 & \\ & & 1 \end{bmatrix} + \sin \phi \begin{bmatrix} 0 & -n_z & n_y \\ n_z & 0 & -n_x \\ -n_y & n_x & 0 \end{bmatrix} \right.$$
$$\left. + (1 - \cos \phi) \begin{bmatrix} n_x^2 & n_x n_y & n_x n_z \\ n_x n_y & n_y^2 & n_y n_z \\ n_x n_z & n_y n_z & n_z^2 \end{bmatrix} \right\} |xyz\rangle. \quad (2)$$

Equation (2) with the condition $\mathbf{n}^2 = 1$, gives (**3**-3.11).

2. Introduce in (**9**-5.4):

$$\lambda = \cos \tfrac{1}{2}\phi, \qquad \mathbf{\Lambda} = \sin \tfrac{1}{2}\phi \mathbf{n}, \qquad \Lambda^2 = \sin^2 \tfrac{1}{2}\phi, \qquad \cos \phi = 1 - 2\Lambda^2. \quad (3)$$

3. Introduce (**9**-5.6) into (**9**-5.5) and subtract.
4. To all orders, from (**9**-4.7), and on using the notation of (**9**-5.6),

$$R_1 R_2 = R(\lambda_1; \mathbf{\Lambda}_1) R(\lambda_2; \mathbf{\Lambda}_2)$$
$$= R(\lambda_1\lambda_2 - \mathbf{\Lambda}_1 . \mathbf{\Lambda}_2; \lambda_1 \mathbf{\Lambda}_2 + \lambda_2 \mathbf{\Lambda}_1 + \mathbf{\Lambda}_1 \times \mathbf{\Lambda}_2) = R_+. \quad (4)$$
$$R_2 R_1 = R(\lambda_2; \mathbf{\Lambda}_2) R(\lambda_1; \mathbf{\Lambda}_1)$$
$$= R(\lambda_1\lambda_2 - \mathbf{\Lambda}_1 . \mathbf{\Lambda}_2; \lambda_1 \mathbf{\Lambda}_2 + \lambda_2 \mathbf{\Lambda}_1 - \mathbf{\Lambda}_1 \times \mathbf{\Lambda}_2) = R_-. \quad (5)$$

Write, with the notation of (**9**-5.7),

$$[R_1, R_2]\mathbf{r} = (R_1 R_2 - R_2 R_1)\mathbf{r} = (R_+ - R_-)\mathbf{r}. \quad (6)$$

4,5|**9**-5.6

$$\lambda = \lambda_1\lambda_2 - \mathbf{\Lambda}_1 . \mathbf{\Lambda}_2, \quad \mathbf{\Lambda} = \lambda_1 \mathbf{\Lambda}_2 + \lambda_2 \mathbf{\Lambda}_1, \quad \mathbf{a} = \mathbf{\Lambda}_1 \times \mathbf{\Lambda}_2. \quad (7)$$

For infinitesimal rotations, to second-order, ($\lambda_i = 1 - \frac{1}{4}\phi_i^2$, $\mathbf{\Lambda}_i = \frac{1}{2}\phi_i \mathbf{n}$), in (6) and thence in (**9**-5.7),

6 $$[R_1, R_2]\mathbf{r} = 4(\mathbf{\Lambda}_1 \times \mathbf{\Lambda}_2) \times \mathbf{r}. \quad (8)$$

Guess from (**9**-5.5) an infinitesimal rotation that *adds* (R8) to **r** to second-order:

$$R(1;2\boldsymbol{\Lambda}_1\times\boldsymbol{\Lambda}_2)\mathbf{r}=\mathbf{r}+4(\boldsymbol{\Lambda}_1\times\boldsymbol{\Lambda}_2)\times\mathbf{r}. \qquad (9)$$

9|R8
$$[R_1, R_2]\mathbf{r}=\{R(1;2\boldsymbol{\Lambda}_1\times\boldsymbol{\Lambda}_2)-E\}\mathbf{r}. \qquad (10)$$

Equation (10) gives (**9**-5.8) and the verification of (**4**-4.10) is straightforward.

PROBLEMS **11**-1

1. Use Fig. **11**-5.1a where the path for C_2, C_2 is class 1.
2. Lengthy way: draw new figure like Fig. **11**-1.1. Short way: In Fig. **11**-1.1a,c rotate the axes (but nothing else) around $\mathbf{x}$ so that $\mathbf{y}$ goes to $\mathbf{z}$ and $\mathbf{z}$ to $-\mathbf{y}$. Figures a and c therein can now be read out for $C_{2y}C_{2x}$. The result is the antipole of C_{2z}. Thus $C_{2y}C_{2x}=\bar{C}_{2z}$, $[C_{2y}, C_{2x}]=-1$, as in Table **8**-1.3.

PROBLEM **11**-5

1. Standard parameters in the form λ, $\boldsymbol{\Lambda}$:

$$E=1,(000);\; C_{2x}=0,(100);\; C_{2y}=0,(010);\; C_{2z}=0,(001). \qquad (1)$$

11-5.2
$$C_{2x}C_{2y}=0,(100)\times(010)=0,(001)=C_{2z}. \qquad (2)$$

Other products similarly.

PROBLEMS **11**-6

1. From (**9**-5.3):

$$R(\phi\mathbf{n})\mathbf{r}=\mathbf{r}+\sin\phi(\mathbf{n}\times\mathbf{r})+(1-\cos\phi)\{(\mathbf{n}\cdot\mathbf{r})\mathbf{n}-n^2\mathbf{r}\} \qquad (1)$$

$$=\mathbf{r}+2\sin\tfrac{1}{2}\phi\cos\tfrac{1}{2}\phi(\mathbf{n}\times\mathbf{r})+2\sin^2\tfrac{1}{2}\phi\{(\mathbf{n}\cdot\mathbf{r})\mathbf{n}-n^2\mathbf{r}\}. \qquad (2)$$

2
$$R(\lambda_g;\boldsymbol{\Lambda}_g)\mathbf{r}=\mathbf{r}+2\lambda_g(\boldsymbol{\Lambda}_g\times\mathbf{r})+2\{(\boldsymbol{\Lambda}_g\cdot\mathbf{r})\boldsymbol{\Lambda}_g-\Lambda_g^2\mathbf{r}\}. \qquad (3)$$

3
$$R(\lambda_g;\boldsymbol{\Lambda}_g)\boldsymbol{\Lambda}_{g_i}=\boldsymbol{\Lambda}_{g_i}+2\lambda_g(\boldsymbol{\Lambda}_g\times\boldsymbol{\Lambda}_{g_i})-2\Lambda_g^2\boldsymbol{\Lambda}_{g_i}+2(\boldsymbol{\Lambda}_g\cdot\boldsymbol{\Lambda}_{g_i})\boldsymbol{\Lambda}_g. \qquad (4)$$

2.

11-6.3|**11**-5.2

$$\lambda_{gg^{-1}}=\lambda_g^2+\Lambda_g^2=1; \qquad \boldsymbol{\Lambda}_{gg^{-1}}=-\lambda_g\boldsymbol{\Lambda}_g+\lambda_g\boldsymbol{\Lambda}_g-\boldsymbol{\Lambda}_g\times\boldsymbol{\Lambda}_g=\mathbf{0}. \qquad (5)$$

Result: $gg^{-1}=E$, $\lambda_E=1$, $\boldsymbol{\Lambda}_E=\mathbf{0}\Rightarrow[g, g^{-1}]=1$, from (**11**-5.4). If $g=C_2$, $g^{-1}=C_2 \Rightarrow [g, g^{-1}]=[C_2, C_2]=-1$, from (**11**-5.9).

3. Operate as in Fig. 1.

4. From Fig. **11**-6.1,

$$\lambda_g = \tfrac{1}{2}, \quad \mathbf{\Lambda}_g = (0, 0, \tfrac{\sqrt{3}}{2}); \quad \lambda_{g_1} = 0, \quad \mathbf{\Lambda}_{g_1} = (100);$$
$$\lambda_{g_2} = 0, \quad \mathbf{\Lambda}_{g_2} = (\tfrac{1}{2}, -\tfrac{\sqrt{3}}{2}, 0), \quad g_2 = g_1^g. \tag{6}$$

11-6.6, **11**-6.7

$$\lambda_{(g_1)^g} = 0, \quad \mathbf{\Lambda}_{(g_1)^g} = (100) - 2\tfrac{3}{4}(100) + (0, 0, \tfrac{\sqrt{3}}{2}) \times (100) \tag{7}$$
$$= (100) - (\tfrac{3}{2}00) + (0, \tfrac{\sqrt{3}}{2}, 0) = (-\tfrac{1}{2}, \tfrac{\sqrt{3}}{2}, 0). \tag{8}$$

From (6), these are the parameters of the antipole of g_1^g.

PROBLEM **11**-7

1.

F**11**-6.1*b*
$$\lambda_{g_1} = 0, \quad \mathbf{\Lambda}_{g_1} = (100); \quad \lambda_g = \tfrac{1}{2}, \quad \mathbf{\Lambda}_g = (0, 0, \tfrac{\sqrt{3}}{2});$$
$$\lambda_{g^{-1}} = \tfrac{1}{2}, \quad \mathbf{\Lambda}_{g^{-1}} = (0, 0, -\tfrac{\sqrt{3}}{2}). \tag{1}$$

Call $g_1^g = gg_1g^{-1} = g'g^{-1} = g''; \quad g' = gg_1.$ (2)

11-5.2 $\lambda_{g'} = 0, \quad \mathbf{\Lambda}_{g'} = (\tfrac{1}{2}00) + (0, \tfrac{\sqrt{3}}{2}, 0) = (\tfrac{1}{2}, \tfrac{\sqrt{3}}{2}, 0).$ (3)

11-5.2 $\lambda_{g''} = 0, \quad \mathbf{\Lambda}_{g''} = \tfrac{1}{2}(\tfrac{1}{2}, \tfrac{\sqrt{3}}{2}, 0) + (-\tfrac{3}{4}, \tfrac{\sqrt{3}}{4}, 0) = (-\tfrac{1}{2}, \tfrac{\sqrt{3}}{2}, 0).$ (4)

F**11**-6.1b $\lambda_{g_2} = 0, \quad \mathbf{\Lambda}_{g_2} = (\tfrac{1}{2}, -\tfrac{\sqrt{3}}{2}, 0).$ (5)

Thus, to form g_1^g in $[g_1^g, g]$, (4) has to be standardized into (5).

$[g_1^g, g] \equiv [g_2, g]$: $\quad \lambda_{g_2g} = 0, \quad \mathbf{\Lambda}_{g_2g} = (\tfrac{1}{4}, -\tfrac{\sqrt{3}}{4}, 0) + (-\tfrac{3}{4}, -\tfrac{\sqrt{3}}{4}, 0)$
$$= (-\tfrac{1}{2}, -\tfrac{\sqrt{3}}{2}, 0). \tag{6}$$

$[g, g_1]$: $\quad \lambda_{gg_1} = 0, \quad \mathbf{\Lambda}_{gg_1} = (\tfrac{1}{2}00) + (0, \tfrac{\sqrt{3}}{2}, 0).$ (7)

From (6) and (7) the characters of g_2 and g_1 will be of opposite signs. Reason: $\mathfrak{H}$ badly chosen. Same work with Fig. **11**-6.1*c* will give correct agreement.

PROBLEM **11**-9

1. $$\sigma_x = C_{2x}\boldsymbol{i}, \quad \sigma_y = C_{2y}\boldsymbol{i}, \quad \sigma_z = C_{2z}\boldsymbol{i}. \tag{1}$$

Pole of σ_x = Pole of C_{2x}, etc. Standard poles from Fig. **2**-5.1 with $\mathfrak{H}$ from Fig. **9**-1.1, listed as λ, $\mathbf{\Lambda}$:

$$\sigma_x = 0, (100); \quad \sigma_y = 0, (010); \quad \sigma_z = 0, (001). \tag{2}$$

11-5.2 $\sigma_x\sigma_y$:

$\lambda = 0, \quad \mathbf{\Lambda} = (100) \times (010) = (001) \quad \Rightarrow \quad \sigma_x\sigma_y = C_{2z}.$ (3)

11-9.17	$[\sigma_x \cdot \sigma_y] = [C_{2x}, C_{2y}][i, i] = -[C_{2x}, C_{2y}].$	(4)
T8-1.3	$[C_{2x}, C_{2y}] = 1 \quad \Rightarrow \quad [\sigma_x, \sigma_y] = -1.$	(5)

Similarly the other products and factors.

PROBLEMS 12-2

1. Use (**12**-2.2) twice on both sides of (**12**-2.3). Write $(\mathbf{A}\times\mathbf{B})\times\mathbf{C} = (\mathbf{C}\,.\,\mathbf{A})\mathbf{B} - (\mathbf{C}\,.\,\mathbf{B})\mathbf{A}$ and similarly $\mathbf{A}\times(\mathbf{B}\times\mathbf{C})$.
2. From (**12**-2.2):

$$[\![0, \mathbf{A}]\!][\![0, \mathbf{B}]\!] = [\![-\mathbf{A}\,.\,\mathbf{B}, \mathbf{A}\times\mathbf{B}]\!] \quad \Rightarrow \quad \mathbf{A}\,.\,\mathbf{B} = 0. \tag{1}$$

3.

$$\mathbb{A}\mathbb{B} = (a + A\mathbb{n})(b + B\mathbb{n}) = ab + AB\mathbb{n}^2 + \mathbb{n}(bA + aB) \tag{2}$$

12-2.13

$$= ab - AB + \mathbb{n}(aB + bA). \tag{3}$$

4. From (**12**-2.2):

$$[\![a, \mathbf{A}]\!][\![0, \mathbf{B}]\!] = [\![-\mathbf{A}\,.\,\mathbf{B}, a\mathbf{B} + \mathbf{A}\times\mathbf{B}]\!] \quad \Rightarrow \quad \mathbf{A}\,.\,\mathbf{B} = 0. \tag{4}$$

PROBLEMS 12-4

1.

12-2.2, 12-4.1

$$(\mathbb{A}\mathbb{B})^* = [\![ab - \mathbf{A}\,.\,\mathbf{B}, -a\mathbf{B} - b\mathbf{A} - \mathbf{A}\times\mathbf{B}]\!], \tag{1}$$

$$\mathbb{B}^*\mathbb{A}^* = [\![b, -\mathbf{B}]\!][\![a, -\mathbf{A}]\!] = [\![ab - \mathbf{A}\,.\,\mathbf{B}, -b\mathbf{A} - a\mathbf{B} + \mathbf{B}\times\mathbf{A}]\!]. \tag{2}$$

2.

$$[\![a, \mathbf{A}]\!][\![b, \mathbf{B}]\!] = [\![c, \mathbf{C}]\!], \qquad a^2 + A^2 = b^2 + B^2 = 1. \tag{3}$$

12-2.2

$$c^2 = (ab - \mathbf{A}\,.\,\mathbf{B})^2 = a^2b^2 + (\mathbf{A}\,.\,\mathbf{B})^2 - 2ab\,\mathbf{A}\,.\,\mathbf{B}. \tag{4}$$

12-2.2

$$C^2 = a^2B^2 + b^2A^2 + (\mathbf{A}\times\mathbf{B})\,.\,(\mathbf{A}\times\mathbf{B}) + 2ab\,\mathbf{A}\,.\,\mathbf{B}. \tag{5}$$

3|4,5

$$c^2 + C^2 = a^2 + b^2A^2 + (\mathbf{A}\,.\,\mathbf{B})^2 + \mathbf{A}\,.\,(B^2\mathbf{A} - \mathbf{B}\,.\,\mathbf{AB})$$
$$= a^2 + A^2 = 1. \tag{6}$$

3.

$$\mathbb{A}^* = -\mathbb{A} \quad \Rightarrow \quad [\![a, -\mathbf{A}]\!] = [\![-a, -\mathbf{A}]\!] \quad \Rightarrow \quad a = 0, \tag{7}$$

etc.

4. From (**12**-4.3), (**12**-4.2):

$$|\mathbb{A}\mathbb{B}|^2 = \mathbb{A}\mathbb{B}(\mathbb{A}\mathbb{B})^* = \mathbb{A}\mathbb{B}\mathbb{B}^*\mathbb{A}^* = |\mathbb{B}|^2\,\mathbb{A}\mathbb{A}^* = |\mathbb{B}|^2\,|\mathbb{A}|^2. \tag{8}$$

5.

12-4.2, 12-4.14

$$\bar{\mathbb{r}}^* = \mathbb{R}\mathbb{r}^*\mathbb{R}^* = -\mathbb{R}\mathbb{r}\mathbb{R}^* = -\bar{\mathbb{r}}. \tag{9}$$

Problem 3 on (9) proves that $\bar{\mathbb{r}}$ is a pure quaternion. From Problem 4, $|\bar{\mathbb{r}}| = |\mathbb{R}|\,|\mathbb{r}|\,|\mathbb{R}^*| = |\mathbb{R}|^2\,|\mathbb{r}|$.

PROBLEMS **12**-5

1. Introduce the products from T**12**-5.1 into (**12**-5.11):

$$[\![a, \mathbf{A}]\!][\![b, \mathbf{B}]\!] =$$

$$= ab - A_xB_x - A_yB_y - A_zB_z + a[\![0, \mathbf{B}]\!] + b[\![0, \mathbf{A}]\!]$$
$$+ (A_xB_y - A_yB_x)\mathbb{k} + (A_yB_z - A_zB_y)\mathbb{i} + (A_zB_x - A_xB_z)\mathbb{j} \quad (1)$$
$$= [\![ab - \mathbf{A}\,.\,\mathbf{B}, \mathbf{0}]\!] + [\![0, a\mathbf{B} + b\mathbf{A}]\!] + (\mathbf{A}\times\mathbf{B})_z[\![0, \mathbf{k}]\!]$$
$$+ (\mathbf{A}\times\mathbf{B})_x[\![0, \mathbf{i}]\!] + (\mathbf{A}\times\mathbf{B})_y[\![0, \mathbf{j}]\!] \quad (2)$$
$$= [\![ab - \mathbf{A}\,.\,\mathbf{B}, a\mathbf{B} + b\mathbf{A} + \mathbf{A}\times\mathbf{B}]\!]. \quad (3)$$

2.

T**12**-5.1 $\quad \mathbb{i}\mathbb{j} = \mathbb{k} \quad \Rightarrow \quad \mathbb{i}\mathbb{j}\mathbb{k} = \mathbb{k}^2 = -1. \quad (4)$

PROBLEMS **12**-6

1. Simple matrix multiplication.
2. Introduce $[\![\lambda, \mathbf{\Lambda}]\!]^* = [\![\lambda, -\mathbf{\Lambda}]\!]$ into (**12**-6.6):

$$\check{R}^{1/2}[\![\lambda, \mathbf{\Lambda}]\!]^* = \check{R}^{1/2}[\![\lambda, -\mathbf{\Lambda}]\!] = \begin{bmatrix} \lambda + i\Lambda_z & \Lambda_y + i\Lambda_x \\ -\Lambda_y + i\Lambda_x & \lambda - i\Lambda_z \end{bmatrix}. \quad (1)$$

This matrix is the transpose of $\{\check{R}^{1/2}[\![\lambda, \mathbf{\Lambda}]\!]\}^*$ as given by (**12**-6.6). Thus: $\check{R}^{1/2}[\![\lambda, \mathbf{\Lambda}]\!]^* = \{\check{R}^{1/2}[\![\lambda, \mathbf{\Lambda}]\!]\}^\dagger = \{\check{R}^{1/2}[\![\lambda, \mathbf{\Lambda}]\!]\}^{-1}$, from the unitary property.

PROBLEMS **12**-8

1.

12-8.8 $\quad \bar{\mathbb{r}} = [\![0, \cos\phi\,\mathbf{r} + \sin\phi\,(\mathbf{n}\times\mathbf{r})]\!]. \quad (1)$

Since $\mathbf{n}\times\mathbf{r}$ is normal to $\mathbf{r}$, this is a rotation by ϕ in the plane $(\mathbf{r}, \mathbf{n}\times\mathbf{r})$.

2.

12-8.8 $\quad \mathbb{R}\mathbb{r}\mathbb{R}^* = [\![0, -\mathbf{r} + 2(\mathbf{n}\,.\,\mathbf{r})\mathbf{n}]\!] \quad \Rightarrow \quad \mathbf{n}\,.\,\mathbf{r} = 0. \quad (2)$

PROBLEM **13**-4

1. $\mathbf{D}_2$ is Abelian; hence it has four classes E, C_{2x}, C_{2y}, C_{2z}. Only the first is regular, thus giving two classes E, $\tilde{E}$ in $\tilde{\mathbf{D}}_2$. C_{2x}, irregular, will give only one class containing C_{2x} and $\tilde{C}_{2x}$ and the same for the

others. Thus the total number of classes in $\tilde{\mathbf{D}}_2$ is five. The number of spinor representations equals the number of regular classes in G; that is, one.

PROBLEMS **14**-2

1. For cyclic groups, (**14**-2.12) obtains; hence all bases are one-dimensional. For dihedral groups, from (**14**-2.14) the largest possible bases are $\langle u^i_j u^i_{-j}|$. For u^i_0 these reduce to a single function, corresponding to R^i_{00}. The result follows from (**14**-2.9), on noticing that k can only take the values 0, 1.
2. Straightforward but tedious application of **14**-2.9. The matrix coincides with (**7**-3.10), as it should, since this matrix transforms $\langle u^1_1 u^1_0 u^1_{-1}|$.
3. You must first choose by convention positive rotation axes $\mathbf{n}$. Choose those listed. Consider the rotation $R(-\frac{2}{3}\pi(1\bar{1}1))$. This has now to be written as $R(\frac{2}{3}\pi(\bar{1}1\bar{1}))$. Remember to normalize the rotation-axis vector: $\mathbf{n}=3^{-1/2}(\bar{1}1\bar{1})$. Also: $\cos\frac{1}{2}\phi=\frac{1}{2}$, $\sin\frac{1}{2}\phi=\frac{1}{2}3^{1/2}$. Thus: $\lambda=\frac{1}{2}$, $\Lambda=(\frac{\bar{1}}{2},\frac{1}{2},\frac{\bar{1}}{2})$. Obtain now ρ and τ from (**14**-2.23) and introduce them into (**14**-2.24). The result agrees with Onodera and Okazaki's Γ_6^+ representation in their Table II.

PROBLEMS **15**-4

1. Form a table like Table **15**-6.1. Your first line should be $\bar{E}_1$ from Table **15**-3.1 and your second and third lines $\bar{E}_2$ and $\bar{E}_3$ guessed from orthogonality. In order to guess the bases reproduce the work done in the lower part of Table **15**-6.1, for $j=\frac{3}{2}$ and $j=\frac{5}{2}$, on using (**15**-6.1) to (**15**-6.3).
2. Extract the matrices for $\mathbf{D}_3$ from Table **15**-3.1. Since $j=\frac{3}{2}$, raise each matrix element to the third power. You will obtain the following matrices.

E	C_3^+	C_3^-	C'_{21}	C'_{22}	C'_{23}	
$\begin{bmatrix}1 & \\ & 1\end{bmatrix}$	$\begin{bmatrix}-1 & \\ & -1\end{bmatrix}$	$\begin{bmatrix}-1 & \\ & -1\end{bmatrix}$	$\begin{vmatrix} & -i \\ -i & \end{vmatrix}$	$\begin{bmatrix} & i \\ i & \end{bmatrix}$	$\begin{bmatrix} & -i \\ -i & \end{bmatrix}$	(1)
1	−1	−1	−i	i	−i	(2)
1	−1	−1	i	−i	i	(3)

In (2) and (3), the matrices (1) are reduced by the rules of eqn (**15**-4.9). The character theorem is not satisfied. Notice that, even changing the sign of the matrix for C'_{21} in (1), as suggested in §**15**-3, would not totally correct the discrepancy.
3. Call A the matrix for C''_{2i} in Table **15**-4.1. It is easy to guess its

eigenvectors as $|1 \quad i\rangle$ and $|1 \quad -i\rangle$. Call

$$U = 2^{-1/2}\begin{bmatrix} 1 & 1 \\ i & -i \end{bmatrix}. \tag{4}$$

Then $U^{-1}AU$ is diagonal with elements $-i$, i respectively. The reduced basis is $\langle u_{3/2+}, u_{3/2-}| U$, the first and second functions corresponding to $-i$, i respectively. This gives the results in Table **15**-4.4. (If you check the orthonormality of these two representations, remember to do so in the Hermitian sense).

PROBLEM **15**-5

1. From Opechowski's Theorem, the classes of $\tilde{\mathbf{D}}_6$ are:

$$(E)(\tilde{E})(2C_6)(2\tilde{C}_6)(2C_3)(2\tilde{C}_3)(C_2\tilde{C}_2)(3C'_{2i}3\tilde{C}'_{2i})(3C''_{2i}3\tilde{C}''_{2i}).$$

Nine classes and order 24 entails four one-dimensional and five two-dimensional representations. Of this, four one-dimensional and two two-dimensional belong to $\mathbf{D}_6$. There are thus only three two-dimensional spinor representations.

REFERENCES

AHARONOV, Y. and SUSSKIND, L. (1967). Observability of the sign change of spinors under 2π rotations. *Phys. Rev.* **158,** 1237–8.

ALTMANN, S. L. (1977). *Induced representations in crystals and molecules. Point, space and nonrigid molecule groups.* Academic Press, London.

—— (1979). Double groups and projective representations. I. General theory. *Molec. Phys.* **38,** 489–511.

—— and DIRL, R. (1984). Symmetrisation of molecular Dirac eigenfunctions. *J. Phys. A: Math. Gen.* **17,** 501–14.

—— and HERZIG, P. (1982). Double groups and projective representations. III. Improper groups. *Molec. Phys.* **45,** 585–604.

—— and PALACIO, F. (1979). Double groups and projective representations. II. Clebsch–Gordan coefficients. *Molec. Phys.* **38,** 513–26.

ANANDAN, J. (1980). On the hypotheses underlying physical geometry. *Foundations of Physics* **10,** 601–29.

ASANO, K. and SHODA, K. (1935). Zur Theorie der Darstellungen einer endlichen Gruppe durch Kollineationen. *Compositio Math.* **2,** 230–49.

BACKHOUSE, N. B. (1973). Projective character tables and Opechowski's theorem. *Physica* **70,** 505–19.

BACRY, H. (1977). La rotation des fermions. *La Recherche* **8,** 1010–1010.

BARFIELD, O. (1972). *What Coleridge thought.* Oxford University Press, London.

BELL, E. T. (1937). *Men of Mathematics.* Gollancz, London.

BERNAL, J. D. (1923). *The analytic theory of point systems.* Occasional Paper No 1, 1981. Department of Crystallography. Birkbeck College, London.

BERNSTEIN, H. J. (1967). Spin precession during interferometry of fermions and the phase factor associated with rotations through 2π radians. *Phys. Rev. Lett.* **18,** 1102–3.

BETHE, H. (1929). Termaufspaltung in Kristallen. *Annalen Phys.* Series 5, **3,** 133–208.

BIEDENHARN, L. C. and LOUCK, J. D. (1981). *Angular momentum in quantum physics. Theory and application.* Addison-Wesley, Reading, Mass.

BOOTH, A. J. (1871). *Saint-Simon and Saint-Simonism, a chapter in the history of socialism in France.* Longmans, Green, Reader & Dyer, London.

BOYLE, E. E. and GREEN, K. F. (1978). The representation groups and projective representations of the point groups and their applications. *Phil: Trans. Roy. Soc.* A **288,** 237–69.

BRADLEY, C. J. and CRACKNELL, A. P. (1972). *The mathematical theory of symmetry in solids. Representation theory for point groups and space groups.* Clarendon Press, Oxford.

BRINK, D. M. and SATCHLER, G. R. (1968). *Angular momentum* (2nd edn). Clarendon Press, Oxford.

BROWN, E. (1970). A simple alternative to double groups. *Am. J. Phys.* **38,** 704–15.

BUERGER, M. J. (1956). *Elementary crystallography. An introduction to the fundamental geometrical features of crystals.* Wiley, New York.

BUTLER, P. H. (1981). *Point group symmetry applications. Methods and tables.* Plenum Press, New York.

CARTAN, E. (1913). Les groupes projectifs qui ne laissent invariante aucune multiplicité plane. *Bull. Soc. Math. de France* **41,** 53–96.

—— (1938). *Leçons sur la théorie des spineurs.* I. *Les spineurs de l'espace a trois dimensions.* Hermann, Paris.

CAYLEY, A. (1843). On the motion of rotation of a solid body. *Cambridge Mathematical Journal* **3,** 224–32.

—— (1845). On certain results relating to quaternions. *Phil. Mag.* **26,** 141–45.

—— (1848). On the application of quaternions to the theory of rotation. *Phil. Mag.* **33,** 196–200.

—— (1888). *The collected mathematical papers.* Vol. I. 13 volumes plus Index (1898).

CHARLÉTY, S. (1936). *Histoire du Saint-Simonisme* (1825–1864) (2nd edn). Hartmann, Paris. (1st edn 1896.)

CLIFFORD, K. (1878). Applications of Grassmann's extensive algebra. *American Journal of Mathematics Pure and Applied* **1,** 350–58.

CONWAY, A. W. (1912). The quaternionic form of relativity. *Phil. Mag.* **24,** 208–8.

COXETER, H. S. M. (1946). Quaternions and reflections. *Am. Math. Monthly* **53,** 136–46.

DIEHL, E. (1940). *Anthologia Lyrica Graeca* (3rd edn). Fasc. 2. Teubner, Leipzig.

EHLERS, J., RINDLER, W. and ROBINSON, I. (1966). Quaternions, bivectors and the Lorentz group. In *Perspectives in geometry and relativity. Essays in honor of Václay Hlavatý* (ed. B. Hoffmann) pp. 134–49. Indiana University Press, Bloomington.

ELLIOTT, R. J. (1954). Spin orbit coupling in band theory – Character tables for some "double" space groups. *Phys. Rev.* **96,** 280–87.

EULER, L. (1775*a*). Formulae generales pro translatione quacunque corporum rigidorum. *Novi Comm. Acad. Sci. Imp. Petrop.* **20,** 189–207.

—— (1775*b*). Nova methodus motum corporum rigidorum determinandi. *Novi. Comm. Acad. Sci. Imp. Petrop.* **20,** 208–38.

FANO, G. (1971). *Mathematical methods of quantum mechanics.* McGraw Hill, New York.

FANO, U. and RACAH, G. (1959). *Irreducible tensorial sets.* Academic Press, New York.

GAUSS, C. F. (1863). *Werke.* 12 vols 1863–1929. Königliche Gesellschaft der Wissenschaften, Göttingen.

GEL'FAND, I. M., MINLOS, R. A. and SHAPIRO, Z. Ya. (1963). *Representations of the rotation and Lorentz groups and their applications* (trans. G. Cummins and T. Boddington). Pergamon Press, Oxford.

GORMLEY, P. G. (1947). Stereographic projection and the linear fractional group of transformations of quaternions. *Proc. Roy. Irish. Acad.* A **51,** 67–85.

GRATTAN-GUINNESS, I. (1983). Personal communication.

GRAVES, R. P. (1882). *Life of Sir William Rowan Hamilton,* vol. I; vol. II, 1885; vol. III, 1889. Hodges, Figgis & Co., Dublin.

GRAY, J. (1980). Olinde Rodrigues' paper of 1840 on transformation groups. *Arch. for History of Exact Sciences* **21,** 375–85.

GREIDER, K. (1980). Relativistic quantum theory with correct conservation laws. *Phys. Rev. Lett.* **44,** 1718–21.

—— (1984). A unifying Clifford Algebra Formalism for relativistic fields. *Foundations of Physics* **14,** 467–506.

Guthrie, W. K. C. (1962). *A history of Greek philosophy* Vol I. *The earlier presocratics and the pythagoreans*. Cambridge University Press, Cambridge.

Haggarty, R. J. and Humphreys, J. F. (1978). Projective characters of finite groups. *Proc. London. Math. Soc.* (3), **36,** 176–92.

Hamermesh, M. (1962). *Group theory and its application to physical problems*. Addison-Wesley, Reading, Mass.

Hamilton, W. R. (1844*a*). On quaternions; or on a new system of imaginaries in algebra. *Phil. Mag.*, 3rd ser, **25,** 489–95.

—— (1844*b*). On a new species of imaginary quantities connected with the theory of quaternions. *Proc. Roy. Irish Acad.* **2,** 424–34.

—— (1846). On symbolical geometry. *Cambridge and Dublin Mathematical Journal* **1,** 45–57.

—— (1853). *Lectures on quaternions*. Hodges & Smith, Dublin.

—— (1899). *Elements of quaternions* (2nd edn, ed. C. J. Jolly). Vol. I. Vol. II, 1901. Longmans, Green and Co., London.

—— (1931). *The mathematical papers of Sir William Rowan Hamilton*. Vol. I. *Geometrical optics* (eds A. W. Conway and J. L. Synge). Vol. II *Dynamics* (eds A. W. Conway and A. J. McConnell) 1940. Vol. III *Algebra* (eds H. Halberstan and R. E. Ingram) 1967. Cambridge University Press, Cambridge.

Hankins, T. L. (1980). *Sir William Rowan Hamilton*. Johns Hopkins University Press, Baltimore.

Herzig, P. (1984). Clebsch–Gordan coefficients for the group chains $\mathbf{O} \supset \mathbf{T} \supset \mathbf{D}_2$ and $\mathbf{T}_d \supset \mathbf{D}_2$ by the method of projective representations. *Int. J. Quantum Chem.* **26,** 297–312.

Hestenes, D. (1966). *Space-time algebra*. Gordon & Breach, New York.

Hilbert, D. and Cohn-Vossen, S. (1952). *Geometry and the imagination* (trans. P. Nemenyi). Chelsea Publishing Co., New York.

Hilton, H. (1903). *Mathematical crystallography and the theory of groups of movements*. Clarendon Press, Oxford.

Hüller, A. and Kroll, D. M. (1975). Rotational tunneling in solids. *J. Chem. Phys.* **63,** 4495–504.

Hurley, A. C. (1966). Ray representations of point groups and the irreducible representations of space groups and double space groups. *Phil. Trans. Roy. Soc.* A **260,** 1–36 (plus one table).

Jansen, L. and Boon, M. (1967). *Theory of finite groups. Applications in physics. Symmetry groups of quantum mechanical systems*. North-Holland, Amsterdam.

Karpilovsky, G. (1985). *Projective representations of finite groups*. Dekker, Basel.

Kim, S. K. (1981*a*). On the projective representations of point groups. *Molec. Phys.* **43,** 1447–50.

—— (1981*b*). A unified theory of the point groups and their general irreducible representations. *J. Math. Phys.* **22,** 2101–7.

—— (1983*a*). A unified theory of the point groups. II. The general projective representations and their applications to space groups. *J. Math. Phys.* **24,** 411–13.

—— (1983*b*). A unified theory of the point groups. III. Classification of basis functions of improper point groups. *J. Math. Phys.* **24,** 414–18.

—— (1983*c*). A unified theory of the point groups. IV. The general corepresentations of the crystallographic and noncrystallographic Shubnikov point groups. *J. Math. Phys.* **24,** 419–23.

—— (1984*a*). A unified theory of the point groups. V. The general projective corepresentations of the magnetic point groups and their applications to magnetic space groups. *J. Math. Phys.* **25,** 189–96.

—— (1984*b*). General irreducibility condition for vector and projective corepresentations of antiunitary groups. *J. Math. Phys.* **25,** 197–9.

KIRK, G. S., RAVEN, J. E. and SCHOFIELD, M. (1983). *The presocratic philosophers* (2nd edn). Cambridge University Press, Cambridge.

KLEIN, F. (1884). *Vorlesungen über das Ikosaeder und die Auflösung der Gleichungen vom fünften Grade.* Teubner, Leipzig. Translated as: *The icosahedron and the solutions of equations of the fifth degree* (trans. G. G. Morrice, 2nd edn). Dover Publications, New York, 1956.

—— (1897). *The mathematical theory of the top.* Scribner, New York. Reprinted in *Congruence of sets and other monographs,* no date, Chelsea Publishing Co., New York.

—— (1926). *Vorlesungen über die Entwicklung der Mathematik im* 19. *Jahrhundert,* vol. I; vol. II, 1927. Springer, Berlin.

—— and SOMMERFELD, A. (1897). *Über die Theorie des Kreisels,* vol. I, vol. II, 1898; vol. III, 1903; vol. IV, 1910. Teubner, Leipzig.

LANCZOS, C. (1967). William Rowan Hamilton – An appreciation. *American Scientist* **55,** 129–43.

LAROUSSE, P. (1866). *Grand dictionnaire universel du XIXᵉ siècle.* Administration du Grand Dictionnaire Universel, Paris. Vol. I, 1866; vol. XIII, no date.

LITTLEWOOD, D. E. (1950). *A university algebra.* Heinemann, London.

MACKAY, A. L. (1977). *The harvest of a quiet eye. A selection of scientific quotations* (ed. M. Ebison). The Institute of Physics, Bristol.

MCCONNELL, A. J. (1943). The Dublin Mathematical School in the first half of the nineteenth century. *Proc. Roy. Irish Acad.* A **50,** 75–92.

MACDUFFEE, C. C. (1944). Algebra's debt to Hamilton. *Scripta Math.* **10,** 25–36.

MCKIE, D. and MCKIE C. (1974). *Crystalline solids.* Nelson, London.

MCWEENY, R. (1979). *Coulson's valence* (3rd edn). Oxford University Press, Oxford.

MESSIAH, A. (1964). *Quantum Mechanics* (trans. J. Potter). Two vols. North-Holland, Amsterdam.

MICHAUD. (1843). *Biographie universelle ancienne et moderne. Nouvelle édition.* Desplaces, Paris. Vol. I, 1843, vol. XXXVI (no date), pp. 288–9.

MISNER, C. W., THORNE, K. S. and WHEELER, J. A. (1970). *Gravitation.* Freeman, San Francisco.

MORRIS, A. O. (1971). Projective representations of finite groups. In *Proceedings of the Conference on Clifford algebra, its generalizations and applications* (ed. A. Ramakrishnan) pp. 43–86. Matscience, Madras.

MURNAGHAN, F. D. (1938). *The theory of group representations.* Johns Hopkins Press, Baltimore.

NEUMANN, C. (1865). *Vorlesungen über Riemann's Theorie der Abel'schen Integrale.* Teubner, Leipzig.

NORMAND, J.-M. (1980). *A Lie group: rotations in quantum mechanics.* North-Holland, Amsterdam.

O'DONNELL, S. (1983). *William Rowan Hamilton. Portrait of a prodigy.* Boole Press, Dublin.

ONODERA, Y. and OKAZAKI, M. (1966). Tables of basis functions for double point groups. *J. Phys. Soc. Japan* **21,** 2400–8.

OPECHOWSKI, W. (1940). Sur les groupes cristallographiques "doubles". *Physica* **7,** 552–62.

PAGE, D. N. and WOOTERS, W. K. (1983). Evolution without evolution: dynamics described by stationary observables. *Phys. Rev.* D **27,** 2885–92.

PONTRYAGIN, L. S. (1966). *Topological groups* (2nd edn, trans. A. Brown). Gordon & Breach, New York.
RAUCH, H., ZEILINGER, A., BADUREK, G., WILFING, A., BAUSPIESS, W. and BOUSE, U. (1975). Verification of coherent spinor rotation of fermions. *Phys. Lett.* **54** A, 427–427.
RIESZ, M. (1946). Sur certaines notions fondamentales en théorie quantique relativiste. *Skandinavisk matematikercongres* (*Dixième Congres des Mathématiciens Scandinaves*) **10,** 123–48.
—— (1958). *Clifford numbers and spinors.* Lecture Series No 38, The Institute for Fluid Dynamics and Applied Mathematics, University of Maryland, Maryland.
RODRIGUES, O. (1840). Des lois géométriques qui régissent les déplacements d'un système solide dans l'espace, et de la variation des coordonnées provenant de ses déplacements considérés indépendamment des causes qui peuvent les produire. *J. de Mathématiques Pures et Appliquées* **5,** 380–440.
ROMAN, P. (1975). *Some modern mathematics for physicists and other outsiders. An introduction to algebra, topology and functional analysis.* (2 vols). Pergamon, New York.
ROSE, M. E. (1957). *Elementary theory of angular momentum.* Wiley, New York.
RUDRA, P. (1965). On projective representations of finite groups. *J. Math. Phys.* **6,** 1273–7.
SCHOENFLIES, A. and GRÜBLER, M. (1902). Kinematik. In *Encyklopädie der mathematischen Wissenschaften mit einschluss ihrer Anwendungen.* Band IV, 1 Teilband pp. 190–278. (Dated 1901–1908). Teubner, Leipzig.
SCHUR, I. (1904). Über die Darstellung der endlichen Gruppen durch gebrochene lineare Substitutionen. *J. reine angew. Math.* **127,** 20–50.
—— (1907). Untersuchungen über die Darstellung der endlichen Gruppen durch gebrochene lineare Substitutionen. *J. reine. angew. Math.* **132,** 85–137.
—— (1910). Über die Darstellung der symmetrischen und der alternierenden Gruppe durch gebrochene lineare Substitutionen. *J. reine angew. Math.* **139,** 155–250. (pp. 165–250 published 1911).
—— (1927). Über die reellen Kollineations-gruppen, die der symmetrischen oder der alternierenden Gruppe isomorph sind. *J. reine angew. Math.* **158,** 63–79.
SCHUTZ, B. (1980). *Geometrical methods of mathematical physics.* Cambridge University Press, Cambridge.
SILBERSTEIN, L. (1912). Quaternionic form of relativity. *Phil. Mag.* **23,** 790–809.
SYLVESTER, J. J. (1850). On the rotation of a rigid body about a fixed point. *Phil. Mag.* **37,** 440–4.
SYNGE, J. L. (1972). Quaternions, Lorentz transformations and the Conway-Dirac-Eddington matrices. *Comm. Dublin Inst. Adv. Studies*, Series A, No. 21, 1–67.
TAIT, P. G. (1890). *An elementary treatise on quaternions* (3rd edn). Cambridge University Press, Cambridge.
TEITLER, S. (1965). "Vector" Clifford algebras and the classical theory of fields. *Nuovo Cimento,* suppl. ser. 1, **3,** 1–14.
—— (1966). The structure of 4-spinors. *J. Math. Phys.* **7,** 1730–8.
TEMPLE, G. (1960). *Cartesian tensors. An introduction.* Methuen, London.
—— (1981). 100 *years of mathematics.* Duckworth, London.
TINKHAM, M. (1964). *Group theory and quantum mechanics.* McGraw Hill, New York.
TUCKERMAN, L. B. (1947). Multiple reflections by plane mirrors. *Quart. Appl. Math.* **5,** 133–48.

VAN DER WAERDEN, B. L. (1932). *Die gruppentheoretische Methoden in der Quantenmechanik.* Springer, Berlin.

WAGNER, H. (1951). Zur mathematischen Behandlung von Spiegelungen. *Optik* **8,** 456–72.

WALLACE, A. H. (1957). *An introduction to algebraic topology.* Pergamon Press, London.

WARREN, J. (1829). On the geometrical representation of the powers of quantities, whose indices involve the square roots of negative quantities. *Phil. Trans. Roy. Soc.* **119,** 339–59.

WEILL, G. (1894). *Saint-Simon et son oeuvre. Un précurseur du socialisme.* Perrin et Cie, Paris.

WERNER, S. A., COLELLA, R., OVERHAUSER, A. W. and EAGEN, C. F. (1975). Observation of the phase shift of a neutron due to precession in a magnetic field. *Phys. Rev. Lett.* **35,** 1053–5.

WEYL, H. (1925*a*). Theorie der Darstellung kontinuierlicher halb-einfacher Gruppen durch lineare Transformationen. I. *Math. Z.* **23,** 271–309.

—— (1925*b*). *Ibid.* II. *Math. Z.* **24,** 328–376.

—— (1925*c*). *Ibid.* III. *Math. Z.* **24,** 377–395.

—— (1926). Nachtrag zu der Arbeit: Theorie der Darstellung kontinuierlicher halb-einfacher Gruppen durch linearen Transformationen. III. *Math. Z.* **24,** 789–91.

—— (1931). *The theory of groups and quantum mechanics* (trans. H. P. Robertson). Methuen, London.

—— (1949). *Philosophy of mathematics and natural science.* Princeton University Press, Princeton.

WHITTAKER, E. T. (1937). *A treatise on the analytical dynamics of particles and rigid bodies* (4th edn). Cambridge University Press, Cambridge.

—— (1940). The Hamiltonian revival. *Math. Gazette* **24,** 153–8.

WIGNER, E. P. (1959). *Group theory and its application to the quantum mechanics of atomic spectra* (trans. J. J. Griffin). Academic Press, New York.

WILSON, E. (1941). *To the Finland station. A study in the writing and acting of history.* Secker & Warburg, London.

ZASSENHAUS, H. (1949). *The theory of groups* (trans S. Kravetz). Chelsea Publishing Co., New York.

INDEX

Entries for expressions such as 'unitary matrices', 'binary rotations', etc, will mainly be found under those headings rather than as sub-entries of 'matrices', 'rotations', etc. A page number in bold-face indicates a definition.

Abbreviations

F	figure
T	table
dim	dimension
rep	representation
rtn	rotation
v-rep	vector representation
p-rep	projective representation

A CATALOG OF SELECTED

DOVER BOOKS

IN SCIENCE AND MATHEMATICS

Astronomy

BURNHAM'S CELESTIAL HANDBOOK, Robert Burnham, Jr. Thorough guide to the stars beyond our solar system. Exhaustive treatment. Alphabetical by constellation: Andromeda to Cetus in Vol. 1; Chamaeleon to Orion in Vol. 2; and Pavo to Vulpecula in Vol. 3. Hundreds of illustrations. Index in Vol. 3. 2,000pp. 6⅛ x 9¼.
Vol. I: 23567-X
Vol. II: 23568-8
Vol. III: 23673-0

EXPLORING THE MOON THROUGH BINOCULARS AND SMALL TELESCOPES, Ernest H. Cherrington, Jr. Informative, profusely illustrated guide to locating and identifying craters, rills, seas, mountains, other lunar features. Newly revised and updated with special section of new photos. Over 100 photos and diagrams. 240pp. 8¼ x 11. 24491-1

THE EXTRATERRESTRIAL LIFE DEBATE, 1750–1900, Michael J. Crowe. First detailed, scholarly study in English of the many ideas that developed from 1750 to 1900 regarding the existence of intelligent extraterrestrial life. Examines ideas of Kant, Herschel, Voltaire, Percival Lowell, many other scientists and thinkers. 16 illustrations. 704pp. 5⅜ x 8½. 40675-X

THEORIES OF THE WORLD FROM ANTIQUITY TO THE COPERNICAN REVOLUTION, Michael J. Crowe. Newly revised edition of an accessible, enlightening book recreates the change from an earth-centered to a sun-centered conception of the solar system. 242pp. 5⅜ x 8½. 41444-2

A HISTORY OF ASTRONOMY, A. Pannekoek. Well-balanced, carefully reasoned study covers such topics as Ptolemaic theory, work of Copernicus, Kepler, Newton, Eddington's work on stars, much more. Illustrated. References. 521pp. 5⅜ x 8½. 65994-1

A COMPLETE MANUAL OF AMATEUR ASTRONOMY: Tools and Techniques for Astronomical Observations, P. Clay Sherrod with Thomas L. Koed. Concise, highly readable book discusses: selecting, setting up and maintaining a telescope; amateur studies of the sun; lunar topography and occultations; observations of Mars, Jupiter, Saturn, the minor planets and the stars; an introduction to photoelectric photometry; more. 1981 ed. 124 figures. 26 halftones. 37 tables. 335pp. 6½ x 9¼. 42820-6

AMATEUR ASTRONOMER'S HANDBOOK, J. B. Sidgwick. Timeless, comprehensive coverage of telescopes, mirrors, lenses, mountings, telescope drives, micrometers, spectroscopes, more. 189 illustrations. 576pp. 5⅝ x 8¼. (Available in U.S. only.) 24034-7

STARS AND RELATIVITY, Ya. B. Zel'dovich and I. D. Novikov. Vol. 1 of *Relativistic Astrophysics* by famed Russian scientists. General relativity, properties of matter under astrophysical conditions, stars, and stellar systems. Deep physical insights, clear presentation. 1971 edition. References. 544pp. 5⅝ x 8¼. 69424-0

Chemistry

THE SCEPTICAL CHYMIST: The Classic 1661 Text, Robert Boyle. Boyle defines the term "element," asserting that all natural phenomena can be explained by the motion and organization of primary particles. 1911 ed. viii+232pp. 5⅜ x 8½. 42825-7

RADIOACTIVE SUBSTANCES, Marie Curie. Here is the celebrated scientist's doctoral thesis, the prelude to her receipt of the 1903 Nobel Prize. Curie discusses establishing atomic character of radioactivity found in compounds of uranium and thorium; extraction from pitchblende of polonium and radium; isolation of pure radium chloride; determination of atomic weight of radium; plus electric, photographic, luminous, heat, color effects of radioactivity. ii+94pp. 5⅜ x 8½. 42550-9

CHEMICAL MAGIC, Leonard A. Ford. Second Edition, Revised by E. Winston Grundmeier. Over 100 unusual stunts demonstrating cold fire, dust explosions, much more. Text explains scientific principles and stresses safety precautions. 128pp. 5⅜ x 8½. 67628-5

THE DEVELOPMENT OF MODERN CHEMISTRY, Aaron J. Ihde. Authoritative history of chemistry from ancient Greek theory to 20th-century innovation. Covers major chemists and their discoveries. 209 illustrations. 14 tables. Bibliographies. Indices. Appendices. 851pp. 5⅜ x 8½. 64235-6

CATALYSIS IN CHEMISTRY AND ENZYMOLOGY, William P. Jencks. Exceptionally clear coverage of mechanisms for catalysis, forces in aqueous solution, carbonyl- and acyl-group reactions, practical kinetics, more. 864pp. 5⅜ x 8½. 65460-5

ELEMENTS OF CHEMISTRY, Antoine Lavoisier. Monumental classic by founder of modern chemistry in remarkable reprint of rare 1790 Kerr translation. A must for every student of chemistry or the history of science. 539pp. 5⅜ x 8½. 64624-6

THE HISTORICAL BACKGROUND OF CHEMISTRY, Henry M. Leicester. Evolution of ideas, not individual biography. Concentrates on formulation of a coherent set of chemical laws. 260pp. 5⅜ x 8½. 61053-5

A SHORT HISTORY OF CHEMISTRY, J. R. Partington. Classic exposition explores origins of chemistry, alchemy, early medical chemistry, nature of atmosphere, theory of valency, laws and structure of atomic theory, much more. 428pp. 5⅜ x 8½. (Available in U.S. only.) 65977-1

GENERAL CHEMISTRY, Linus Pauling. Revised 3rd edition of classic first-year text by Nobel laureate. Atomic and molecular structure, quantum mechanics, statistical mechanics, thermodynamics correlated with descriptive chemistry. Problems. 992pp. 5⅜ x 8½. 65622-5

FROM ALCHEMY TO CHEMISTRY, John Read. Broad, humanistic treatment focuses on great figures of chemistry and ideas that revolutionized the science. 50 illustrations. 240pp. 5⅜ x 8½. 28690-8

Engineering

DE RE METALLICA, Georgius Agricola. The famous Hoover translation of greatest treatise on technological chemistry, engineering, geology, mining of early modern times (1556). All 289 original woodcuts. 638pp. 6¾ x 11. 60006-8

FUNDAMENTALS OF ASTRODYNAMICS, Roger Bate et al. Modern approach developed by U.S. Air Force Academy. Designed as a first course. Problems, exercises. Numerous illustrations. 455pp. 5⅜ x 8½. 60061-0

DYNAMICS OF FLUIDS IN POROUS MEDIA, Jacob Bear. For advanced students of ground water hydrology, soil mechanics and physics, drainage and irrigation engineering, and more. 335 illustrations. Exercises, with answers. 784pp. 6⅛ x 9¼. 65675-6

THEORY OF VISCOELASTICITY (Second Edition), Richard M. Christensen. Complete, consistent description of the linear theory of the viscoelastic behavior of materials. Problem-solving techniques discussed. 1982 edition. 29 figures. xiv+364pp. 6⅛ x 9¼. 42880-X

MECHANICS, J. P. Den Hartog. A classic introductory text or refresher. Hundreds of applications and design problems illuminate fundamentals of trusses, loaded beams and cables, etc. 334 answered problems. 462pp. 5⅜ x 8½. 60754-2

MECHANICAL VIBRATIONS, J. P. Den Hartog. Classic textbook offers lucid explanations and illustrative models, applying theories of vibrations to a variety of practical industrial engineering problems. Numerous figures. 233 problems, solutions. Appendix. Index. Preface. 436pp. 5⅜ x 8½. 64785-4

STRENGTH OF MATERIALS, J. P. Den Hartog. Full, clear treatment of basic material (tension, torsion, bending, etc.) plus advanced material on engineering methods, applications. 350 answered problems. 323pp. 5⅜ x 8½. 60755-0

A HISTORY OF MECHANICS, René Dugas. Monumental study of mechanical principles from antiquity to quantum mechanics. Contributions of ancient Greeks, Galileo, Leonardo, Kepler, Lagrange, many others. 671pp. 5⅜ x 8½. 65632-2

STABILITY THEORY AND ITS APPLICATIONS TO STRUCTURAL MECHANICS, Clive L. Dym. Self-contained text focuses on Koiter postbuckling analyses, with mathematical notions of stability of motion. Basing minimum energy principles for static stability upon dynamic concepts of stability of motion, it develops asymptotic buckling and postbuckling analyses from potential energy considerations, with applications to columns, plates, and arches. 1974 ed. 208pp. 5⅜ x 8½. 42541-X

METAL FATIGUE, N. E. Frost, K. J. Marsh, and L. P. Pook. Definitive, clearly written, and well-illustrated volume addresses all aspects of the subject, from the historical development of understanding metal fatigue to vital concepts of the cyclic stress that causes a crack to grow. Includes 7 appendixes. 544pp. 5⅜ x 8½. 40927-9

Mathematics

FUNCTIONAL ANALYSIS (Second Corrected Edition), George Bachman and Lawrence Narici. Excellent treatment of subject geared toward students with background in linear algebra, advanced calculus, physics, and engineering. Text covers introduction to inner-product spaces, normed, metric spaces, and topological spaces; complete orthonormal sets, the Hahn-Banach Theorem and its consequences, and many other related subjects. 1966 ed. 544pp. 6⅛ x 9¼. 40251-7

ASYMPTOTIC EXPANSIONS OF INTEGRALS, Norman Bleistein & Richard A. Handelsman. Best introduction to important field with applications in a variety of scientific disciplines. New preface. Problems. Diagrams. Tables. Bibliography. Index. 448pp. 5⅜ x 8½. 65082-0

VECTOR AND TENSOR ANALYSIS WITH APPLICATIONS, A. I. Borisenko and I. E. Tarapov. Concise introduction. Worked-out problems, solutions, exercises. 257pp. 5⅝ x 8¼. 63833-2

THE ABSOLUTE DIFFERENTIAL CALCULUS (CALCULUS OF TENSORS), Tullio Levi-Civita. Great 20th-century mathematician's classic work on material necessary for mathematical grasp of theory of relativity. 452pp. 5⅝ x 8¼. 63401-9

AN INTRODUCTION TO ORDINARY DIFFERENTIAL EQUATIONS, Earl A. Coddington. A thorough and systematic first course in elementary differential equations for undergraduates in mathematics and science, with many exercises and problems (with answers). Index. 304pp. 5⅜ x 8½. 65942-9

FOURIER SERIES AND ORTHOGONAL FUNCTIONS, Harry F. Davis. An incisive text combining theory and practical example to introduce Fourier series, orthogonal functions and applications of the Fourier method to boundary-value problems. 570 exercises. Answers and notes. 416pp. 5⅜ x 8½. 65973-9

COMPUTABILITY AND UNSOLVABILITY, Martin Davis. Classic graduate-level introduction to theory of computability, usually referred to as theory of recurrent functions. New preface and appendix. 288pp. 5⅜ x 8½. 61471-9

ASYMPTOTIC METHODS IN ANALYSIS, N. G. de Bruijn. An inexpensive, comprehensive guide to asymptotic methods–the pioneering work that teaches by explaining worked examples in detail. Index. 224pp. 5⅜ x 8½ 64221-6

APPLIED COMPLEX VARIABLES, John W. Dettman. Step-by-step coverage of fundamentals of analytic function theory–plus lucid exposition of five important applications: Potential Theory; Ordinary Differential Equations; Fourier Transforms; Laplace Transforms; Asymptotic Expansions. 66 figures. Exercises at chapter ends. 512pp. 5⅜ x 8½. 64670-X

INTRODUCTION TO LINEAR ALGEBRA AND DIFFERENTIAL EQUATIONS, John W. Dettman. Excellent text covers complex numbers, determinants, orthonormal bases, Laplace transforms, much more. Exercises with solutions. Undergraduate level. 416pp. 5⅜ x 8½. 65191-6

Math–Geometry and Topology

ELEMENTARY CONCEPTS OF TOPOLOGY, Paul Alexandroff. Elegant, intuitive approach to topology from set-theoretic topology to Betti groups; how concepts of topology are useful in math and physics. 25 figures. 57pp. 5⅜ x 8½. 60747-X

COMBINATORIAL TOPOLOGY, P. S. Alexandrov. Clearly written, well-organized, three-part text begins by dealing with certain classic problems without using the formal techniques of homology theory and advances to the central concept, the Betti groups. Numerous detailed examples. 654pp. 5⅜ x 8½. 40179-0

EXPERIMENTS IN TOPOLOGY, Stephen Barr. Classic, lively explanation of one of the byways of mathematics. Klein bottles, Moebius strips, projective planes, map coloring, problem of the Koenigsberg bridges, much more, described with clarity and wit. 43 figures. 210pp. 5⅜ x 8½. 25933-1

CONFORMAL MAPPING ON RIEMANN SURFACES, Harvey Cohn. Lucid, insightful book presents ideal coverage of subject. 334 exercises make book perfect for self-study. 55 figures. 352pp. 5⅝ x 8¼. 64025-6

THE GEOMETRY OF RENÉ DESCARTES, René Descartes. The great work founded analytical geometry. Original French text, Descartes's own diagrams, together with definitive Smith-Latham translation. 244pp. 5⅜ x 8½. 60068-8

PRACTICAL CONIC SECTIONS: The Geometric Properties of Ellipses, Parabolas and Hyperbolas, J. W. Downs. This text shows how to create ellipses, parabolas, and hyperbolas. It also presents historical background on their ancient origins and describes the reflective properties and roles of curves in design applications. 1993 ed. 98 figures. xii+100pp. 6½ x 9¼. 42876-1

THE THIRTEEN BOOKS OF EUCLID'S ELEMENTS, translated with introduction and commentary by Thomas L. Heath. Definitive edition. Textual and linguistic notes, mathematical analysis. 2,500 years of critical commentary. Unabridged. 1,414pp. 5⅜ x 8½.. Three-vol. set. Vol. I: 60088-2 Vol. II: 60089-0 Vol. III: 60090-4

GEOMETRY OF COMPLEX NUMBERS, Hans Schwerdtfeger. Illuminating, widely praised book on analytic geometry of circles, the Moebius transformation, and two-dimensional non-Euclidean geometries. 200pp. 5⅝ x 8¼. 63830-8

DIFFERENTIAL GEOMETRY, Heinrich W. Guggenheimer. Local differential geometry as an application of advanced calculus and linear algebra. Curvature, transformation groups, surfaces, more. Exercises. 62 figures. 378pp. 5⅜ x 8½. 63433-7

CURVATURE AND HOMOLOGY: Enlarged Edition, Samuel I. Goldberg. Revised edition examines topology of differentiable manifolds; curvature, homology of Riemannian manifolds; compact Lie groups; complex manifolds; curvature, homology of Kaehler manifolds. New Preface. Four new appendixes. 416pp. 5⅜ x 8½. 40207-X

History of Math

THE WORKS OF ARCHIMEDES, Archimedes (T. L. Heath, ed.). Topics include the famous problems of the ratio of the areas of a cylinder and an inscribed sphere; the measurement of a circle; the properties of conoids, spheroids, and spirals; and the quadrature of the parabola. Informative introduction. clxxxvi+326pp; supplement, 52pp. 5⅜ x 8½. 42084-1

A SHORT ACCOUNT OF THE HISTORY OF MATHEMATICS, W. W. Rouse Ball. One of clearest, most authoritative surveys from the Egyptians and Phoenicians through 19th-century figures such as Grassman, Galois, Riemann. Fourth edition. 522pp. 5⅜ x 8½. 20630-0

THE HISTORY OF THE CALCULUS AND ITS CONCEPTUAL DEVELOPMENT, Carl B. Boyer. Origins in antiquity, medieval contributions, work of Newton, Leibniz, rigorous formulation. Treatment is verbal. 346pp. 5⅜ x 8½. 60509-4

THE HISTORICAL ROOTS OF ELEMENTARY MATHEMATICS, Lucas N. H. Bunt, Phillip S. Jones, and Jack D. Bedient. Fundamental underpinnings of modern arithmetic, algebra, geometry, and number systems derived from ancient civilizations. 320pp. 5⅜ x 8½. 25563-8

A HISTORY OF MATHEMATICAL NOTATIONS, Florian Cajori. This classic study notes the first appearance of a mathematical symbol and its origin, the competition it encountered, its spread among writers in different countries, its rise to popularity, its eventual decline or ultimate survival. Original 1929 two-volume edition presented here in one volume. xxviii+820pp. 5⅜ x 8½. 67766-4

GAMES, GODS & GAMBLING: A History of Probability and Statistical Ideas, F. N. David. Episodes from the lives of Galileo, Fermat, Pascal, and others illustrate this fascinating account of the roots of mathematics. Features thought-provoking references to classics, archaeology, biography, poetry. 1962 edition. 304pp. 5⅜ x 8½. (Available in U.S. only.) 40023-9

OF MEN AND NUMBERS: The Story of the Great Mathematicians, Jane Muir. Fascinating accounts of the lives and accomplishments of history's greatest mathematical minds–Pythagoras, Descartes, Euler, Pascal, Cantor, many more. Anecdotal, illuminating. 30 diagrams. Bibliography. 256pp. 5⅜ x 8½. 28973-7

HISTORY OF MATHEMATICS, David E. Smith. Nontechnical survey from ancient Greece and Orient to late 19th century; evolution of arithmetic, geometry, trigonometry, calculating devices, algebra, the calculus. 362 illustrations. 1,355pp. 5⅜ x 8½. Two-vol. set. Vol. I: 20429-4 Vol. II: 20430-8

A CONCISE HISTORY OF MATHEMATICS, Dirk J. Struik. The best brief history of mathematics. Stresses origins and covers every major figure from ancient Near East to 19th century. 41 illustrations. 195pp. 5⅜ x 8½. 60255-9

Physics

OPTICAL RESONANCE AND TWO-LEVEL ATOMS, L. Allen and J. H. Eberly. Clear, comprehensive introduction to basic principles behind all quantum optical resonance phenomena. 53 illustrations. Preface. Index. 256pp. 5⅜ x 8½. 65533-4

QUANTUM THEORY, David Bohm. This advanced undergraduate-level text presents the quantum theory in terms of qualitative and imaginative concepts, followed by specific applications worked out in mathematical detail. Preface. Index. 655pp. 5⅜ x 8½. 65969-0

ATOMIC PHYSICS: 8th edition, Max Born. Nobel laureate's lucid treatment of kinetic theory of gases, elementary particles, nuclear atom, wave-corpuscles, atomic structure and spectral lines, much more. Over 40 appendices, bibliography. 495pp. 5⅜ x 8½. 65984-4

A SOPHISTICATE'S PRIMER OF RELATIVITY, P. W. Bridgman. Geared toward readers already acquainted with special relativity, this book transcends the view of theory as a working tool to answer natural questions: What is a frame of reference? What is a "law of nature"? What is the role of the "observer"? Extensive treatment, written in terms accessible to those without a scientific background. 1983 ed. xlviii+172pp. 5⅜ x 8½. 42549-5

AN INTRODUCTION TO HAMILTONIAN OPTICS, H. A. Buchdahl. Detailed account of the Hamiltonian treatment of aberration theory in geometrical optics. Many classes of optical systems defined in terms of the symmetries they possess. Problems with detailed solutions. 1970 edition. xv+360pp. 5⅜ x 8½. 67597-1

PRIMER OF QUANTUM MECHANICS, Marvin Chester. Introductory text examines the classical quantum bead on a track: its state and representations; operator eigenvalues; harmonic oscillator and bound bead in a symmetric force field; and bead in a spherical shell. Other topics include spin, matrices, and the structure of quantum mechanics; the simplest atom; indistinguishable particles; and stationary-state perturbation theory. 1992 ed. xiv+314pp. 6⅛ x 9¼. 42878-8

LECTURES ON QUANTUM MECHANICS, Paul A. M. Dirac. Four concise, brilliant lectures on mathematical methods in quantum mechanics from Nobel Prize–winning quantum pioneer build on idea of visualizing quantum theory through the use of classical mechanics. 96pp. 5⅜ x 8½. 41713-1

THIRTY YEARS THAT SHOOK PHYSICS: The Story of Quantum Theory, George Gamow. Lucid, accessible introduction to influential theory of energy and matter. Careful explanations of Dirac's anti-particles, Bohr's model of the atom, much more. 12 plates. Numerous drawings. 240pp. 5⅜ x 8½. 24895-X

ELECTRONIC STRUCTURE AND THE PROPERTIES OF SOLIDS: The Physics of the Chemical Bond, Walter A. Harrison. Innovative text offers basic understanding of the electronic structure of covalent and ionic solids, simple metals, transition metals and their compounds. Problems. 1980 edition. 582pp. 6⅛ x 9¼. 66021-4